The HANDBOOK of
ENVIRONMENTAL
COMPLIANCE in ONTARIO

Third Edition

JOHN-DAVID PHYPER,

M.A.Sc., M.B.A.

BRETT IBBOTSON,

M. Eng., P.Eng.

Toronto Montréal Boston Burr Ridge, IL Dubuque, IA Madison, WI
New York San Francisco St. Louis Bangkok Bogotá Caracas Kuala Lumpur
Lisbon London Madrid Mexico City Milan New Delhi Santiago Seoul
Singapore Sydney Taipei

McGraw-Hill Ryerson

ISBN: 0-07-090795-1

1234567890 TRI 098765432
Printed and bound in Canada.

National Library of Canada Cataloguing in Publication

Phyper, John-David
    The handbook of environmental compliance in Ontario / John-David Phyper, Brett Ibbotson. — 3rd ed.

Includes bibliographical references and index.
ISBN 0-07-090795-1

1. Environmental law—Ontario. I. Ibbotson, Brett II. Title.

| KEO717.P59 2002 | 344.713′046 | C2002-905139-8 |
| KF3775.ZB3P59 2002 | | |

Publisher: **Julia Woods**
Editorial Co-ordinator: **Catherine Leek**
Art Director: **Dianna Little**
Production Co-ordinator: **Mary Pepe**
Editor: **Katherine Coy**
Electronic Page Design and Composition: **Computer Composition of Canada Inc.**
Cover Design: **Andrew Gileno**

To my wife Christine and daughters Megan and Madison.

J.-D.P.

To my parents for their unfailing support and to Debra for her patience and encouragement.

B.G.I.

Acknowledgments

This *Handbook* represents the efforts of more than the two authors. During its preparation, and even before, many others shared information, experience and opinions with the authors.

To ensure that the material contained in the *Handbook* has been presented properly and is current, draft versions of chapters were reviewed by individuals active in that particular area of environmental management. All of the following people freely gave of their time and ideas and the authors are indebted to each of them for their assistance:

Dr. Ron Brecher – GlobalTox International Consultants Inc.

Mr. Bill Gastmeier – HGC Engineering

Dr. Neil Hutchinson – Gartner Lee Limited

Mr. David Hopper - Angus Environmental Limited

Mr. Gary Loftus – Canadian Environmental Auditors Association

Professor Donald Mackay – Canadian Environmental Modelling Centre, Trent University

Mr. P. Douglas Petrie - Willms & Shier, Environmental Lawyers

Mr. L. Piccioni – City of Hamilton/Region of Hamilton-Wentworth

Mr. Alex Safonsky – Resource Environmental Associates

Mr. J. Sinnige – Golder Associates Limited

Ms. Jeanette Southwood - Angus Environmental Limited

Mr. Tony Tarsitano – Integrated Management Solutions Limited

Mr. Gary Zikovitz - Ontario Ministry of the Environment

The authors also thank the staff at Angus Environmental Limited who assisted with the researching and collecting of information to update many of the chapters: Gayle Giesbrecht, Vera Lusney, Grant Piraine and Sean Rayman.

Table of Contents

List of Figures

List of Tables

Preface

In the preface to the first edition, we stated that the *Handbook* had been written to assist people responsible for ensuring that businesses and organizations in Ontario comply with environmental regulations and requirements — individuals who, whether they look after a small shop or direct the environmental affairs for a large corporation, have a personal stake in the environmental management of those companies.

Since the release of the first edition early in 1991, we have received comments from many of those environmental managers and engineers that we thought would be interested in the *Handbook*. We also have been pleasantly surprised to hear from lawyers, students, regulators, teachers and consultants. Their comments have been overwhelmingly positive, and we are grateful for their support and encouragement.

We expected that the changing nature of environmental management would make it necessary to revise parts of the *Handbook* after a few years, but we did not anticipate some of the landmark events and broad changes that have taken place since the second edition was published in 1994.

Some of the more important examples include:

- major shifts occurring at both the federal and provincial level in the philosophical underpinnings of legislation and in the roles these agencies want to pursue;
- substantial reductions in budgets and staff at the Ontario Ministry of the Environment;
- events such as the drinking water contamination at Walkerton and the subsequent intensive review of government policies and practices; and
- proliferation of Internet sites where environmental information can be obtained.

Descriptions of the shifts that are occurring in regulatory philosophy can be found in many documents, but for Ontario, the 2001 report entitled *Managing the Environment — A Review of Best Practices* (also known as the Gibbons report) outlines these changes and provides the supporting rationale.

Not as important, but nevertheless a complicating factor in revising the *Handbook* has been the frequent combining and separating of the Ministry of the Environment and the Ministry of Energy. We use the abbreviation MOE throughout the text except when referring to documents released when the two ministries were not combined. Our records show that the MOE was a separate entity from 1972 to 1993, and then from 1997 until April, 2002, and then again since August, 2002. It was the Ministry of the Environment and Energy (MOEE) from 1993 to 1997, and then from April to August, 2002. Since these name changes seem to be happening fairly frequently, we can only suggest that readers try to keep pace.

We also are aware that there are proper formats for naming statutes and regulations. For example, the correct citation is the *Environmental Protection Act*, R.S.O. 1990, c. E. 19 [as amended by S.O. 1992, c. 1; 1993, c. 27; 1994, c. 5, c. 23, c. 27; 1997, c. 6, c. 7, c. 19, c. 30, c. 37; 1998, c. 15, c. 35; 2000, c. 22, c.26; and 2001, c. 9]; however we prefer to use the simpler and less daunting "EPA" and make similar simplifications when referring to other statutes and regulations.

Although it has been necessary to make substantial changes to the *Handbook,* we have tried to remain true to its original goal of providing information succinctly and directly that responds to the following questions:

What are the applicable regulatory requirements for a specific situation?

What should be monitored or tested to evaluate compliance?

What methods are there for achieving compliance?

What are the penalties for not complying?

Are things likely to change in the near future? If so, how?

The material in this book is based on the experience and opinions of the authors, as well as written information available at the time that the book was being prepared. Every effort has been made to ensure that the information is current and reflects recent developments in environmental management. The authors welcome suggestions for ways to improve the *Handbook* and information on any errors or misinterpretations that may have been included.

CHAPTER 1

The Evolving Nature of Environmental Compliance

1.1 Overview

Compliance can be thought of as a two-step process. It starts with legislation, which represents the regulator's concept of what is necessary or acceptable. It ends with a response or reaction from the person or organization being regulated. Missing from this simple description is a forewarning of how complicated the process can be.

In the context of environmental compliance, it can be argued that there has always been environmental legislation in Canada (the federal Fisheries Act pre-dates confederation), but the mid-1950s is often considered as the time when the first environmental legislation was passed (such as the Ontario Water Resources Act of 1957) and the first environmental departments were established (such as the Ontario Water Resources Commission, the predecessor of the current Ministry of the Environment or MOE).

From the mid-1950s to the mid-1970s, the amount of environmental legislation grew at a slow but steady pace. By the early 1970s, major pieces of environmental legislation were being enacted. In 1971, Canada became the second country (after France) to establish a national environmental agency (Environment Canada). From the mid-1970s to the mid-1990s, this pace quickened. Many acts and regulations were introduced or expanded.

Around 1985, the nature of environmental compliance went through a distinct period of change. After three decades of compliance largely as a result of voluntary action and negotiated persuasion, regulatory agencies — including the MOE — decided that a less lenient approach was necessary. As a result, the years 1985 to 1995 were marked by increased penalties for non-compliance, stronger enforcement by regulatory agencies, and greater efforts by some organizations to ensure that they were in compliance.

Sometime around 1995, many aspects of environmental regulation and management entered a quiet period. Government budget and program cuts prompted some to complain that the monitoring and enforcement of environmental regulations were suffering and that compliance was slipping.

Criticism grew in the wake of events such as the drinking water contamination that occurred in Walkerton in 1999.

The first few years of the new millennium are showing signs of renewed vigour in many aspects of environmental legislation and management. Six of these indicators are described below.

1.2   Evolution of Legislation and Regulation

In the past, regulators relied on the "command-and-control" approach. Regulatory agencies set the rules for others to follow. Limits or prescribed actions were set for specific facilities or sectors. The annoying phrase "dilution is the solution to pollution" was an important consideration in setting those rules. Many of the current regulations reflect this approach.

After several decades of experience, it is clear that this approach should be improved. Many of the larger issues that confront us today are regional, even international in scope. Often, political boundaries are incompatible limits for addressing environmental issues. Sources of emissions and effluents need to be evaluated individually and cumulatively.

The conventional approach of appointing one agency or department as the prime source of environmental regulation is also recognized as being artificial. The challenges of environmental management are broader than the mandate of one department. A broader government-wide vision is needed along with alignment of resources to avoid both duplication and gaps.

While there have always been examples of regulators and those being regulated working co-operatively on some specific issues, there is a growing sense that effective environmental management is a shared responsibility. In 1999, the MOE announced its Recognizing and Encouraging Voluntary Action initiative (REVA). While not implemented, philosophical aspects of REVA are present in a growing number of MOE programs introduced since 1999. In this mode, the broad goals of the MOE are to foster an ethic that encourages the regulated community to go beyond regulatory compliance, and to provide assistance in reaching for that goal. At the heart of the program is the belief that actions taken to eliminate or reduce the creation of pollutants (often referred to as pollution prevention, or P2) are better than the traditional approach of investing equipment and capital in pollution control technology.

1.3   Frequent Changes to Regulatory Framework

Even during periods when interest drops in environment issues or the resources allocated by government agencies are reduced, there is a framework of environmental acts, regulations, policies, standards and guidelines that regulate virtually any activity in Ontario that has the potential to degrade the natural environment or make it unacceptable for people. Furthermore, there are stiff penalties for non-compliance.

Environmental regulations address more parameters than ever before, and the acceptable concentrations of many parameters continue to become more protective of the environment.

The environmental framework has become increasingly complex as regulations are put in place that address more aspects of environmental management, try to mesh with other regulations, are constructed or revised to withstand legal challenges, and try to fit with national and international policies and initiatives.

Merely keeping pace with changes in the regulatory framework has become a major challenge. Table 1.1 lists many of the major pieces of federal and provincial environmental legislation.

1.4   Changing Roles of Regulators

All three levels of government in Canada have roles and responsibilities in creating the environmental regulatory regime.

Federal responsibilities include environmental management at federal facilities (such as airports and international harbours), transportation (other than on roads within provinces), discharges that may be deleterious to fisheries, the import and export of goods and wastes and issues deemed to be in the national interest (global warming) or that involve the management of special materials (such as polychlorinated biphenyls). Federal responsibilities also can be triggered by projects with federal funding, that involve First Nations, or that require a federal licence or permit.

The Canadian Environmental Protection Act, 1999 (CEPA, 1999) is the main federal environmental statute. Key aspects of CEPA, 1999 are summarized in Table 1.2.

Provincial responsibilities include regulating emissions to air, discharges to water, waste management, and intraprovincial transportation on roads. While federal requirements can be adopted by provincial agencies if province-specific requirements have not been set, the regulatory framework in Ontario is extensive and this seldom is necessary.

The Environmental Protection Act (EPA) is the main environmental statute in Ontario. Key aspects of EPA are summarized in Table 1.3.

Municipal governments can impose environmental requirements on activities within their jurisdictions or on those who use facilities owned or operated by municipalities. Many municipalities in Ontario have established noise by-laws and sewer-use by-laws. Most participate in site selection and operation of wastewater treatment plants and solid waste management facilities. Some have offices to evaluate environmental aspects of local planning and transportation activities. Municipal agencies often act in a co-operative and/or support role to provincial agencies.

Since the mid-1990s, the MOE has indicated the intention of moving away from "local issues" and transferring responsibility for these to municipalities. This became evident in 1996 with some issues pertaining to the redevelopment of contaminated property. The Municipal Act, 2001 — in effect January, 2003 — authorizes municipalities to prohibit or regulate odour and dust, two areas overseen by the MOE in the past.

1.5   Changes in Regulator Investigation and Enforcement Efforts

Another aspect of environmental management that undergoes substantial changes from time to time is the amount of effort spent identifying and prosecuting offenders. Until the mid-1980s, prosecuting offenders was seen as a last resort and seldom was pursued. From the mid-1980s to the mid-1990s, various steps were taken in Ontario that showed a braced resolve to investigate polluters and prosecute environmental offenders. Fines for environmental offences in 1992 totalled $3.6 million, a 500% increase over 1986. Similar statistics show total fines down to less than $1 million in 1997 and 1998. Then in 2000, the MOE introduced the Environmental SWAT Team, which conducts strategic sector-by-sector inspections for compliance with environmental requirements. Team members can issue "tickets" (relatively small fines), orders, or refer cases to the Investigation and Enforcement Branch for possible prosecutions. According to the MOE, the numbers of charges laid and the total fines levied by the courts increased substantially in the first half of 2001 relative to 2000. The MOE has clearly stated that tough and aggressive investigation and enforcement is essential for the other anticipated shifts in regulatory philosophy noted above.

The liabilities and maximum possible penalties for non-compliance continue to grow (such as the round of increases announced in 2001 in Ontario). Legislation squarely places responsibility not only on corporations and organizations, but also on their officers and managers. Both federal and provincial legislation make company officers who participate, authorize or acquiesce to an offence, guilty of the offence and liable for punishment. Stiff fines and jail terms can be imposed.

1.6   Changing Public Attitudes and Expectations

In the long term, public attitudes and expectations will be as influential as the regulatory framework (and perhaps even more so) in establishing the importance of environmental management at businesses and organizations.

Attitudes about environmental management are beginning to influence the marketplace and reach into corporate boardrooms. Within the past few years, anyone with a computer connected to the Internet has access to environmental information once found only in the files of regulators, environmental practitioners and a few public organizations. Many Web sites have been created that provide information on environmental emissions, waste generation, government approvals, charges and convictions.

1.7   Changing Attitudes of Regulated Community

Corporate attitudes and expectations are changing. It is not important to determine which came first: the awakening of environmental awareness or environmental policy-making. What is important is that they support one another. In response to growing concern about the environment, the increased costs of non-compliance and the increased liabilities, growing numbers of companies are seeing a "greening" of the boardroom. Many companies are realizing that pollution is bad for business and that sound environmental management can protect and enhance the value of physical assets as well as corporate reputations. In some cases, a product that is certified as environmentally responsible can be a marketing advantage or even a requirement for sales to some customers.

Companies that do not have acceptable environmental records will struggle for credibility with the public. Each company will have to wrestle with what is the most appropriate way for it to respond to the changes that are occurring. One component of corporate response could be to adopt principles such as those listed in Table 1.4, which are based on the CERES Principles (formerly the "Valdez Principles"). These require a company to curb pollution, reduce waste, offer compensation for environmental damage and report every year about their operations with respect to the environment.

1.8   Purpose of the *Handbook*

The Handbook of Environmental Compliance in Ontario is intended to serve as a guide that plant managers, environmental engineers, advisors and students can consult to understand regulatory requirements, the way(s) to achieve compliance and the liabilities of not complying. Each chapter addresses a different facet of environmental management in terms of current and proposed legislation, obtaining approvals and permits, methods used to assess compliance, sampling and modelling techniques, penalties and liabilities, and reporting requirements.

Chapter 2: Air Quality and Atmospheric Emissions presents current regulatory requirements as defined by federal and provincial acts, regulations, policies and initiatives. This aspect of environmental management is in transition as the regulatory approach evolves from one of command-and-control to one with an emphasis on pollution prevention; emissions caps and trading; regional issues such as ground level ozone; and changes in the methods used to issue emission permits, set standards, and model the dispersion of atmospheric emissions.

Chapter 3: Water Quality and Liquid Discharges focuses on the Municipal, Industrial Strategy for Abatement (MISA) regulations implemented under the Ontario EPA. MISA is technology driven and requires that dischargers install the best available technology that is economically achievable. MISA regulations have been issued for direct dischargers in nine industrial sectors. For indirect liquid dischargers (those who discharge to sewer systems), there are municipal sewer use by-laws.

Chapter 4: Waste Management and Transportation presents the requirements of EPA and its associated regulations (notably Ontario Regulation 347), which set out a comprehensive system for monitoring hazardous and liquid industrial wastes from point of generation to ultimate disposal. The chapter also provides information on the provincial and federal statutes that describe the requirements for packing, labelling, etc. of dangerous goods, including wastes, during transportation.

Chapter 5: Assessing and Remediating Contaminated Property addresses the environmental issues that arise whenever contaminated properties are being considered for sale or redevelopment. Like most jurisdictions, Ontario is faced with the difficult task of insuring that contaminated properties are adequately cleaned up prior to rezoning, re-use or redevelopment. While this falls under guidelines issued by the MOE in 1996, the Brownfields Statute Law Amendment Act of 2001 and anticipated regulations could see substantial changes. Other pertinent statutes and restrictions under common law and contract law also are presented.

Chapter 6: Noise and Vibration are defined as contaminants under the EPA. The regulation of noise for the most part, however, is managed through municipal by-laws, which in turn are influenced by the MOE Model Municipal Noise Control By-Law and numerous MOE guidelines. A brief overview of the noise guidelines in Ontario is presented.

Chapter 7: Special Materials addresses polychlorinated biphenyls (PCBs), one of the most regulated groups of chemicals in Canada. Various provincial and federal regulations govern the use, storage, handling, transportation and disposal of PCBs. This chapter also presents the regulatory requirements for managing materials that contain asbestos.

Chapter 8: Enforcement addresses this key part of environmental regulation. In Ontario, the Investigation and Enforcement Branch of the MOE investigates and makes recommendations regarding appropriate legal remedies against pollution and polluters, as well as supplies information in support of prosecutions. Since 2000, inspections for compliance also have been made by the Environmental SWAT Team. This chapter discusses the various levels of government response to violations and describes both corporate and individual responses to site visits and interviews. Liabilities and penalties under various federal and provincial acts and regulations are summarized.

Chapter 9: Environmental Fate is a primer to understanding the behaviour of chemicals in the environment, which can be a valuable aid to managing the production, storage, handling, transportation and disposal of chemicals. Information is presented on key environmental parameters, partition coefficients, reaction rates and transport and transformation processes.

Chapter 10: Environmental Audits illustrates how audits have become a standard component of environmental management systems. This chapter discusses the objectives and potential applications for various types of audits and addresses the issue of confidentiality.

Chapter 11: Risk Assessment and Management demonstrates how these practices have emerged from being implicit to now being explicitly part of some environmental management practices. Different approaches to risk assessment and the situations where risk assessment can be applied are presented.

Chapter 12: Emergency Planning and Spills describes the types of procedures that need to be in place to respond to unscheduled releases of materials. In Ontario, Part X of the EPA outlines the responsibilities and requirements for the notification and restoration following a spill. This chapter discusses spill prevention and the federal and provincial requirements following a spill.

Chapter 13: Toxicity Testing describes how toxicity testing techniques are incorporated into the regulation of liquid effluents and their potential role in evaluating soil quality. Both acute and chronic toxicity tests are examined.

Table 1.1

Summary of Major Environmental Acts

Federal Acts
- Canadian Environmental Protection Act, 1999
- Transportation of Dangerous Goods Act, 1992
- Fisheries Act
- Canadian Water Act
- Arctic Water Pollution Prevention Act
- Northwest Territories Waters Act
- International Boundary Waters Treaty Act
- Canadian Environmental Assessment Act
- Canada Shipping Act
- Pest Control Products Act

Ontario Acts
- Environmental Protection Act
- Dangerous Goods Transportation Act
- Ontario Water Resources Act
- Environmental Assessment Act
- Environmental Statute Amendment Act
- Pesticides Act
- Planning Act
- Conservation Authorities Act
- Municipal Act
- Brownfields Statute Law Amendment Act
- Technical Standards and Safety Act
- Nutrient Management Act
- Toughest Environmental Penalties Act

Table 1.2

Key Aspects of the Canadian Environmental Protection Act, 1999

The main goal of the renewed Canadian Environmental Protection Act, 1999 (CEPA, 1999) is to contribute to sustainable development through pollution prevention and to protect the environment, human life and health from the risks associated with toxic substances.

CEPA, 1999 is divided into 11 Parts and six Schedules. In the context of the *Handbook*, the key aspects of CEPA, 1999 include:

Part 2 gives citizens better access to environmental information and encourages public participation in environmental decision-making and regulating. It establishes an environmental registry for posting notices, approvals, objections to approvals, policies, proposed regulations, orders and documents submitted to a court by the Minister. It also gives citizens the right to sue where a violation results in significant harm to the environment, and the federal government fails to take appropriate action, and it provides expanded "whistle blower" protection.

Part 3 authorizes the Environment Minister to conduct environmental studies, establish a national inventory of releases of pollutants (see Chapter 2, section 2.2.3), and issue objectives, guidelines and codes or practice related to preserving the quality of the environment.

Part 4 authorizes the Environment Minister to request Pollution Prevention (P2) Plans to control substances specified in the List of Toxic Substances or identified in other sections of the Act. Pollution prevention is described as the use of process, practices, materials, products and substances or energy that avoid or minimize the creation of pollutants and waste, and reduce the overall risk to the environment or human health (see Chapter 3).

Part 5 organizes substances into various lists. The Domestic Substance List (DSL) includes approximately 23,000 substances that are manufactured in, imported into, or used in Canada on a commercial basis. (There also is a much longer Non-Domestic Substances List.) From the DSL, Environment Canada has created the Priority Substance List. These are the first substances to be assessed to determine whether or not they are toxic or capable of becoming toxic. A **toxic substance** is defined as one that can have an immediate or long-term harmful effect on the environment or its biological diversity; constitute a danger to the environment on which life depends; or constitutes a danger in Canada to human life or health. Once various types of information are collected and assessed, a toxic substance from the DSL can be placed on the List of Toxic Substances in Schedule 1 of CEPA, 1999. The regulation of toxic substances is a central goal of CEPA, 1999. If a toxic substance is assessed as being persistent, bioaccumulative, and present in the environment primarily due to human activity, then it is targeted for virtual elimination under the federal government's Toxic Substances Management Policy. **Virtual elimination** is described as reducing environmental releases of a substance to a level that cannot be measured accurately. Substances to be virtually eliminated currently include aldrin, chlordane, DDT, dieldrin, endrin, heptachlor, hexachlorobenzene, mirex, PCBs, dioxins, furans, and toxaphene. All are persistent organic pollutants (and occasionally referred to as POPs).

Part 7 addresses the control of pollution and the management of waste. It specifically addresses concentration of nutrients in products; the protection of marine environments from land-based sources of pollution; disposal at sea; the issuing of permits for waste disposal; requirements for fuels; emissions from vehicles, engines and equipment; international air pollution issues (see Chapter 2); international water pollution issues (see Chapter 3); and the import and export of hazardous waste (see Chapter 4).

Part 8 provides Environment Canada additional regulatory powers related to spills, explosions and leaks. It can request emergency plans for accidental releases of toxic substances (see Chapter 12).

Table 1.3
Key Aspects of the Environmental Protection Act

Originally promulgated in 1971, the Environmental Protection Act (EPA) is directed toward the protection and conservation of the natural environment.

The EPA is divided into 17 parts. In the context of the *Handbook*, key aspects of EPA include:

Part II has several key provisions. Section 14(1) requires that no person shall discharge a contaminant or cause or permit the discharge of a contaminant into the natural environment that causes or is likely to cause an adverse effect. Adverse effect is defined as:

- impairment of the natural environment for any use that can be made of it
- injury or damage to property or plant or animal life
- harm or material discomfort to any person
- an adverse effect on the health of any person
- impairment of the safety of any person
- rendering any property or plant or animal life unfit for use by man
- loss of enjoyment of normal use of property
- interference with the normal conduct of business

Section 9(1) of Part II states that no person shall construct, alter, extend or replace anything or alter the production rate of any process that may discharge a contaminant into any part of the natural environment, except water, without a Certificate of Approval (C of A). The process of obtaining a C of A is described in Chapter 2 for air emissions, in Chapter 3 for liquid effluents, and in Chapter 4 for waste management activities.

Other sections of Part II authorize other types of orders including control orders, stop orders and remedial orders. These and other aspects of enforcement are the focus of Chapter 8.

Part V addresses waste management. This is the subject of Chapter 4.

Part VI deals with ozone-depleting substances. It prohibits the transport, storage, use or display of anything that contains an ozone-depleting substance, with the exception of prescription drugs containing an ozone depleting substance that acts as a propellant (metered dose inhalers).

Part X of the EPA deals with spills of pollutants. This is the subject of Chapter 12.

Table 1.4
Environmental Principles for Corporations

Protection of the Biosphere
Minimize and strive to eliminate the release of any substance that may damage the environment or its inhabitants. Safeguard wildlife habitat and open spaces while preserving biodiversity. Minimize contributions to global concerns such as the greenhouse effect, depletion of the ozone layer and acid rain.

Sustainable Use of Natural Resources
Use natural resources such as water, soils and forests in ways that are sustainable. Conserve non-renewable natural resources through efficient use and careful planning.

Reduction and Disposal of Waste
Minimize the creation of waste, especially hazardous waste. Wherever possible, recycle materials. Dispose of wastes safely and responsibly.

Efficient Use of Energy
Invest in improved energy efficiency and conservation. Maximize the energy efficiency of products and services.

Risk Reduction
Minimize the environmental, health and safety risks to employees and local communities by employing safe technologies and operating procedures, and by being constantly prepared for emergencies.

Market Safe Products and Services
Offer products or services that minimize adverse environmental impacts. Inform consumers of the environmental benefits and/or impacts of products or services.

Damage Compensation

Take responsibility for harm caused to the environment. Be prepared to restore the environment and compensate persons who are adversely affected.

Disclosure

Disclose incidents that cause environmental harm or pose undue hazards to employees and to the public. Disclose potential environmental, health or safety hazards posed by operations. Do not take action against employees who report any condition that creates a danger to the environment or poses health and safety hazards.

Environmental Directors and Managers

Appoint at least one senior officer who is qualified to represent environmental interests. Have that officer report directly to the chief executive officer. Commit resources to implement these principles and report on a regular basis as to their implementation.

Assessment and Annual Audit

Make public an annual evaluation of progress in implementing these principles and in complying with all applicable laws and regulations.

Incorporate the Importance of Environmental Management into Corporate Attitudes

Recognize that environmental management is not something to be avoided or dismissed. Realize that proper environmental management can protect and enhance physical assets and corporate reputations.

CHAPTER 2

Air Quality and Atmospheric Emissions

2.1 Overview

The regulation of air quality in Ontario is in transition. From the early 1970s to the early 1990s, a framework of regulations, policies and guidelines was established. Sources of emissions were allowed or tolerated as long as limits were not exceeded. Limits were based on either the available control technologies or the assumptions that sufficient dilution and degradation would occur in the atmosphere. Relatively little thought was given to the weighing of costs and benefits of control technologies, cumulative effects from multiple sources, interactions between the air and other environmental media such as the water and soil, or the potential for impacts to occur at locations remote from the point of release.

In recent years, perspectives on air quality have become broader and now include issues that are regional (such as the transboundary migration of smog precursors) and even global (such as greenhouse gas emissions and global warming) in scope. This changing perspective is influencing many aspects of regulating air quality and emissions to the atmosphere. Some examples include:

- Greater emphasis is being placed on pollution prevention rather than emission control.
- Risk assessment and risk management are beginning to play a larger role in the processes used to set numerical limits such as acceptable concentrations of chemicals in air.
- Models used to predict how emissions behave in the atmosphere are becoming more sophisticated and more representative of actual conditions.
- Monitoring and public reporting of the results is becoming more comprehensive and frequent.
- New concepts are emerging such as emissions trading and permit-by-rule regulations.

2.2   Federal Acts and Regulations

2.2.1   Canadian Environmental Protection Act, 1999

Prior to 1988, the Clean Air Act was the only federal legislation that addressed air pollution. In 1988, the Clean Air Act, the Environmental Contaminants Act and the Ocean Dumping Control Act were consolidated into the Canadian Environmental Protection Act (CEPA).

In the current version of the Act (CEPA, 1999), the key elements that are relevant to air quality and atmospheric emissions include:

- provisions to control all aspects of the life cycle of toxic substances including development, manufacturing, storage, transportation, use and disposal
- the regulation of fuels and emissions from vehicles and engines
- the regulation of emissions from federal departments, boards, agencies and Crown corporations
- provisions to create guidelines and environmentally safe codes of practice
- provisions to control sources of air pollution in Canada where a violation of international agreement would otherwise result.

Table 1.2 provides a summary of the key aspects of CEPA, 1999.

2.2.2   CEPA, 1999 Regulations

While the powers to regulate air emissions are fairly broad in CEPA, 1999, the regulating of most types of air pollution sources largely is left to provincial regulators. As a result, there are relatively few regulations under CEPA, 1999 related to air emissions and these are directed at either specific types of industrial facilities or at specific chemicals:

- Chloro-Alkali Mercury Release Regulations (SOR/90-130)
- Federal Halocarbon Regulations (SOR/99-255)
- Federal Mobile PCB Treatment and Destruction Regulations (SOR/90-5)
- Ozone-depleting Substances Regulations, 1998 (SOR/99-7)
- Secondary Lead Smelter Release Regulations (SOR/91-155)
- Vinyl Chloride Release Regulations (SOR/92-631)

In addition, there are several regulations under CEPA, 1999 that address the quality of various petroleum fuels. These mostly pertain to sulphur content of fuels and are intended to reduce or control emissions of sulphur dioxide to the atmosphere.

The federal government takes the lead in international initiatives (such as transboundary migration of air pollutants), directs research on issues relevant across the country (such as greenhouse gases), and establishes numerical limits that provincial and territorial governments can adopt as binding standards. Details for these types of activities are provided in several of the following sections.

Listings of CEPA, 1999 documents often include emission guidelines for various types of sources. Many of the guidelines date back to the 1970s and are of limited value. Some of the more recent guidelines (those developed in the 1990s) relate to the management of halons (1996), hazardous waste incinerators (1992), stationary combustion turbines (1992) and thermal power generating facilities (1993). Table 2.1 lists these and other Environment Canada documents that pertain to air quality and emissions to the atmosphere.

2.2.3   National Pollutant Release Inventory

As noted in Table 1.2, Part 3 of CEPA, 1999 authorizes the Environment Minister to establish a national inventory of releases of pollutants. The resulting National Pollutant Release Inventory (NPRI) was initiated in 1993 (with the first reports submitted in 1994).

Any facility that uses or processes a substance on the NPRI list in excess of the reporting thresholds is required to report the emissions of those substances into the Canadian environment. In recent years, the deadline for reporting has been June 1 of the following year.

Since 1993, the number of substances on the NPRI list has grown and therefore more facilities have had to file reports. For most substances, reporting is necessary if a facility has manufactured, processed or otherwise used (MPO) 10 tonnes or more, on an annual basis, of materials that contain chemicals of interest in concentrations greater than 1%, and whose employees collectively work 20,000 or more person-hours a year. For some activities (hazardous waste incineration, wood preservation and others) the 20,000-hour employee threshold does not apply. A few substances have lower MPO thresholds and/or lower concentration thresholds.

In 2002, criteria air contaminants were added to the NPRI list of reportable substances. These include nitrogen oxides (NOx), sulphur dioxide, volatile organic compounds, carbon monoxide and particulate matter (with various thresholds based on annual releases). The NPRI reportable substance list changes frequently. To ensure compliance, facilities must re-examine their reporting status annually.

The report format and reporting procedures are similar to those used in the United States. Environment Canada provides an electronic version of the forms to generate the reports and to submit the results electronically. The reports are then entered into a database for public access. Environment Canada also prepares an annual printed report that summarizes the largest emitters by location, by material emitted, by industry, etc.

NPRI information must be kept on file at the facility and be available to Environment Canada inspectors for at least three years.

2.3   CCME and Setting Canada-Wide Standards

Through the Canadian Council of Ministers of the Environment (CCME), the federal government develops limits (or "standards") that provincial and territorial governments can adopt. In general, the CCME examines issues that are relevant across most or all of the country. CCME members agree on priorities and undertake the studies necessary to set the limits.

Under the 1998 Canada-Wide Accord for Environmental Harmonization and the subsequent Canada-Wide Environmental Standards Sub-Agreement, the CCME has set Canada-Wide Standards (CWS) for various chemicals in air. These include benzene, ground-level ozone, mercury emissions from incinerators and from base-metals smelting, dioxins and furans from incineration and boilers at pulp and paper mills that burn salt-laden wood, particulate matter (PM) — notably the fraction smaller than 10 mm in diameter (PM_{10}) and the fraction smaller than 2.5 mm in diameter ($PM_{2.5}$). Work is underway on CWS for dioxins and furans emissions from iron sintering, steel manufacturing and conical waste burners, and mercury emissions from electric power generation facilities. Some of these activities began well before the Accord and Sub-Agreement.

In general, a CWS document contains numeric limits, a timetable for implementation or attainment, and a framework for monitoring progress and reporting to the public. Each CCME member (the Government of Canada, each province and territory) does what makes sense in their jurisdiction so that the issue or problem is addressed by the collective actions. Each jurisdiction decides whether or not to adopt the numerical limits.

In addition to developing CWS documents, the CCME has prepared other guidelines for emissions from specific types of sources. Examples include emission guidelines for commercial/industrial boilers and heaters, cement kilns, the plastics processing industry, the use of surface coating products and degreasing facilities. More information about CCME publications can be obtained at **www.ccme.ca.**

For owners and operators of facilities that have atmospheric emissions, the CWS documents and other CCME guidelines are useful descriptions of best available technology and the time of publication.

2.4   NTREE and Climate Change

The National Round Table on the Environment and Economy (NTREE) was created in 1994 as an advisory council to the prime minister. Members develop recommendations on how best to integrate environmental, economic and social considerations into decision-making. One issue that NTREE has examined is climate change and the possible role of emission trading. In 2002, it concluded that emission trading is an important tool for minimizing the cost of responding to climate change. This conclusion has encouraged the use of emission trading (see Section 2.15.2). More about NTREE can be obtained at **www.ntree-trnee.ca.**

2.5   International Agreements

Canada is a signatory to several international agreements that pertain to air quality. One of these is the Canada-U.S. Air Quality Agreement of 1991. The Ozone Agreement of December 2000 addresses transboundary issues, notably those pertaining to NOx and volatile organic compounds (VOCs). Ground level ozone is produced when NOx and VOCs react in the presence of sunlight. The Agreement specifies reductions in both countries for industry, transportation and electricity generating facilities. For Ontario, the electricity generating sector will have a NOx cap of approximately 25 kilotonnes, which is approximately 50% less than emissions in 1998.

In 2001, Canada became the first country to sign and ratify the Stockholm Convention. This agreement is aimed at virtually eliminating releases of the 12 persistent organic pollutants (often referred to as POPs). The "dirty dozen" chlorinated compounds include PCBs, DDT, chlordane, dioxins and furans.

Through the National Contaminants Program, concentrations of POPs are monitored in the Canadian north. The results are shared with eight countries in the circumpolar community.

Through the North America Free Trade Agreement (NAFTA) Commission on Environmental Cooperation, Canada (in conjunction with the United States and Mexico) will develop regional action plans for PCBs, DDT and chlordane. Consideration is being given to expanding that list to include lindane, dioxins, furans and hexachlorobenzene.

Under the Kyoto Protocol, countries are to reduce greenhouse gases (primarily carbon dioxide, methane and nitrous oxide) in response to concerns about global warming and climate change. Canada's target is to reduce these emissions by 6% relative to 1990 levels by 2008 to 2012. As of late 2002, it was unclear whether, when, or how the protocol will be implemented in Canada.

2.6   Other Federal Programs and Initiatives

The National Air Pollution Surveillance (NAPS) Network was established in 1969 as a joint project of the federal and provincial governments to monitor the quality of ambient air in urban areas across Canada. Samples are collected for 24 hours once every six days. The monitored chemicals include total suspended particulate matter, PM_{10}, $PM_{2.5}$ and sulphate.

In 1990, the CCME adopted a Management Plan for Nitrogen Oxides (NOx)and Volatile Organic Compounds (VOCs) to identify ways to protect the Canadian environment and to ensure that Canada fulfilled its international obligations. The objective of the plan is to reduce ground-level ozone concentrations to be less than the one-hour ambient air quality objective of 82 ppb (0.16 mg/m³) across Canada.

Elements of the plan called for initiatives such as energy conservation measures, reduced emissions from industrial sources and product modifications. Some of these initiatives have been implemented and most have evolved into actions such as the addition of these parameters to NPRI reporting described in Section 2.2.3, the recent CWS noted in Section 2.3, and the Ozone Agreement noted in Section 2.5.

In 1998, federal, provincial, and territorial governments signed the Canada-Wide Acid Rain Strategy for Post-2000. The primary goal of the strategy is to reduce acid rain loadings across Canada. The strategy sets targets and timelines for sulphur dioxide emissions in Ontario, Quebec, New Brunswick and Nova Scotia. It also includes seeking further reductions from sources in the eastern United States. More information can be obtained at **www.ec.gc/acidrain**.

2.7   Provincial Acts and Regulations

2.7.1   Environmental Protection Act

Table 1.3 summarizes key aspects of the EPA. While much of the text of the EPA does not specifically address air quality or atmospheric emissions, the broad intent and general language of the EPA clearly encompasses air quality and emissions to the atmosphere. For example, s. 9(1) of the EPA states that no person shall construct, alter, extend or replace anything or alter the production rate of any process that may discharge a contaminant into any part of the natural environment, except water, without a Certificate of Approval (C of A). Obtaining approvals for air emissions is addressed in detail in Section 2.9.

Similarly, the general prohibition in s. 14(1) of the EPA that no person shall discharge a contaminant or cause or permit the discharge of a contaminant into the natural environment that causes or is likely to cause an adverse effect clearly includes effects that can be caused by emissions to the atmosphere. For example, the loss of enjoyment of normal use of property can include nuisances such as odour and dust.

Other parts of the EPA can apply to air quality issues under certain conditions. For example, Part VI prohibits the transport, storage, use or display of anything that contains an ozone-depleting substance, with the exception of prescription drugs containing an ozone depleting substance that acts as a propellant (metered dose inhalers).

Part X of the EPA deals with spills of pollutants. This can include spills to the atmosphere. (See Chapter 12 for additional information about spills.)

2.7.2   EPA Regulations

Numerous regulations have been promulgated under the EPA that pertain to air quality or emissions to the atmosphere:

- Air Contaminants from Ferrous Foundries (R.R.O. 1990, Reg. 336)
- Airborne Contaminant Discharge Monitoring and Reporting (O. Reg. 127/01)
- Ambient Air Quality Criteria (R.R.O. 1990, Reg. 337)
- Boilers Regulation (R.R.O. 1990, Reg. 338)
- Certificate of Approval Exemptions – Air (O. Reg. 524/98)
- Dry Cleaners (O. Reg. 323/94)
- Emissions Trading (O. Reg. 397/01)
- Fees – Certificates of Approval (O. Reg. 363/98)
- Gasoline Volatility (O. Reg. 271/91)
- General – Air Pollution (R.R.O. 1990, Reg. 346)
- Ground Source Heat Pumps (O. Reg. 177/98)
- Hot Mix Asphalt Facilities (R.R.O. 1990, Reg. 349)
- Lakeview Generating Station (O. Reg. 396/01)
- Lambton Industry Meteorological Alert (R.R.O. 1990, Reg. 350)
- Mobile PCB Destruction Facilities (R.R.O. 1990, Reg. 352)
- Motor Vehicles (O. Reg. 361/98)
- Ontario Power Generation Inc. (O. Reg. 153/99)
- Ozone-Depleting Substances – General (R.R.O. 1990, Reg. 356)
- Recovery of Gasoline Vapour in Bulk Transfers (O. Reg. 455/94)
- Refrigerants (O. Reg. 189/94)
- Solvents (O. Reg. 717/94)
- Sterilants (O. Reg. 718/94)
- Sulphur Content of Fuels (R.R.O. 1990, Reg. 361)

Details of several of the regulations are discussed in subsequent sections of this chapter.

As its name implies, Regulation 346 is the primary air pollution regulation in Ontario. It sets numerical limits that should not be exceeded, a shortened version of the adverse effects definition that appears in the EPA, a discussion of visible emissions, restrictions on fuels used in fuel-burning appliances, and an appendix that describes atmospheric dispersion modelling. The fundamental philosophy that guides Regulation 346 is that if there is sufficient dilution of pollutants in the atmosphere, then the overall impact will be acceptable. The numerical limits are addressed in Section 2.10. Current models and proposed changes to the models are described in Section 2.11.

The process of obtaining a C of A for an air emission source is presented in the EPA, while O. Reg. 363/98 describes the fees associated with approvals, and O. Reg. 524/98 identifies sources that are exempt from C of A requirements. The process of applying for a C of A and the associated fees are described below in Section 2.9.

O. Reg. 127/01 is relatively recent and sets out reporting requirements for many types of emission sources. Details are described in Section 2.12.

Regulation 356 reiterates federal legislation regarding pressurized containers containing chlorofluorocarbons. The regulation also details restrictions on the manufacture of flexible polyurethane foam and various rigid insulation foams. The restrictions include the percentage content of the chlorofluorocarbon in the product and the schedule for its reduction.

Many of the other regulations pertain to specific types of activities (mostly industrial activities) or to specific industries. O. Reg. 153/99 sets an emission cap of 215 kilotonnes/year of NOx and sulphur dioxide from Ontario Power Generation Inc. This is a component of the Countdown Acid Rain Program (see Section 2.14.1). Similarly, O. Reg. 396/01 requires that the Lakeview generating station stop burning coal by 2005.

O. Reg. 397/01 is relatively recent and introduces the concepts of emission caps and trading to the regulatory framework in Ontario. Details about emission trading are presented in Section 2.15.2.

Regulation 352 addresses emission from PCB destruction facilities. This is described further in Section 7.1.13.

2.7.3   Occupational Health and Safety Act

Sections of the Ontario Health and Safety Act should be reviewed prior to placing an emission point near a fresh air intake vent. Section 132(2) requires that the replacement air shall be free from contamination with any hazardous dust, vapour, smoke, fume, mist or gas. Section 132(3) requires that the discharge of air from any exhaust system shall be in such a manner so as to prevent the return of contaminants to any work place.

2.8   Municipal Requirements

Air quality is not usually covered by municipal by-laws, however, new powers in the Municipal Act, 2001 (in effect January, 2003) give Ontario municipalities the authority to prohibit or regulate odour and dust. This slight shift in regulatory involvement reflects efforts by the MOE to step back from local issues. Where a municipal by-law conflicts with a provincial Act or regulation, the provincial legislation often has mechanisms to override the by-law.

2.9   Obtaining Approvals for Emissions

2.9.1   EPA Requirements

Section 9(1) of the EPA requires that a C of A be obtained by any person who intends to:

- construct, alter, extend or replace any plant, structure, equipment, apparatus, mechanism or thing that may discharge, or from which may be discharged, a contaminant into any part of the natural environment other than water; or
- alter the process or rate of production with the result that a contaminant may be discharged into any part of the natural environment other than water or the rate or manner of discharge of a contaminant into any part of the natural environment other than water may be altered.

Section 9(2) requires an applicant for a C of A to provide whatever information is needed by the MOE to assess the emission source properly. The application must include:

- the name of the owner and operator of the source
- a complete description of the process generating the contaminant including a process flow sheet, operating schedule, production data and raw materials
- a description of the systems, if any, to be used to reduce emissions
- an emission inventory for all contaminants that could be discharged to the atmosphere
- surrounding land use and points of impingement
- general information concerning toxic materials, waste storage and noise
- an assessment of the compliance with Regulation 346

The MOE has prepared several documents to help in the preparation of a C of A for air emissions. There is a blank C of A application, the *Guide for Applying for Approval-Air*, the *Procedure for Preparing an Emission Summary and Dispersion Modelling Report* (see Section 2.9.4 for details) and the *Protocol of Updating Certificates of Approval for Air Emissions* (see Section 2.9.5).

There also are guides for specific types of air emission equipment or activities. These include new biomedical waste incinerators, stationary combustion turbines, new municipal waste incinerators, boilers and heaters, and new cement plants. The titles of these and other MOE documents related to air are provided in Table 2.2.

Until recently, an emission source that existed unchanged since June 30, 1988 did not require a C of A, and could not be issued one, however, in 2001, the MOE reinterpreted this rule, and will now issue certificates for all sources at a facility. This allows the types of Cs of A described in Section 2.9.5 to be issued.

Applicants are required to complete dispersion modelling to determine the point of impingement (POI) concentrations of contaminants and submit the results as part of a C of A application.

Once an application is submitted, it is first checked for completeness. Incomplete applications or those that do not follow MOE requirements may be returned without a technical review or held while the applicant responds to requests for additional information. For applications that are complete, the facility owner/operator will receive a standard acknowledgement letter from the Ministry indicating that the technical review is underway.

Most delays in processing C of A applications result from incomplete or confusing information. It is important to respond completely and accurately to the MOE information requirements. This is particularly true in cases where unusual or new and innovative processes or control systems are proposed, or where difficult or contentious pollutants could potentially be emitted.

It is important to remember that having and complying with the terms and conditions of a C of A does not exclude the possibility of being prosecuted. For example, a C of A might describe maximum emission rates of specific chemicals. The rates could be met, but if the emission still causes odours or other adverse effects, the emitter may not be in compliance.

2.9.2   Exemptions

Section 9(3) of the EPA identifies several exemptions from the requirement to obtain a C of A:

- routine maintenance carried out on any plant, structure, equipment, apparatus, mechanism or thing;
- equipment for the combustion of fuel, other than waste incinerators, in buildings or structures designed for the housing of not more than three families;
- any equipment, apparatus, mechanism or thing in or used in connection with a building or structure designed for the housing of not more than three families where the only contaminant produced by such equipment, apparatus, mechanism or thing is sound or vibration;
- any plant, structure, equipment, apparatus, mechanism or thing that may be a source of contaminant of a class exempted from the regulation;
- any plant, structure, equipment, apparatus, mechanism or thing used in agriculture; and
- any motor or motor vehicle that is subject to the provisions of EPA, Part III.

Additional exemptions are provided in O. Reg. 524/98. Some are paraphrased below, but the regulation should be checked for other exemptions and complete descriptions.

- Equipment that burns fuel for comfort heating if the total thermal output is less than 1.58 million joules per hour (approximately 1.5 million BTU/h).
- Equipment, other than a waste incinerator, associated with a dwelling in a building that contains one or more dwellings and is used by the occupants of not more than three of the dwellings.
- Equipment used for building construction, demolition, drilling or blasting.
- A fireplace or wood stove if the only fuel is natural gas, untreated wood, or manufactured fire logs.
- Equipment that is used to transfer outdoor air into a building that is not a cooling tower.
- Equipment used to vent indoor air out of a space other than a lab; parking garage; building that handles or bales aerosol cans; and a building used to produce, process, repair or store waste.
- Equipment used to vent emissions from a motor vehicle or locomotive used in a warehouse.
- Equipment used in areas for the preparation of food or beverages such as a restaurant, snack bar, cafeteria or banquet hall if the primary business at the location is not wholesale distribution or sale to retail facilities.
- Equipment at cleaning operations that use only aqueous detergent solutions.
- Equipment used for fire fighting exercises other than a fuel-fired generator set.
- Mobile equipment used to make snow; clean ducts, carpets and upholstery; remove asbestos; or crush or screen aggregate.
- A lagoon, clarifier or pond that is used to treat or detain sewage.
- Any source of light used for illumination or advertising.
- A racecourse if the only contaminants emitted are noise, vibration, odour and dust attributable to the races.
- Equipment used during an outdoor entertainment, artistic or sporting event, including a fair, parade, fireworks display, art show, air show or car show.
- Equipment used solely to mitigate the effects of an emergency.

Section 9(6) of the EPA is important but often overlooked. It allows a proponent (the owner and/or operator of a source) to make changes to a system without obtaining a C of A or an amendment to an existing C of A by notifying the MOE Director. The section only applies when the changes will result in a decrease of emissions

2.9.3   C of A Fees

A processing and handling fee is payable upon submission of an application for a C of A. The fee provisions are contained in O. Reg. 363/98. The total cost to process an application typically consists of the following items:

1. Administrative processing – $200 ($100 for administrative applications).
2. Fixed cost for the general technical review – the minimum is $400 but the total depends on equipment covered by application and can be much higher.
3. Emission summary review cost – depends on number of sources and is not always necessary but can be as high as $3,000 if the application includes more than 20 sources not previously reviewed.
4. Noise assessment review cost – depends on equipment, number of sources and location of nearest residential building and is not always necessary but can be as high as $2,500 for each large industrial facility such as an arc furnace, asphalt plant, gas turbine or blow-down device.

The MOE has prepared two documents that pertain to C of A fees. There is a blank form for determining the application fee, and the *Guide: Application Costs for Air Emissions*. These are listed in Table 2.2.

2.9.4   Emission Summary and Dispersion Modelling Report

Since 1998, a complete facility emission summary and dispersion model (ESDM) report must be submitted with the C of A application. An ESDM report must include the following:

1. Facility description – should include a general description, process flow diagram, scaled site plan, elevation view and locations of nearby receptors.
2. Completed source summary table – should include source data — including an identifier and a listing of general information for each emission source — estimate the maximum emission rate for every contaminant emitted, assess data quality, describe the emission estimating techniques used and provide percentage of overall emission.
3. Completed Emission Summary Table – should provide the contaminant name, its Chemical Abstract Service (CAS) number, the aggregate half-hour emission rate for each contaminant, the aggregate maximum point of impingement (POI) concentration, the half-hour POI limit, the percentage of the POI limit that the maximum POI represents, the output of the dispersion modelling and a description of the maximum emission scenario on which emissions are reported.

The document entitled *Procedure for Preparing an Emission Summary and Dispersion Modelling Report* (MOE, 1998) describes this in greater detail.

ESDM reports are reviewed by the MOE under the Selected Targets for Air Compliance (STAC) program. The detailed information is examined to determine if the aggregate emissions from a facility comply with all of the air quality limits set out in Regulation 346. (These limits are described below in Section 2.10.) Facilities that do comply may be requested to apply for the types of approvals described in Section 2.9.5.

2.9.5   Multiple Emission Sources

For facilities or properties with multiple emission sources, applying for, revising and administering Cs of A have been cumbersome. In an effort to improve the managing of multiple emission sources, the MOE announced two new types of approvals (MOE, 2002).

For a facility with multiple Cs of A, these can be replaced with a single Consolidated C of A. The Consolidated C of A also can include new or historically unapproved sources. A Consolidated C of A can include all sources at a facility or be limited to those that emit the same contaminants. The owner/operator of the facility is required to apply for an amendment before making any modifications requiring approval.

Alternatively, a facility with multiple sources can apply for a Basic Comprehensive C of A. It replaces all existing Cs of A and includes all new or historically unapproved sources. It provides owner/operators with "limited" operational flexibility to make modifications up to an approved maximum production rate without the need to apply for an amendment. In return the facility must demonstrate that it complies with performance limits and monitoring requirements, most notably the requirements set out in Regulation 346. A log must be kept of all modifications made under the limited operational flexibility. The flexibility conditions expire in five years at which time the owner/operator will need to renew or apply for an amended C of A. If no application is received before the end of the five years, the operational flexibility ends and the C of A becomes a Consolidated C of A.

A deficiency that the MOE has identified in the current version of Regulation 346 is that there is no provision for a C of A to be reviewed or to expire. The Basic Comprehensive C of A is an attempt to address that deficiency.

2.9.6   Public Access

It is a goal of the MOE to provide opportunities for public comment on decision-making activities. Under the Environmental Bill of Rights (EBR) Act, the public can apply to review an existing C of A. In some cases, a C of A will be posted on the EBR registry for public comment. In those same cases, the final MOE decision regarding the C of A also is posted at the end of the application review process.

2.10   Numerical Limits

Two basic types of numerical limits have been established by the MOE for assessing air quality and emissions to the atmosphere.

In Regulation 346, s. 5 forbids a person from causing or permitting the concentration of a contaminant to exceed the values set in Schedule 1 at prescribed locations called points of impingement (POI).

Schedule 1 contains the half-hour, POI standards for approximately 90 substances. Those standards and additional POI guidelines (for chemicals not listed in Schedule 1) are presented in Table 2.1. Some of the POI limits are identified as "interim". Additional information will be considered by the MOE prior to these being finalized.

The second type of limit is the ambient air quality criterion (AAQC). These criteria are used for assessing general air quality and the potential for causing adverse effects. AAQCs are listed in Table 2.3.

The MOE has indicated that approaches used in the past to set POI limits and AAQC need to be improved. The MOE has proposed to develop a framework that uses risk assessment and risk management to set air quality standards in the future. Once risks to human health and the environment have been determined, risk management would be used to examine practical considerations (MOE, 2001c).

Table 2.1 contains a few "future, effects-based" POI limits that the MOE feels are consistent with the risk assessment component. Subject to the risk management component, these will replace current POI limits.

2.11   Air Dispersion Models

2.11.1   General Types of Models

Mathematical models can be used to predict how atmosphere pollutants behave and the concentrations that result at specific times and locations. Over the last three decades, the sophistication of such models has grown rapidly. Many of the latest models attempt to address various factors that influence atmospheric dispersion such as atmospheric stability, irregular or complex terrain, multiple-emission sources, local building wake effects, and short-term events.

Downwind locations of interest (such as the nearest house, a place where people spend time or a garden) often are called POI. The models are used to predict the concentration at the POI of a chemical emitted from a source.

Although many air dispersion models have been developed, they usually fall into a few general categories. In Gaussian plume models, it is assumed that the emitted plume spreads both laterally and vertically. The concentrations of a substance in such a plume are described in mathematical terms as having a "normal" or Gaussian distribution. Downwind concentrations are calculated based on the height of release, the emission rate, exit velocity and temperature from the source, wind speed, and general atmospheric conditions. Gaussian plume models are generally considered to be capable of predicting annual average concentrations at a point of exposure within a factor of two to four for pollutants released continuously over flat terrain (Cohrssen and Covello, 1989). More complex conditions contribute to greater uncertainty in predictions.

Puff models are based on the same principles as Gaussian models, but they are used to simulate the transport of emissions after episodic or short-duration releases, such as explosions or accidental releases.

Long-range atmospheric transport models attempt to predict pollutant concentrations over geographical regions as large as entire continents. Trajectories that released pollutants might follow are based on historical wind data from weather stations within the region. Generally, these models are thought to predict annual average concentrations within a factor of three to five (Cohrssen and Covello, 1989).

In addition to these mathematical models, physical models can be used to assess air dispersion patterns, notably in urban settings where building sizes and configurations make complex dispersion patterns. Due to the steep costs of constructing and conducting tests with physical models, they typically are reserved situations not easily modelled such as the complex wind patterns that can be created around buildings in urban settings.

2.11.2   Models in Regulation 346

The models described in the Appendix of Regulation 346 are relatively simple Gaussian plume dispersion models. They are used to calculate half-hour POI concentrations.

Different models are identified for various configurations of emission sources, receptor locations and wake effects of adjacent buildings. The configuration of the building on which the emission point is located and the adjacent buildings determines which algorithm is most appropriate for a specific situation.

To assist in the selection of the appropriate model, figures are provided in Regulation 346 that illustrate various configurations of stack and building heights and location of adjacent buildings. Four different scenarios are described.

Scenario 1

The receptor is located within 5 m of the source and the emission is caught in either the wake of the building or an adjacent building (that is, less than 100 m).

Scenario 2

The receptor is located greater than 5 m from the source and the emission is caught in either the wake of the building or an adjacent building (that is, less than 100 m).

Scenario 3

The receptor is located greater than 5 m from the source and the emission is not being caught by the wake of the building or an adjacent building (that is, less than 100 m).

Scenario 4

There are multiple sources. For the receptor, worst-case concentrations from each individual source are to be added together.

Various scenario characteristics need to be known or estimated to use the models. These can include stack height, emission exit velocity and temperature, emission rate of the pollutant, and the distance and height of closest receptor.[1]

Most models take into account atmospheric stability, which represents the turbulence of the transporting wind and its ability to disperse emitted materials (Bowne, 1984). Stability categories can be semi-quantitatively specified in terms of wind speed, incoming solar radiation during the day, and cloud cover during the night (Pasquill and Smith, 1983).

Surface Wind Speed (m/s)	DAY			NIGHT	
	Incoming Solar Radiation			Thinly Overcast or	
	Strong	Moderate	Slight	≥4/8 low cloud	≤3/8 cloud
<2	A	A-B	B	—	—
2-3	A-B	B	C	E	F
3-5	B	B-C	C	D	E
5-6	C	C-D	D	D	D
>6	C	D	D	D	D

Stability is least for Class A and highest for Class F. The neutral class (D) is assumed to occur for overcast conditions during day or night.

The MOE models for Scenarios 2 and 3 implicitly use atmospheric stability classes C and D in the modelling.

---

1 Software for dispersion modelling may be requested from the MOE office at 125 Resources Rd., Rexdale, ON, M4V 1K6 (telephone: 416-235-5772 or 416-235-5764).

While the models in Regulation 346 are relatively simple, the results produced by the models often are key considerations in decisions for issuing or denying a C of A.

2.11.3   Recent Advances

The United States EPA recommends several atmospheric dispersion models in its *Guidelines on Air Quality Models* (U.S. EPA, 2001). Some of these models may be better suited to evaluating some conditions than the models in Regulation 346, however, permission from the MOE should be obtained prior to using the models for assessment of compliance. (Also see Section 2.11.4.)

RAM Gaussian Plume Multiple-Source Air Quality Model

This is a steady-state model for estimating concentrations of stable pollutants for average times of an hour to a day from point and area sources.

Climatological Dispersion Model

The climatological dispersion model (CDA) is a climatological steady-state Gaussian Plume Model for determining long-term (seasonal or annual) arithmetic average pollution concentrations at a ground level receptor in an urban area.

Industrial Source Complex Models

Industrial source complex models (ISC) have both a short-term (ISCST3) and a long-term version of this Gaussian plume dispersion model. The models account for settling, dry deposition of particles and downwash. In addition, the models can handle area, line or volume sources, plume rise as a function of downwind distance, and separation of point sources. Meteorological conditions are assumed to be constant. Limited adjustment can be made for terrain. This model has been widely used in the United States since the 1970s.

A simplified version of ISC (called SCREEN3) has been used for many years by the U.S. EPA as a permit screening tool. It can accommodate only one emission source at a time and assumes that meteorological conditions are constant.

Urban Airshed Model

The urban airshed model (UAM) is a three-dimensional, grid-type numerical simulation model which incorporates a condensed photo-chemical kinetics mechanism for urban atmospheres. It is designed for computing ozone concentrations under short-term, episodic conditions (one or two days) resulting from oxides of nitrogen, volatile organic compounds and carbon monoxide.

Complex Terrain Dispersion Model plus Algorithms for Unstable Situations

The complex terrain dispersion model plus algorithms for unstable conditions (CTDMPLUS) is a refined, point source Gaussian air quality model for use in all stability conditions for complex terrain applications.

The next generation of air dispersion models to be used by the U.S. EPA will include AERMOD and ISC-PRIME.

AERMOD is a new platform for steady-state plume modeling. This model includes air dispersion fundamentally based on planetary boundary layer turbulence structure, scaling and concepts and can be used for both surface and elevated sources and for simple or complex terrain. It is used in conjunction with AERMAP (AERMOD Terrain Preprocessor), and AERMET (AERMOD Meteorological Preprocessor). A screening level version of AERMOD is being developed.

Industrial Source Complex Plume Rise Model Enhancements (ISC-PRIME) is a Gaussian plume model widely used to assess pollution concentration and/or deposition flux on receptors from a wide variety of sources. Meteorological data can be changed hourly. It can accommodate multiple emission sources. The model is similar to ISCST3 but contains enhanced building downwash analysis.

2.11.4   Proposed Changes to Ontario's Dispersion Models

The science of air dispersion modelling has also advanced considerably since Regulation 346 was first issued. Some of the deficiencies that the MOE has identified in the current models include the inability to address multiple sources, long-range transport and deposition, very short-term effects, very long-term effects and synergistic effects of pollutants.

In 2001, the MOE issued a proposal to replace the models in Regulation 346 with a tiered approach that would use SCREEN3 and "worst case" conditions in Tier one, then use more refined models such as AERMOD and ISC-PRIME along with regional or local meteorological data. (SCREEN3 subsequently would be replaced by a screening level version of AERMOD when that becomes available.) The MOE also has indicated that it may consider interpreting model results using a statistical technique. For example, the xth highest concentration would be used to assess compliance where the potential effect results from chronic exposure over a lifetime to odours (MOE, 2001d).

The proposal outlines the phasing-in of the new models over three to five years, since a facility may be in compliance using the Regulation 346 models and be out of compliance using the new models. The phase-in period will allow facilities to assess emissions using the new models and address compliance issues where necessary. Facilities approved under the old model will not be grandparented.

During the phase-in period, the MOE proposes to accept results of old and new models. Existing Cs of A are valid. As before, the MOE can amend or revoke a C of A at any time. For abatement and/or compliance purposes, the MOE will continue to exercise its discretion on which models to use.

2.12   Reporting

2.12.1   General Requirements

Section 13 of the EPA requires that every person who discharges into the natural environment or who is responsible for a source that discharges to the environment any contaminant in an amount, concentration or level in excess of that prescribed by the regulations shall forthwith notify the Ministry. Accordingly, anyone who emits a substance into the atmosphere in concentrations that exceed the conditions of an order or C of A must notify the MOE.

2.12.2   Monitoring and Reporting for O. Reg. 127/01

The Airborne Contaminant Discharge Monitoring and Reporting Regulation (O. Reg. 127/01) came into effect in 2001. Unlike the federal National Pollutant Release Inventory (see Section 2.2.3) that primarily targets industry, O. Reg. 127/01 includes industry with commercial facilities (universities, shopping centres, office buildings) and instructs the owner and operator of a facility to ensure compliance, imposing legal and due diligence obligations on owners who lease property.

O. Reg. 127/01 applies to all facilities located in Ontario and listed in the associated guideline entitled *Step by Step Guideline for Emission Calculation, Record Keeping and Reporting for Airborne Contaminant Discharge* (MOE, 2001a).

Facilities are organized into three classes: Class A – electricity generation, Class B – large sources, and Class C – small sources. Facilities in Classes A and B started reporting in June, 2002. Class C facilities are scheduled to start reporting in June, 2003.

The substances to be reported are organized in three tables: Table 2A, Table 2B and Table 2C. Figure 2.1 illustrates how the three facility classes and the three tables of substances determine the reporting requirements for a facility. These requirements also are described below.

Table 2A contains criteria air contaminants (NOx, sulphur dioxide, carbon monoxide, VOCs, and particulate matter) and greenhouse gases. A facility will need to report if it can reasonably be expected to meet any of the following screening criteria:

- use coal, refuse, wood or waste oil as a fuel at any time of the year;
- have a name plate capacity of more than 3 million BTU per hour;
- use 3,000 kg or more of solvent during the year;

- use 3,000 kg or more of coating materials during the year;
- use 3,000 kg of printing ink during the year; or use 5,000 kg or more of welding rods or wires during the year.

The reporting thresholds are based on quantities emitted during smog season (May 1 to September 30).

If a facility satisfies any of the screening criteria, the annual emissions of all the CAC chemicals must be calculated. If the emission of a CAC chemical exceeds the reporting threshold listed in Table 2A, the emitted quantity must be reported. If the emission of a CAC chemical is less than the threshold listed in Table 2A, the facility must use the MOE codes to indicate whether the emission is below the reporting threshold or whether there is no reportable emission.

While much of O. Reg. 127/01 is focussed on annual emissions, there are exceptions. For CAC chemicals, the amounts emitted during the year and the amounts during the smog season (May 1 to September 30) must be reported separately. For large dischargers of sulphur dioxide and NOx, these emissions must be monitored and reported quarterly. This applies to facilities where a discharge unit's name plate capacity exceeds 73 megawatts, and its annual emissions exceed 20 tonnes of sulphur dioxide or 14 tonnes of NOx.

Table 2B contains 79 manufactured, processed or otherwise used chemicals. A facility will need to report if it can be reasonably expected to meet any of the following screening criteria:

- employ or engage people who work a total of 20,000 hours or more during the year; and
- manufacture, process or otherwise use at any time a substance listed in Table 2B in an amount equal to or greater than the corresponding threshold amount for that substance.

If a facility meets the screening criteria, emissions must be monitored and the annual emitted quantities reported for all MPO chemicals that exceed the thresholds in Table 2B.

Table 2C contains the 268 substances on the NPRI substance list. Any facility required to submit a NPRI report is required to submit to the MOE that portion of the NPRI report that contains the details of atmospheric emissions.

To determine the quantities of chemicals emitted, the MOE recommends that the following types of information should be considered:

- map and/or drawing of the facility
- list of exhaust stacks/vents
- list of fuel combustion equipment
- annual fuel consumption for each fuel type
- process flow charts and production quantities

- solvent usage and type
- list of fugitive emission sources
- list of related internal environmental reports

O. Reg. 127/01 requires that all reports and records be kept for at least seven years.

Industries are responsible for making the information available to the public either at their place of business or via the Internet. In addition, the MOE has indicated that it will post the information online.

In response to industry's request for one-window reporting, Environment Canada and the MOE are conducting a one-year pilot study to harmonize reporting systems, criteria for reporting, reporting thresholds, and lists of substances for Ontario industry air pollutant releases, as well as providing software that generates both reports.

For those emission sources that need to be monitored, O. Reg. 127 refers to a guideline for continuous emission monitoring (CEM) systems (MOE, 2001b).

2.12.3   Incident Reporting

There are several sections of the EPA that impose requirements to notify various parties when a contaminant is released into the natural environment that is out of the course of normal events. For example, section 15(1) of the EPA requires that every person who discharges a contaminant or causes or permits the discharge of a contaminant into the natural environment out of the normal course of events that causes or is likely to cause an adverse effect shall forthwith notify the Ministry.

Similarly, s. 92(1) of Part X requires every person having control of a pollutant that is spilled and every person who spills or causes or permits a spill of a pollutant that causes or is likely to cause an adverse effect shall forthwith notify the MOE, the municipality or the regional municipality where the spill occurred, and the owner of the pollutant. Additional information on spills and associated reporting requirements is provided in Chapter 12.

Directly relevant to incidents that involve air quality is s. 9 of Regulation 346, which requires notification where a failure to operate in the normal manner or a change in operating conditions occurs, or a shutdown of the source or part thereof, results in the emission of air contaminants that exceeds the POI limits specified in Schedule 1 or causes an adverse effect, or exceeds visible emission criteria. Under such circumstances, the owner or operator of the source of air pollution shall immediately notify a provincial officer and provide details of such failures, change or shutdown. A written account shall be provided as soon as is practicable.

A provincial officer may authorize, in writing, the continuance of such operation for a reasonable period of time given the circumstances and may impose upon the owner or operator such terms and conditions for continued operation as the officer considers necessary.

2.13   Monitoring

2.13.1   Monitoring Objectives

Monitoring air emissions or ambient air quality may be needed or advisable for various reasons. Routine monitoring may be a requirement of a control order, C of A or other directive. An emitter may monitor as part of an internal environmental management program. Special or one-time efforts may be undertaken to investigate specific conditions. Whatever the reason, it is vital that the individuals who collect samples — whether for regulatory purposes or as part of an internal sampling program — be properly trained, follow appropriate sampling protocols and have samples analyzed using techniques with appropriate detection limits.

2.13.2   Source Testing Technologies and Frequency

There are two broad types of emission testing: continuous emission monitoring and campaign or compliance testing. Continuous emission monitoring (CEM) involves the real-time or near real-time measurement of the concentration of a pollutant in an emission stream, and the simultaneous measurement of flow rate parameters.

Instrumentation to measure concentration in real time in industrial environments is limited to a small range of pollutants: opacity, NOx, sulphur dioxide, total hydrocarbons, carbon monoxide, carbon dioxide, hydrogen sulphide and some organic compounds. CEM systems are complex and expensive to operate and maintain, and therefore are used only on the largest or more important sources. The MOE has prepared a guideline for CEM systems (MOE, 2001b).

Campaign or compliance testing refers to test programs carried out over short time periods to monitor emissions from an operating facility.

The protocols for source testing are lengthy and comprehensive. The Ontario Source Testing Code (Version 2) describes the methodologies preferred by the MOE (MOE, 1980). These include:

Method 1 – location of sampling site and sampling points

Method 2 – determination of stack gas velocity and volumetric flow rate

Method 3 – determination of molecular weight of dry stack gas

Method 4 – determination of moisture content of stack gases

Method 5 – determination of particulate matter emissions from stationary sources

There are several techniques that can be used to sample emission points such as stacks and exhausts. Particulate matter is sampled using isokinetic sample trains. Isokinetic sampling involves matching the sampling velocity with the gas velocity through alterations in the volumetric flow rate of the sample and the probe nozzle diameter.

Many organic compounds are sampled using non-isokinetic sample trains that include appropriate absorption solutions or adsorbents. Metals may be present in several forms simultaneously (adsorbed to particles, as vapours or fumes) and therefore should be sampled using a combination of isokinetic sampling and adsorbents. Dioxins, furans and trace organic compounds are usually sampled using a modified, particulate matter sampling train.

Emissions from many combustion processes are now believed to be sources of dioxins and furans. Because of concerns about potential human health effects and environmental distribution of these compounds, considerable attention is being directed at quantifying and reducing combustion emissions. If the MOE suspects that dioxins and furans may be formed during an industrial process, especially one involving high temperatures, a dioxin and furan sampling program may be required.

In addition to the Ontario Source Testing Code, the MOE frequently refers to the sampling methodologies required by the U.S. EPA. These include test methods for specific chemicals (such as NOx, volatile organic compounds, particulate matter and individual metals), and for specific types of emission points (such as facilities that use surface coatings, sewage sludge incinerators and non-ferrous smelters). Details can be found at **www.epa.gov/ttn/emc**.

2.13.3   Visible Emissions

Many emissions are visible due to the presence of liquid sprays or mists, solid particles or coloured gases. Section 8 of Regulation 346 describes how the acceptability of a visible emission is determined based on its opacity (the degree to which it obstructs the passage of light):

1. Subject to ss. 8(2), no person shall cause or permit to be caused a visible emission that obstructs the passage of light to a degree greater than 20% at the point of emission for a period of not more than four minutes in the aggregate in any 30-minute period.
2. A visible emission from a source of combustion employing solid fuel may obstruct the passage of light to a degree greater than 20% but no greater than 40% at the point of emission for a period of not more than four minutes in the aggregate in any 30-minute period.

Opacity is the most commonly used parameter for evaluating emissions from stationary sources. The most common way of determining the opacity of visible emissions is through the observations of certified observers.

The current version of Regulation 346 only recognizes observations made by trained provincial officers during the course of enforcement procedures if the results are to be used to assess compliance in a legal context. Such observations must be made in accordance with s. 7(3) of Regulation 346.

In the past, the MOE offered courses to certify people outside the MOE as visible emission observers. The classroom instruction component of the training addressed types of sources, legislation and factors such as sun angle, plume angle and point of observation. Individuals were familiarized with plumes of various colours and opacities. A certificate, valid for six months, was awarded if a sufficient number of randomly selected emissions were properly evaluated during a test at the end of the course.

Three key parameters used to assess visible emissions are duration of emission, opacity and the type of fuel involved (if relevant). At facilities where opacity is a concern, online meters can be installed that continuously record the opacity of the emission stream as it passes through a stack.

2.13.4   Ambient Air Sampling

Ambient air monitoring may be undertaken by regulatory agencies to evaluate general conditions, establish the concentrations typically present at a location, identify trends in air quality and identify regional differences or similarities. Ambient air monitoring can also be used to collect data for enforcement purposes.

The MOE maintains a network of air monitoring stations at approximately 80 locations for monitoring concentrations of various parameters including gases such as sulphur dioxide, nitrous oxide and particulate matter and its constituents, such as lead and dustfall. The MOE has recently added mercury to this list.

Annual air quality reports have been issued by the MOE since 1971. The latest report describes conditions in 2000 (MOEE, 2001e). It notes that over the past 30 years, carbon monoxide concentrations have declined by 81%, NOx concentrations have declined by 49% and nitrous oxide concentrations have declined by 23%. These declines have occurred despite increases in population, economic activity and vehicular traffic. Ground level ozone and particulate matter remain concerns, notably in the Windsor to Ottawa corridor. Both occasionally exceed AAQCs and have been on an upward trend over the past 20 years. It also is noted that more than 50% of the ground level ozone in Ontario originates in the eastern United States.

Ambient air quality surveys also can be undertaken prior to a source becoming operational and the data can be subsequently used to evaluate the effects that the source has on local air quality.

At industrial facilities, ambient air monitoring may be undertaken to assess the potential off-site migration that could occur in the event of a spill or unscheduled release. Such information can be valuable during an emergency response situation.

In most instances, ambient air concentrations are relatively low compared to concentrations in emission sources. The preferred sampling method is to draw air through a loop, cell or reaction chamber where the concentration(s) is measured and the result is available within a few moments of taking the sample. If such real-time analysis is impractical, there are several other approaches that can be employed:

- whole air samples can be pumped into bags or containers often made of special plastics that are later transported to a laboratory.
- air can be passed through a filter that traps particles. The filter can be weighed to determine the mass of material collected or the filter can be dissolved and the residue analyzed.
- air can be passed through a trap that is maintained at a relatively low temperature. The condensed material that collects in the trap can be analyzed.
- air can be bubbled through a liquid medium or passed through a packed bed of charcoal granules, synthetic material or a polyurethane plug to absorb gaseous compounds.

2.13.5   Odours

Two approaches are used in the assessment of odour. One is to monitor individual substances being emitted and compare their concentrations to odour thresholds. The other is to consider the odour potential of an entire emission. The latter is determined by an odour panel, a group of trained people who smell samples to determine if an odour is discernible. Typically, a series of dilutions is created by mixing samples of the emission with odour-free air. A positive response is produced when 50% or more of the members of the panel can detect an odour in a sample. The findings of an odour panel can be used to describe the total number of odour units associated with an emission.

While the term "odour threshold" implies that detection occurs at a specific concentration, detectability, in fact, is highly variable. An individual's sensitivity can be influenced by many factors and, as a result, odour thresholds may be reported as specific values, ranges or as geometric means and standard deviations.

To obtain a C of A where odour is a concern, the MOE requires the use of models that consider all dispersion conditions. This is not possible with the models in Regulation 346, therefore models used by the U.S. EPA must be used (see Section 2.11.3).

Table 2.4 presents a list of odour thresholds for selected chemicals. The odour thresholds for some compounds such as hydrogen sulphide are extremely low. Up to 50% of complaints made by the public to the MOE are associated with odour.

2.13.6   The Need for Quality Assurance and Quality Control

A rigorous quality assurance/quality control (QA/QC) program is an essential part of any air quality and emission testing effort. The following aspects of sample collection, handling and analysis should be addressed in a QA/QC program:

- sampling procedures (including the cleaning of equipment between sampling efforts)
- calibration of field equipment
- chain-of-custody procedures for samples
- standard reference methods for laboratory standard solutions
- method blank samples and method blank samples spiked with a standard solution be included during the course of an analytical run
- assessments of precision, accuracy and completeness

2.14   Other MOE Programs

2.14.1   Countdown Acid Rain Program

This program set emission caps for NOx and sulphur dioxide for the electricity generating sector at plants where fossil fuels are burned. These caps are set out in O. Reg. 153/99.

2.14.2   Clean-Air Plan for Industry

This program was announced late in 2001. It sets emission caps for NOx and sulphur dioxide for sectors other than electricity generation. The targeted sectors include pulp and paper, cement and concrete manufacturers, iron and steel, petroleum refining, chemical, and non-iron metal smelting.

2.14.3   Ontario's Drive Clean

This vehicle inspection program is aimed at reducing emissions of NOx, carbon monoxide and VOCs, by improving vehicle maintenance.

2.14.4   Smog Alert Program

This program is a joint effort of the MOE and Environment Canada. Smog advisories are issued when elevated ozone levels are forecast. In 2000, Ontario enhanced the program to include the issuing of a "smog watch" when there is a 50% chance of a smog day within 72 hours, and issuing a "smog advisory" when there is the strong likelihood of a smog day within the next 24 hours.

2.14.5   Air Pollution Index

The Air Pollution Index (API) is described in s. 4 of Regulation 346. It is based on the 24-hour running averages of sulphur dioxide, suspended particulate matter and the coefficient of haze. If the API is anticipated to exceed 50 for the next six hours or longer, major sources can be ordered to curtail emissions. If the API is anticipated to exceed 100 for the next six hours or longer, activities non-essential to the public can be ordered to cease operating. The API has seldom exceeded 30 since 1996 and has not reached 50 or 100.

2.14.6   Air Quality Index

The Air Quality Index (AQI) is based on hourly measurements of sulphur dioxide, ozone, nitrous oxide, total reactive sulphur, carbon monoxide, suspended particulate matter and the coefficient of haze. AQI values are released to the public as they are determined. An AQI of less than 32 is considered to be good.

2.15   Future Directions

2.15.1   Permit-by-Rule Regulations

Initially introduced as "standardized approvals regulations", permit-by-rule regulations would provide automatic approvals for routine activities that have predictable, controllable and well-understood impacts on the environment. These approval conditions would be set out in the regulations. Rather than applying for a C of A, a professional engineer or other designated professional could determine which emission sources meet the conditions in the regulations. Written documentation (in a format specified by the MOE) would take the place of and be the equivalent of a C of A. This would greatly reduce the time and effort spent preparing, submitting and processing C of A applications.

2.15.2   Emission Capping and Trading

Since 1996, the Ontario government has participated in the Pilot Emissions Reduction Trading (PERT) project to become familiar with emission capping and trading, and study the U.S. experience where these approaches have been used since the mid-1990s. Convinced that emission capping and trading will result in cost-effective reductions, the government has started to incorporate these concepts into regulations. O. Reg. 397/01 sets the caps for the electricity generating sector. These caps began in 2002 and will be fully implemented by 2007 for NOx and sulphur dioxide. Ontario emitters can trade with emitters in the eastern United States or any other sources that have reduced emissions. An emission trading code is being developed by the MOE to set out the rules for trading. It is anticipated that eventually the Ontario government will set emission caps for other types of chemicals and then distribute allowances among groups of emitters. The emitters will use whatever mechanisms they feel will allow them to comply with the cap in the most cost-effective manner. This could include improved controls, process modifications and market mechanisms (mostly trading). Emission trading also is identified as a viable alternative to reducing NOx emissions in a CCME guideline for commercial and industrial boilers and heaters.

2.15.3   Emission Performance Standards

Rather than setting limits as concentrations, the MOE is considering setting limits as a function of output. Using electricity generation as an example, an emission performance standard would be expressed as a mass of chemical emitted per megawatt-hour of power generated. The standard would apply to any facility that provides electricity to Ontario consumers.

2.15.4   Effects-based Air Quality Limits

Using a tiered approach, traditional one-hour POI limits might be used in an initial tier but subsequent tiers would use effects-based POI limits and likely reflect emissions over longer periods of time such as 24 hours. The MOE has published effects-based POI limits for a few chemicals (see Table 2.3) but these have not yet replaced the current limits.

2.15.5   Continuous Emission Monitoring

Continuous emission monitoring (CEM) will become more commonplace. MOE guidelines already are supporting the use of CEM for applications such as monitoring required by O. Reg. 127/01.

2.16   Summary

The regulatory framework for air quality and emissions to the atmosphere is in a state of transition. The changes reflect a greater emphasis being placed on pollution prevention rather than on emission control; the use of new, more sophisticated air dispersion models; and the use of risk assessment and risk management to set numerical limits. In addition, relatively new concepts such as emissions capping and trading, and permit-by-rule regulations are being given important roles.

Both the MOE and Environment Canada have substantially increased requirements for monitoring and reporting emission information. This encompasses more facilities and more chemicals. The assembled information is used to update emission inventories and databases that are publicly accessible.

References

Bowne, N.E. 1984. "Atmospheric Dispersion." In *Handbook of Air Pollution Technology*. S. Calvert and H.M. Englund, eds. Toronto: John Wiley and Sons.

Canadian Council of Ministers of the Environment (CCME). 1998. *National Emission Guidelines for Commercial/Industrial Boilers and Heaters*.

Clayton, G., and F. Clayton. 1981. *Patty's Industrial Hygiene and Toxicology*. Toronto: John Wiley and Sons.

Cohrssen, J.J., and V.T. Covello. 1989. *Risk Analysis: A Guide to Principles and Methods for Analyzing Health and Environmental Risks*. Office of the President of the United States. ISBN 0-934213-20-8.

Fazzalari, F. 1978. *Compilation of Odour and Taste Threshold Values Data*. American Society for Testing and Materials, DS 48A.

Leonardos, G. 1984. "Odour Sampling and Analysis." In *Handbook of Air Pollution Technology*. S. Calvert and H.M. Englund, eds. Toronto: John Wiley and Sons.

Ontario Ministry of the Environment (MOE). November 1980. *Ontario Source Testing Code (Version 2)*. Report ARB-66-80. Reprinted in 1991 and 1998. Now PIBS 1310.

Ontario Ministry of the Environment (MOE). June 1998. *Procedure for Preparing an Emission Summary and Dispersion Modelling Report*.

Ontario Ministry of the Environment (MOE). 2001a. *Step by Step Guideline for Emission Calculation, Record Keeping and Reporting for Airborne Contaminant Discharge*.

Ontario Ministry of the Environment (MOE). 2001b. *Guideline for the Installation and Operation of Continuous Emission Monitoring (CEM) Systems and Their Use in Reporting*.

Ontario Ministry of the Environment (MOE). March 2001c. *Discussion Paper: A Proposed Risk Management Framework for the Air Standard Setting Process in Ontario*.

Ontario Ministry of the Environment (MOE). March 2001d. *Updating Ontario's Air Dispersion Models – A Discussion Paper*.

Ontario Ministry of the Environment (MOE). 2001e. *Air Quality in Ontario, 2000 Report*. PIBS 4226e.

Ontario Ministry of the Environment (MOE). September 2001f. *Summary of Point of Impingement Standards, Point of Impingement Guidelines, and Ambient Air Quality Criteria (AAQCs)*. Air Resources Branch.

Ontario Ministry of the Environment (MOE). April 2002. *Protocol for Updating Certificates of Approval for Air Emissions*.

Pasquill, F., and F.B. Smith. 1983. *Atmospheric Diffusion*. 3rd ed. Chichesten: Ellis Horwood Ltd. ISBN 0-85312-587-2.

Stahl, W.H. 1973. *Compilation of Odor and Taste Threshold Values Data.* American Society for Testing and Materials (ASTM), Data Series DS48, Philadelphia, PA.

U.S. Environmental Protection Agency. 2001. *Guideline on Air Quality Models.* Appendix W, 40CFR, Part 51.

Table 2.1
Environment Canada Guidelines

Arctic Mining Industry Emission Guidelines, 1976

Asphalt Paving Industry National Emission Guidelines, 1975

Cement Industry National Emission Guidelines, 1974

Code of Good Operating Practice for Vinyl Chloride and Polychloride Manufacturing Operations, 1991

Environmental Code of Practice for Elimination of Fluorocarbon Emissions from Refrigeration and Air Conditioning Systems, 1996

Environmental Code of Practice for the Reduction of Solvent Emissions from Dry Cleaning Facilities, 1992

Environmental Code of Practice on Halons, 1996

Hazardous Waste Incineration Facilities Guideline, 1992

Metallurgical Coke Manufacturing Industry Emission Guidelines, 1975

National Emission Guidelines for Stationary Combustion Turbines, 1992

Packaged Incinerators National Emission Guidelines, 1978, updated 1982

Thermal Power Generation Emissions – National Guidelines for New Stationary Sources, 1993

Wood Pulping Industry National Emission Guidelines, 1979

Table 2.2
MOE Guides and Forms

Application for Approval for Air, February 2000 [blank form]

Guide for Applying for Approval (Air), January 2000

Protocol for Updating Certificates of Approval for Air Emissions, April 2002

Costs for EPA s. 9 (Air) Applications – October 1998 [blank form]

Guide: Application Costs for Air Emissions, August 1998

Summary of Point of Impingement Standards, Point of Impingement Guidelines, and Ambient Air Quality Criteria (AAQCs), September 2001

Procedure for Preparing an Emission Summary and Dispersion Modelling Report, June 1998

Step by Step Guideline for Emission Calculation, Record Keeping and Reporting for Airborne Contaminant Discharge [referenced in O. Reg. 127/01]

Guideline for the Installation and Operation of Continuous Emission Monitoring (CEM) Systems and Their Use in Reporting [referenced in O. Reg. 127/01]

Guideline A-1: Combustion and Air Pollution Control Requirements for New Biomedical Waste Incinerators

Guideline A-3: Operation of the Air Pollution Index (API), The Air Quality Index (AQI) and the Air Quality Advisory

Guideline A-5: Atmospheric Emissions from Stationary Combustion Turbines

Guideline A-7: Combustion and Air Pollution Requirements for New Municipal Waste Incinerators

Guideline A-9: NOx Emissions from Boilers and Heaters, March 2001

Guideline F-1: Particulate Emissions at New Cement Plants

Most of these documents are available from the MOE. Many can be downloaded from www.ene.gov.on.ca/envision/gp/index.htm.

Table 2.3
Point of Impingement (POI) Limits and Ambient Air Quality Criteria (AAQC)

Contaminant Name	Contaminant Code or CAS No.	Half-hour POI Limit (µg/m³)	POI Limiting Effect	Status	Annual (µg/m³)	24-Hour (µg/m³)	1-Hour (µg/m³)	10-Minute (µg/m³)	AAQC Limiting Effect
Acetaldehyde	75-07-0	500	Health	G		500			Health
Acetic acid	64-19-7	2500	Odour	S		2500			Odour
Acetone	67-64-1	48000	Odour	S		48000			Odour
Acetophenone	98-86-2	625	Odour	G			1167	850	Health and Odour
Acetylene	74-86-2	56000	Odour	S		56000	23.3		Odour
Acrolein	107-02-8	28	Health	G		15			Health
Acrylamide	79-06-1	45	Health	S		0.6			Health
Acrylonitrile	107-13-1	180	Interim[a]	S[a]	0.12				Health
Adipic acid	124-04-9	3500	Health	G		1167			Particulate
Alkyltoluene sulphonamide, N-	N/A	100	Health	G		120			Health
Allyl glycidyl ether	106-92-3	180	Health	G		60			Health
Aluminum distearate	300-92-5	100	Particulate	G		2180			Particulate
Aluminum oxide	1344-28-1	100	Particulate	G		120			Health
Aluminum stearate	7047-84-9	100	Particulate	G		2180			Health
Aluminum tristearate	637-12-7	100	Particulate	G		2180			Particulate
Ammonia	7664-41-7	3600	Odour[a]	G		100			Health and Odour
Ammonium chloride	12125-02-9	100	Particulate	S[a]		120			Health and Odour
Amyl acetate, iso-	123-92-2			G		53200			Health and Odour
Amyl acetate, n-	628-63-7			S		53200			Health
Amyl acetate, secondary	626-38-0			G		66500			(A) Health
Antimony and compounds	7440-36-0	75	Health	S		25			Health
Arsenic and compounds	7440-38-2	1	Health	G		0.3			Health
Arsine	7784-42-1	10	Health	S		5			Health
Asbestos (fibres > 5 µm in length)	1332-21-4			G		0.04 fibres/cm³			Health
Asbestos (total)	1332-21-4	see "Part VI/EPA"		G					Health
Barium - total water soluble	7440-39-3	30	Health	G		10			Health
Benzene	71-43-2	0.0033	Health	CARC					Health
Benzo(a)pyrene - single source	50-32-8			G	0.00022	0.0011			Health
Benzo(a)pyrene - all sources	50-32-8			G	0.0003				Health
Benzoic acid	68-85-0	2100	Health	G		700			Health
Benzothiazole	95-16-9	200	Health	G		70			Health
Benzoyl chloride	98-88-4	350	Health	G		125			Corrosion and Health
Benzyl alcohol	100-51-6	2640	Health	S		880			Health
Beryllium and compounds	7440-41-7	0.03	Health	G		0.01			Health
Biphenyl	92-52-4	60	Odour	G			60		Odour
Borax	1303-96-4	100	Health	S		33			Health
Boric acid	10043-35-3	100	Health	G		33			Health
Boron	7440-42-8	100	Particulate	S		120			Particulate
Boron tribromide	10294-33-4	100	Corrosion	S		35			Corrosion
Boron trichloride	10294-34-5	100	Corrosion	S		35			Corrosion
Boron trifluoride	7637-07-2	5	Health	S		2			Vegetation
Bromacil	314-40-9	30	Health	S		10			Health
Bromine	7726-95-6	70	Health	G		20			Health
Bromochlorodifluoromethane (Halon 1211)	N/A			S					Ozone depleting
Bromoform	75-25-2	165	Health	G		55			Health

Contaminant Name	Contaminant Code or CAS No.	Point of Impingement (POI) Limit			Ambient Air Quality Criteria (AAQC)				
		Half-hour POI Limit (µg/m³)	POI Limiting Effect	Status	Annual (µg/m³)	24-Hour (µg/m³)	1-Hour (µg/m³)	10-Minute (µg/m³)	AAQC Limiting Effect
Bromotrifluoromethane (Halon 1301)	75-63-8	see	"Part VI/EPA"						Ozone depleting
Butanol, iso-	78-83-1	1940	Odour	G		655	15000	2640	Odour; Health; Odour
Butanol, n-	71-36-3	2278	Odour	G		770	15000	3100	Odour; Health; Odour
Butanol, tertiary	75-65-0			UD		30300			Health
Butoxy-2-propanol, 1-	5131-66-8	9900	Health	G		3300			Health
Butyl acetate, n-	123-86-4	735	Odour	G		248	15000	1000	Odour, Health; Odour
Butyl acrylate	141-32-2	100	Particulate	G		120			Particulate
Butyl benzene sulphonamide, N-	3622-84-2	105	Health	G		35			Health
Butyl benzyl phthalate	85-68-7	450	Health	G		150			Health
Butyl stearate	123-95-5	100	Particulate	G		120			Particulate
Cadmium and compounds	7440-43-9	5	Health	S		2			(A) Health
Calcium carbide	75-20-7	20	Corrosion	G		10			Corrosion
Calcium cyanide (as total salt)	592-01-8	100	Particulate	G		120			Particulate
Calcium hydroxide	1305-62-0	27	Corrosion	S		13.5			Corrosion
Calcium oxide	1305-78-8	20	Corrosion	S		10			Corrosion
Calcium stearate	1592-23-0	100	Particulate	G		35			Particulate
Captan	133-06-2	75	Health	S		25			Health
Carbon black	1333-86-4	25	Soiling	S		10			Soiling
Carbon disulphide	75-15-0	330	Odour	S		330			Odour
Carbon monoxide	630-08-0	6000	Health	S		15700 (8 hr average)	36200		(A) see note below
Carbon tetrachloride	56-23-5	7.2	Health	G		2.4			Health
Chloramben	133-90-4	100	Particulate	G		120			Particulate
Chlordane	57-74-9	15	Health	G		5			Health
Chlorinated dibenzo-p-dioxins (CDDs) (See Table 1)	N/A	15 pgTEQ/m³	Health	G		5 pgTEQ/m³			Health
Chlorine	7782-50-5	300	Interim#	S*		10		230	Health; Odour
Chlorine dioxide	10049-04-4	85	Health	S		30			Health
Chlorodifluoromethane (Freon 22)	75-45-6	1050000	Health	G		350000			Health
Chloroform	67-66-3	300	Interim#	S*	0.2	1			Health
Chloropentafluoroethane (CFC-115)	76-15-3	see	"Part VI/EPA"	G					Ozone depleting
Chromium -di, tri- and hexavalent forms	7440-47-3	5	Health	G		1.5			Health
Citric acid	77-92-9	100	Particulate	G		120	300		Health and Particulate
Coal tar pitch volatiles - soluble fraction	8007-45-2	3	Health	G	0.2	1			Health
Cobalt	7440-48-4	0.3	Health	G		0.1			Health
Copper	7440-50-8	100	Health	S		50			Health
Cresols	1319-77-3	230	Health	S		75			Health
Cyanogen chloride	506-77-4	15	Health	G		12			Health
Cyclohexane	110-82-7	300000	Health	G		100000			Health
Dalapon sodium salt	127-20-8	100	Health	G		50			Health
Decaborane	17702-41-9	50	Health	S		25			Health
Decane, n	124-18-5			UD					Health and Odour
Decene, 1-	872-05-9	180000	Health	G		60000	60000		Health

| Contaminant Name | Contaminant Code or CAS No. | Point of Impingement (POI) Limit | | | Ambient Air Quality Criteria (AAQC) | | | | |
		Half-hour POI Limit (µg/m³)	POI Limiting Effect	Status	Annual (µg/m³)	24-Hour (µg/m³)	1-Hour (µg/m³)	10-Minute (µg/m³)	AAQC Limiting Effect
Detergent enzyme (Subtilisin)	1395-21-7	0.2	Health	G		0.06			Health
Diacetone alcohol	123-42-2	990	Odour	G		335		1350	Odour
Diazinon	333-41-5	9	Health	G		3			Health
Diborane	19287-45-7	20	Health	S		10			Health
Dibromotetrafluoroethane (Halon 2402)	124-73-2	see "Part VI/EPA"	"Part VI/EPA"	S					Ozone depleting
Dibutyl amine	111-92-2	100		UD			2645		Health
Dibutyl phthalate (DBP, di-n-butyl phthalate)	84-74-2		Health	G		50			Health
Dibutyltin dilaurate	77-58-7	100	Health	G		30			Health
Dicapryl phthalate	131-15-7	100		S		120			Particulate
Dichloro-1 1,2,2,-tetrafluoroethane, 1,2, (Freon 114)	76-14-2	2100000	Health	G		700000	see "Part VI/EPA"		Health
Dichlorobenzene,ortho-	95-50-1	37000	Health	G			30500		Health
Dichlorobenzene, para-	106-46-7	285	Health	G		95			Health
Dichlorobenzidine 3,3-	91-94-1			CARC					Health
Dichloroethane, 1,1-	75-34-3	600	Health	G		200			Health
Dichloroethylene, cis-1,2-	156-59-2	315	Health	G		105			Health
Dichloroethylene, sym-1,2-	540-59-0	315	Health	G		105			Health
Dichloroethylene, trans-1,2-	156-60-5	315	Health	G		105			Health
Diethyl amine	109-89-7			UD					Health
Diethyl phthalate (DEP)	84-66-2	100	Health	G		125	2910		Health
Diethylene glycol monobutyl ether	112-34-5					65			Health
Diethylene glycol monobutyl ether acetate	124-17-4					85			Health
Diethylene glycol monoethyl ether	111-90-0	800	Odour	G		273		1100	Odour
Diethylene glycol monoethyl ether acetate	112-15-2					1800			Health
Diethylene glycol monomethyl ether	111-77-3	800	Odour	G		1200			Health
Diethylhexyl phthalate (DEHP)	117-81-7	100	Health	G		50	see "Part VI/EPA"		Health
Difluorodichloromethane (Freon 12)	75-71-8	1500000	Health	G		500000			Health
Dihexyl phthalate (DHP)	84-75-3	100	Health	G		50			Health
Disobutyl ketone	108-83-8	470	Odour	G		3500		649	Health; Odour
Dimethyl acetamide, N,N-	127-19-5	900	Health	G		300	1840		Health
Dimethyl amine	124-40-3			UD			40		Health and Odour
Dimethyl disulphide	624-92-0	40	Odour	S					Odour
Dimethyl ether	115-10-6	2100	Odour	G		2100			Odour
Dimethyl methylphosphonate	756-79-6					875			Health
Dimethyl phthalate (DMP)	131-11-3	100	Health	G		125			Health
Dimethyl sulfoxide	67-68-5	6300	Health	G		2100			Health
Dimethyl sulphide	75-18-3	30	Odour	S			30		Odour
Dimethyl-1,3-diamino propane, N,N-	109-55-7	60	Health	G		20			Health
Dioctyl phthalate	117-84-0	100	Particulate	S		120			Particulate
Dioxane	123-91-1			UD		3500			Health

Contaminant Name	Contaminant Code or CAS No.	Point of Impingement (POI) Limit			Ambient Air Quality Criteria (AAQC)				
		Half-hour POI Limit (µg/m^3)	POI Limiting Effect	Status	Annual (µg/m^3)	24-Hour (µg/m^3)	1-Hour (µg/m^3)	10-Minute (µg/m^3)	AAQC Limiting Effect
Dioxolane-1,3	646-06-0	30	Health	G		10			Health
Diphenylamine	122-39-4	50	Health	G		17.5			Health
Diquat dibromide -respirable	85-00-7	0.096	Health	G		0.032			Health
Diquat dibromide -total in ambient air	85-00-7	0.48	Health	G		0.16			Health
Dodecyl benzene sulphonic acid	1886-81-3	100	Particulate	G		120			Particulate
Dodine	2439-10-3	30	Health	G		10			Health
Droperidol	548-73-2	3	Health	G		1			Health
Dustfall	N/A	8000 (µg/m^2)	Soiling	S	4.6 g/m^2 (annual)	7 g/m^2/30 day)			(A) Soiling
Ethanol (Ethyl alcohol)	64-17-5	19000	Odour	G			19000		Odour
Ethyl acetate	141-78-6	19000	Odour	S			19000		Odour
Ethyl acrylate	140-88-5	4.5	Odour	S			4.5		Odour
Ethyl benzene	100-41-4	3000	Health*	S*		1000		1900	Health; Odour
Ethyl ether	60-29-7	7000	Interim*	S*		8000		950	Health; Odour
Ethyl hexanol, 2-	104-76-7	600	Odour	G			600		Odour
Ethyl-3-ethoxy propionate	763-69-9	147	Odour	G		50		200	Odour
Ethylanthraquinone, 2-	84-51-5	30	Health	G		10			Health
Ethylene	74-85-1			UD		40			Vegetation
Ethylene dibromide	106-93-4	9	Health	G		3			Health
Ethylene dichloride	107-06-2	6	Health	G	0.4	2			Health
Ethylene glycol	107-21-1	350	Odour	G		12700			Health
Ethylene glycol butyl ether (Butyl cellosolve)	111-76-2	500	Odour	G		2400		500	Health;Odour
Ethylene glycol butyl ether acetate (But.cell.ace)	112-07-2	10	Odour	G		3250		700	Health;Odour
Ethylene glycol dinitrate	628-96-6	800	Health	G		3			Health
Ethylene glycol ethyl ether (Cellosolve)	110-80-5	220	Odour	G		380		1100	Health;Odour
Ethylene glycol ethyl ether acetate (Cell.ace.)	111-15-9	15	Odour	G		540		300	Health;Odour
Ethylene glycol monohexyl ether	112-25-4	100	Health	G		2500			Health
Ethylene oxide	75-21-8	0.06	Health	G		5			Health
Ethylenediaminetetra acetic acid	60-00-4	75	Health	G		120			Particulate
Fentanyl citrate	990-73-8		Health	G		0.02			Health
Ferric oxide	1309-37-1		Soiling	S		25			Soiling
Fluoridation -as total fluorides, total GS	7664-39-3					40 µg/100 cm^2/30 day			(A) Vegetation
Fluoridation -as total fluorides, total NGS	7664-39-3					80 µg/100 cm^2/30 day			(A) Vegetation
Fluorides (as HF) - gaseous – growing season GS	1664-39-3					0.34 µg/m^3			(A) Vegetation
Fluorides (as HF) - gaseous – growing season GS	7664-39-3	4.3	Vegetation	S		0.86			(A) Vegetation
Fluorides (as HF) - total, growing season GS	7664-39-3	8.6	Vegetation	S		1.72			(A) Vegetation

Contaminant Name	Contaminant Code or CAS No.	Point of Impingement (POI) Limit			Ambient Air Quality Criteria (AAQC)				
		Half-hour POI Limit (µg/m³)	POI Limiting Effect	Status	Annual (µg/m³)	24-Hour (µg/m³)	1-Hour (µg/m³)	10-Minute (µg/m³)	AAQC Limiting Effect
Fluorides (as HF) - total, growing season GS	7664-39-3					0.69 µg/m³/30 day			(A) Vegetation
Fluorides (as HF)- total, non growing season NGS	7664-39-3	17.2	Vegetation	S		3.44			(A) Vegetation
Fluorides (as HF)- total non-growing season NGS	7664-39-3					1.38 µg/m³/30 day			(A) Vegetation
Fluorides in dry forage-dry weight	7664-39-3					35 ppm/30 day ave.,* 80 ppm/30 day ave.** 60 ppm/60 day ave.***			(A) Effects on animals (A) Effects on animals (A) Effects on animals
Fluorinert 3M-FC-70	N/A	100	Particulate	G		120			Particulate
Formaldehyde	50-00-0	65	Odour	S		65			Health
Formic acid	64-18-6	1500	Health	S		500			Health
Furfural	98-01-1	1000	Odour	S					Odour
Furfuryl alcohol	98-00-0	3000	Health	S		1000	1000		Health
Glutaraldehyde	111-30-8	42	Health	G		14	35		Health
Haloperidol	52-86-8	0.3	Health	G		0.1			Health
n-Heptane	142-82-5	33000	Health	G		11000			Health
Hexachlorocyclopentadiene	77-47-4	6	Health	G		2			Health
Hexamethyl disilazane	999-97-3	5	Health	G		2			Health
Hexamethylene diisocyanate monomer	822-06-0	1.5	Health	G		0.5			Health
Hexamethylene diisocyanate trimer	4035-89-6	3	Health	G		1			Health
Hexamethylenediamine	124-09-4	48	Health	G		16			Health
Hexamethyleneimine	111-49-9	945	Health	G		315			Health
Hexane	110-54-3	35000	Health	G		12000	12000		Health
Hexylene glycol	107-41-5	14400	Health	G			668		Health
Hydrogen bromide	10035-10-6	800	Corrosion[a]	S[a]		20			Health
Hydrogen chloride	7647-01-0	100	Health	G		575			Health
Hydrogen cyanide	74-90-8	1150	Health	S		30			Health
Hydrogen peroxide	7722-84-1	90	Health	S					Health
Hydrogen sulphide	7783-06-4	30	Odour	G			30		(A) Odour
Iron - metallic	15438-31-0	10	Soiling	S		4			Soiling
Isabutyl acetate	110-19-0	1220	Odour	G		412		1660	Odour; Odour
Isopropyl ether	108-20-3	220	Odour	G		110000			Health
Isopropyl acetate	108-21-4	1470	Odour	G		500		2000	Odour; Odour
Isopropyl benzene	98-82-8	100	Odour	S		400			Health
Lead	7439-92-1	6	Health	S		2 0.7 µg/m³/30 day +			(A) Health (A) Health

Contaminant Name	Contaminant Code or CAS No.	Half-hour POI Limit (µg/m³)	POI Limiting Effect	Status	Annual (µg/m³)	24-Hour (µg/m³)	1-Hour (µg/m³)	10-Minute (µg/m³)	AAQC Limiting Effect
		Point of Impingement (POI) Limit			**Ambient Air Quality Criteria (AAQC)**				
Lead - in dustfall	7439-92-1					0.1 g/m²/30 day			Health
Lindane (Hexachlorocyclohexane)	58-89-9	15	Health	G		5			Health
Lithium -other than hydrides	7439-93-2	60	Health	S		20			Health
Lithium hydrides	7580-67-8	7.5	Health	S		2.5			Health
Magnesium oxide	1309-48-4	100	Particulate	S		120			Particulate
Magnesium stearate	557-04-0	100	Particulate	G		35			Health
Maleic anhydride	121-75-5	100		G		120			Particulate
Malathion	108-31-6	100	Health	G		30			Health
Manganese compounds (including permanganates)	7439-96-5	7.5	Health	G		2.5			Health
Mercaptans (as Methyl mercaptan) -total	74-93-1	20	Odour	S			20		(A) Odour
Mercaptobenzothiazole disulphide	120-78-5	100	Particulate	G		120			Particulate
Mercury	7439-97-6	5	Health	S		2			(A) Health
Mercury (as Hg) - alkyl compounds	7439-97-6	1.5	Health	S		0.5			Health
Metaldehyde (Acetaldehyde tetramer)	108-62-3	100	Particulate	G		120			Particulate
Methacrylic acid	79-41-4	2000	Odour	G		2000			Odour
Methane diphenyl diisocyanate (MDI)	101-68-8	3	Health	G		1			Health
Methanol (Methyl alcohol, Wood alcohol)	67-56-1	12000	Health	S		4000			Health
Metboxy-1-propyl acetate,2-	70657-70-4	4600	Health	G		1530			Health
Methoxychlor	72-43-5	100	Particulate	G		120			Particulate
Methyl acrylate	96-33-3	4	Odour	S		4			Odour
Methyl bromide	74-83-9	4000	Health	G		1350			Health
Methyl chloride	74-87-3	20000	Health	G		7000			Health
Methyl ethyl ketone (2-Butanone)	78-93-3	30000	Interim*	S*		1000			Health; Health
Methyl ethyl ketone peroxide	1338-23-4	250	Health	S		80	200		Odour
Methyl isobutyl ketone	108-10-1	1200	Odour	S		1200			Odour
Methyl mercapto aniline	2987-53-3			UD					Odour
Methyl methacrylate	80-62-6	860	Odour	S		860			Health
Methyl salicylate	119-36-8	300	Health	G		100			Health
Methyl styrene, alpha	98-83-9			UD			24000		Health
Methyl tert-butyl ether	1634-04-4	2200	Odour	G		7000			Odour
Methyl-2-hexanone, 5-	110-12-3	460	Odour			160			Health
Methyl-2-pyrrolidone, N-	872-50-4			UD			40000	630	Odour
Methyl-n-amyl ketone	110-43-0	18000	Health			4600			Health
Methylal	109-87-5		Health	G		6200			Health
Methylcyclopentadienyl manganese tricarbonyl (MMT)	12108-13-3	30	Health	G		10			Health
Methylene chloride	75-09-2	5300	Interim*	G*	44	220			Health;Health
Methylene dianiline	101-77-9	30	Health	G		10			Health
Methylene iodide	75-11-6	195	Health	G		65			Health
Methylene-bis-2-chloroaniline, 4,4-	101-14-4	30	Health	G		10			Health
Miconazole nitrate	22832-87-7	15	Health	G		5			Health

Contaminant Name	Contaminant Code or CAS No.	Half-hour POI Limit (µg/m³)	POI Limiting Effect	Status	Annual (µg/m³)	24-Hour (µg/m³)	1-Hour (µg/m³)	10-Minute (µg/m³)	AAQC Limiting Effect
			Point of Impingement (POI) Limit		**Ambient Air Quality Criteria (AAQC)**				
Milk powder	N/A	20	Soiling	S		20			Soiling and Odour
Mineral Spirits[2]	N/A	7800	Health#	S#		2600			Health
Molybdenum	7439-98-7	100	Particulate	G		120	3500	4500	Particulate
Monochlorobenzene	108-90-7	4200	Health	S					Health; Odour
Monomethyl amine	74-89-5	25	Odour	S		25			Odour
Naphthalene	91-20-3	36	Odour	G		22.5		50	Health; Odour
Naphthol, alpha-	90-15-3	100	Health	G		100			Health
Nickel	7440-02-0	5	Vegetation	S		2			(A) Vegetation
Nickel carbonyl	13463-39-3	1.5	Health	S		0.5			
Nitric acid	7697-37-2	100	Corrosion	S		35			Corrosion
Nitrilotriacetic acid	139-13-9	100	Health	S		120			Particulate
Nitrogen oxides[3]	10102-44-0	500	Health	S		200	400		(A) Health; Health
Nitroglycerin	55-63-0	10	Health	G		3			Health
Nitrosodiethylamine, N-	55-18-5			CARC					Health
Nitrosodimethylamine, N-	62-75-9			CARC					Health
Nitrous oxide	10024-97-2	27000	Health	G		9000			Health
Octane, 1-	111-65-9	45400	Odour	G		15300			Odour; Odour
Octene, 1-	25377-83-7	150000	Health	G		50000		61800	Health
Oleic acid	112-80-1	6	Health	G			5		Health
Oxalic acid	144-62-7	75	Health	G		25			Health
Oxo-heptyl acetate	90438-79-2	255	Health	G		85			Health
Oxo-hexyl acetate	88230-35-7	255	Health	G		85			Health
Ozone	10028-15-6	200	Health	S		85	165		(A) Health and Vegetation
Palladium - water soluble compounds	7657-10-1	30	Health	G		10			Health
Paraquat dichloride - respirable	1910-42-5	0.009	Health	G		0.003			Health
Paraquat dichloride - total in ambient air	1910-42-5	0.045	Health	G		0.015			Health
Penicillin	1406-05-9	0.3	Health	G		0.1			Health
Pentaborane	19624-22-7	3	Health	S		1			Health
Pentachlorophenol	87-86-5	60	Health	G		20			Health
Perchloroethylene	127-18-4	10000	Interim#	G#		360			Health
Phenol	108-95-2	100	Health	S		100			Health
Phosgene	75-44-5	130	Health	S		45			Health
Phosphine	7803-51-2	30	Particulate	S		10			Health
Phosphoric acid (as P_2O_5)	7664-38-2	100	Health	G		120			Particulate
Phosphorus oxychloride	10025-87-3	40	Health	G		12			Health
Phosphorus pentachloride	10026-13-830	30	Health	S		10			Health
Phthalic anhydride	85-44-9	100	Particulate	S		120			Particulate
Pimozide	2062-78-4	3	Health	G		1			Health
Platinum - water soluble compounds	7440-06-4	0.6	Health	G		0.2			Health
Polybutene -1-sulphone	N/A	100	Particulate	G		120			Particulate
Polychlorinated biphenyls (PCBs)	1336-36-3	0.45	Health	G	0.035	0.15			Health
Polychloroprene	25267-15-6	100		G		500			Particulate
Potassium cyanide	151-50-8	100		G		120			Particulate

Contaminant Name	Contaminant Code or CAS No.	Point of Impingement (POI) Limit			Ambient Air Quality Criteria (AAQC)				
		Half-hour POI Limit (µg/m³)	POI Limiting Effect	Status	Annual (µg/m³)	24-Hour (µg/m³)	1-Hour (µg/m³)	10-Minute (µg/m³)	AAQC Limiting Effect
Potassium hydroxide	1310-58-3	28	Corrosion	G		14			Corrosion
Potassium nitrate	7757-79-1	100		G		120			Particulate
Propanol, iso- (Isopropyl alcohol, Isopropanol)	67-63-0	24000	Odour	G		24000			Odour
Propanol, n- (Propyl alcohol)	71-23-8	48000	Health	G		16000			Health
Propionaldehyde	123-38-6	7	Odour	G		2.5		10	Odour; Odour
Propionic Acid	79-09-04	100	Odour	G			100		Odour
Propionic anhydride (as Propionic acid)	123-62-6	100	Odour	G			100		Odour
Propyl acetate n-	109-60-4	900	Odour	G		6600			Health
Propylene dichloride	78-87-5	2400	Odour	S		2400			Odour
Propylene glycol	57-55-6	100	Health	G		120			Health
Propylene glycol methyl ether	107-98-2	89000	Odour	G		30000		121000	Odour, Odour
Propylene glycol monomethyl ether acetate	108-65-6	5000	Odour	G		5000			Odour
Propylene oxide	75-56-9	450	Interim#	S#	0.3	1.5			Health; Health
Pyridine	110-86-1	60	Odour	G		150		80	Health; Odour
Quinone	106-51-4	45	Health	G		15			Health
Selenium	7782-49-2	20	Health	G		10			Health
Silane	7803-62-5	450	Health	G		150			Health
Silica -respirable (<10 µm diameter), cristabolite	14464-46-1	15	Health	G		5			Health
Silica -respirable (<10 µm diameter), quartz	14808-60-7	15	Health	G		5			Health
Silica -respirable (<10 µm diameter), tridymite	15468-32-3	15	Health	G		5			Health
Silver	7440-22-4	3	Health	S		1			Health
Sodium bisulphite	7631-90-5	100	Particulate	G		120			Particulate and Hea
Sodium chlorate	7775-09-9	18	Health	G		6			Health
Sodium chlorite	7758-19-2	60	Health	G		20			Health
Sodium cyanide	143-33-9	100	Particulate	G		120			Particulate
Sodium hydroxide	1310-73-2	20	Corrosion	G		10			Corrosion
Sodium nitrate	7631-99-4	100	Health	G		7000			Health
Stannous chloride (as Sn)	7772-99-8	30	Particulate	G		10			Health
Strontium	7440-24-6	100	Particulate	G		120			Particulate
Strontium carbonate	1633-05-2	100	Particulate	G		120			Particulate
Strontium hydroxide	18480-07-4	100	Particulate	G		120			Particulate
Strontium oxide	1314-11-0	100	Particulate	G		120			Particuate
Styrene	100-42-5	400	Odour	S		400			Health
Sulfamic acid	5329-14-6	100	Particulate	G		120			Particulate
Sulphur dioxide	7446-09-5	830	Health	S	55	275	690		(A) Health and Vegetation
Sulphur hexafluoride	2551-62-4	1800000	Health	G		600000			Health
Sulphuric acid	7664-93-9	100	Corrosion	S		35			Corrosion
Suspended particulate matter < 44 µm aero. dia.	N/A	100	Visibility	S	60++	120			(A) Visibility

Contaminant Name	Contaminant Code or CAS No.	Point of Impingement (POI) Limit			Ambient Air Quality Criteria (AAQC)				
		Half-hour POI Limit (µg/m³)	POI Limiting Effect	Status	Annual (µg/m³)	24-Hour (µg/m³)	1-Hour (µg/m³)	10-Minute (µg/m³)	AAQC Limiting Effect
Talc - fibrous	14807-96-6	5	Health	G		2			Health
Tellurium - excluding hydrogen telluride	13494-80-9	30	Health	S		10			Health
Tetrabutylurea	4559-86-8	30	Health	G		10			Health
Tetrahydrofuran	109-99-9	93000	Odour	S		93000			Odour
Tetramethyl thiuram disulphide	137-26-8	30	Health	G		10			Health
Thiourea	62-56-6	60	Health	S		20			Health
Tin	7440-31-5	30	Health	S		10			Health
Titanium	7440-32-6	100	Particulate	G		120			Particulate
Titanium dioxide	13463-67-7	100	Health	G		34			Health
Tolmetin sodium	35711-34-3	15	Health	S		5			Health
Toluene	108-88-3	2000	Odour	S		2000			Odour
Toluene diisocyanate	584-84-9	1	Health	S		0.5			Health
Total reduced sulphur (as hydrogen sulphide)	N/A	40	Odour	G			40		Odour
Tributyltin oxide	56-35-9	0.42	Health	G		0.14			Health
Trichlorobenzene, 1,2,4-	120-82-1	100	Health	G		400			Health
Trichloroethane, 1,1,1,- (Methyl chloroform)	71-55-6	350000	Health	S		115000			Health
Trichloroethylene	79-01-6	3500	Interim[#]	S[#]	23	115			Health
Trichlorofluoromethane	75-69-4	18000	Health	G		6000	see "Part VI/EPA"		Health
Trifluoroacetic acid	76-05-1	45	Health	G		15			Health
Trifluorotrichloroethane	76-13-1	2400000	Health	S		800000	see "Part VI/EPA"		Health
Trimethyl amine	75-50-3	0.5	Odour	G			0.5		Odour
Trimethylbenzene, 1,2,4-	95-63-6	500	Odour	G		1000			Health and Odour
Trimethylol propane	77-99-6	100	Health	G		1250			Health
Tripropyltin methacrylate	N/A	3	Health	G		1			Health
Vanadium	7440-62-2	5	Health	G		2			(A) Health
Vinyl chloride	75-01-4	3	Health	G	0.2	1			Health
Vinylidene chloride (1,1-Dichloroethene)	75-35-4	30	Health	S		10			Health
Warfarin	81-81-2	30	Health	G		10			Health
Whey powder	N/A	100	Particulate	G		120			Particulate
Xylenes	1330-20-7	2300	Odour	S		2300			Odour
Zinc	7440-66-6	100		S		120			Particulate
Zinc chloride	7646-85-7	12	Health	G			10		Health
Zinc stearate	557-05-1	100	Particulate	G		35			Health

Table 3# - Future Effects-based POI limits with current interim values subject to RM Framework for Air Standards (currently under development)

Contaminant Name (µg/m³)	Contaminant CAS No.	Future Effects-based POI Limit	Limiting Effect
Acrylonitrile	107-13-1	1.8	Health
Ammonia	7664-41-7	300	Health
Chlorine	7782-50-5	30	Health
Chloroform	67-66-3	3	Health
Ethyl benzene	100-41-4	1400	Odour
Ethyl ether	60-29-7	700	Odour
Hydrogen chloride	7647-01-0	60	Health
Methyl ethyl ketone (2-Butanone)	78-93-3	3000	Health
Methylene chloride	75-09-2	660	Health
Mineral spirits	N/A	3000	Odour
Perchloroethylene	127-18-4	1080	Odour
Propylene oxide	75-56-9	4.5	Health
Trichloroethylene	79-01-6	350	Health

TERMS:
[1] = Carbon monoxide AAQC is for an 8-hour average based on high background levels from automobiles
[2] = Mineral spirits are petroleum distillate mixtures of C_1-C_{12} hydrocarbons, with boiling points ranging from 130-220 °C and flash points ranging from 21-60 °C. Please see Rationale document: "Ontario Air Standards for Mineral Spirits" for further detail.
[3] = NOx (Nitrogen Oxides) are assumed to be the sum of nitrogen dioxide and nitrogen monoxide. AAQCs are based on nitrogen dioxide.
S = Air Quality Standard, G = Guideline, CARC = Carcinogen, UD = Under Development, or odour threshold review.
A = AAQC Chemicals listed in Regulation 337 (formerly Regulation 296) under the Environmental Protection Act.
Part VI/EPA ="Part VI/EPA" refers to Part VI of the Ontario Environmental Protection Act R.S.O. 1990, C. E-19, which addresses the manufacture, use, storage, disposal, etc., of ozone depleting substances.
N/A = Not Available
GS =Growing Season   May 1 - September 30- Northern Ontario,
                          Mid-Ontario & N Regions
                          April 1 - October 31 - Southern Ontario, SW, WC,
                          E & C Regions
NGS = Non Growing Season October 1 - April 30 - Northern Ontario, Mid Ontario & N
                          Regions
                          November 1 -March 31 - Southern Ontario, SW,
                          WC, E & C Regions

* average monthly results for growing season
** average results for any single month
*** average of 2 consecutive months
+ = arithmetic mean, ++ = geometric mean
# = Status of Standard/Guideline is interim, pending the outcome of the Risk Management (RM) Framework for Air Standards (currently under development). See Table 3 for list of pending future Effects-based limits.

Reference: MOE, 2001f.

Table 2.4

Some Odour Recognition Thresholds

Chemical	Recognition Odour Threshold (ppm)	Reference
acetone	100.0	Leonardos, 1989
acrolein	0.21	Leonardos, 1989
ammonia	47.0	Leonardos, 1989
benzene	4.7	Leonardos, 1989
cellosolve	0.550	Fazzalari, 1978
cellosolve acetate	0.14	Fazzalari, 1978
cyclohexanone	0.120	Fazzalari, 1978
hydrogen sulfide	0.00047	Leonardos, 1989
methyl isobutyl ketone (MIBK)	0.28	Fazzalari, 1978
nitrobenzene	0.0047	Leonardos, 1989
n-butanol	1.0	Fazzalari, 1978
phenol	0.047	Stahl, 1973
toluene	5.0	Fazzalari, 1978
trichloroethylene	21.4	Leonardos, 1989
V.M. and P. naphtha	0.86	Clayton and Clayton, 1981

Figure 2.1

Quick Reference for Reporting under O. Reg. 127

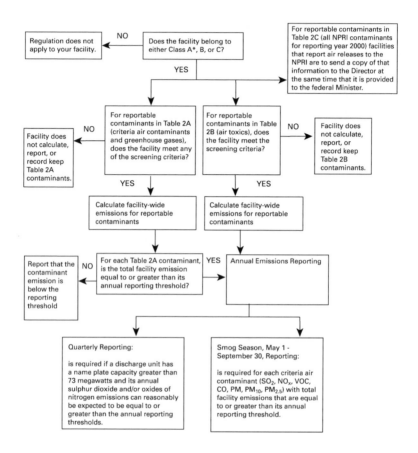

*   For Class A, this regulation does not apply to a generation facility that has a generating capacity of
1 megawatt or less or that sells 10 percent or less of its total electricity generated to the IMO-administered
markets.

Reference: MOE, 2001a.

Chapter 3

Water Quality and Liquid Discharges

3.1 Overview

Environmental legislation in Canada has addressed water quality issues longer than any other aspect of environmental management; the federal Fisheries Act predates confederation. Environmental legislation in Ontario started with the Ontario Water Resources Act in 1956, and the origins of the MOE can be traced back to the Ontario Water Resources Commission.

Today, all three levels of government are involved in water quality management and the regulation of liquid discharges in Ontario. Regulatory requirements are laid out in a relatively comprehensive combination of legislation, policies, guidelines, numerical limits and municipal by-laws.

While the earliest legislation was aimed predominantly at protecting fisheries, the current regulatory regime addresses the quality of effluents discharged to receiving waters and sewers, protecting drinking water supplies and ensuring that water resources are suitable for a wide range of uses.

3.2 Federal Acts, Regulations, Agreements and Initiatives

3.2.1 Canadian Environmental Protection Act

The goal of the renewed Canadian Environmental Protection Act (CEPA, 1999) is to contribute to sustainable development through pollution prevention and to protect the environment, human life and health from the risks associated with toxic substances.

Table 1.2 provides a summary of the key aspects of CEPA, 1999. Those related to water quality and liquid discharges include:

- the provisions in Part 4 to require Pollution Prevention (P2) Plans for substances specified in the List of Toxic Substances in Schedule 1 or identified in other sections of the Act;
- the provisions in Part 5 to regulate toxic substances and to virtually

eliminate releases to the environment of those toxic substances that are persistent and bioaccumulative;
- the provisions in Part 7 to protect marine environments from land-based sources of pollution (not as relevant for Ontario) and to address international water pollution issues; and
- the provisions in Part 8 that provide Environment Canada with additional regulatory powers related to spills, explosions and leaks. Environment Canada can request emergency plans for accidental releases of toxic substances.

3.2.2   CEPA Regulations

Numerous regulations under CEPA, 1999 pertain to liquid effluents and releases to aquatic environments:

- Chlor-Alkali Mercury Release Regulations (SOR/90-130)
- Disposal at Sea Regulations (SOR/2001-275)
- Persistence and Bioaccumulation Regulations (SOR/2000-107)
- Phosphorus Concentration Regulations (SOR/89-501)
- Prohibition of Certain Toxic Substances Regulations (SOR/96-237)
- Pulp and Paper Mill Defoamer and Wood Chip Regulations (SOR/92-268)
- Pulp and Paper Mill Effluent Chlorinated Dioxins and Furans Regulations (SOR/92-267)
- Regulations respecting Applications for Permits for Disposal at Sea (SOR/2001-276)

Two are highlighted below because recent amendments are indicators of how regulatory expectations under CEPA 1999 are evolving.

SOR/92-268 restricts the content of dibenzofurans and dibenzo-para-dioxins (both appear on the CEPA, 1999 List of Toxic Substances) in defoamers that are manufactured, imported, offered for sale or sold for use in Canadian pulp and paper mills. Reporting requirements include quarterly reports for any person who manufactures, imports, offers for sale or sells defoamers for use in mills in Canada. In addition, five years' transaction records of all defoamer products must be maintained by anyone who manufactures, imports, offers for sale or sells defoamers for use in mills in Canada.

SOR/92-267 requires that all pulp and paper mills in Canada shall not discharge into the environment any final effluent that contains measurable concentrations of 2,3,7,8-tetrachlorodibenzo-paradioxins or 2,3,7,8-tetra-chlorodibenzofurans. This is consistent with the CEPA, 1999 goal of virtually eliminating these persistent, bioaccumulative substances.

In August, 2002, the Environment Minister announced proposed Environmental Emergency (E2) regulations. The regulations would have an initial target list of 174 substances and encompass an estimated 1,500 facilities. It is anticipated that the regulations will be finalized by 2003.

3.2.3   Fisheries Act

The Fisheries Act prohibits the depositing (or permitting the deposit) of a deleterious substance in any type of waters frequented by fish or in any other place under circumstances where the substance could enter the water. The requirements of the Fisheries Act apply to all fish habitat, not just places where fish are present. A deleterious substance is defined by the Act as:

- any substance that, if added to water, would degrade or alter the quality of that water so that it is rendered harmful to fish; and
- any water that contains a substance in which quantity or concentration would, if added to water, degrade the quality of the water and, therefore, cause harm to fish.

Since the early 1990s, considerable effort has been directed toward developing appropriate methodologies for sampling and testing effluents, and interpreting the results of those tests. This has culminated in the development of Environmental Effects Monitoring (EEM) and regulatory requirements to conduct EEM studies for some industrial sectors. Since 1992, all new effluent regulations under the Fisheries Act have required regulated sites to conduct EEM studies. Chapter 13 presents additional information on federal requirements concerning the aquatic toxicity tests used to identify deleterious substances.

3.2.4   Fisheries Act Regulations and Guidelines

For certain industries, regulations have been developed under the Fisheries Act to allow the deposit of certain substances in prescribed amounts or concentrations into waters frequented by fish.

- Chlor-Alkai Mercury Liquid Effluent Regulations (C.R.C. 1978, c. 811)
- Meat and Poultry Products Plant Liquid Effluent Regulations (C.R.C. 1978, c. 818)
- Metal Mining Liquid Effluent Regulations (C.R.C. 1978, c. 819)
- Metal Mining Effluent Regulations (SOR/2002-222)
- Petroleum Refinery Liquid Effluent Regulations (C.R.C. 1978, c. 828)
- Port Alberni Pulp and Paper Effluent Regulations (SOR/92-638)
- Potato Processing Plant Liquid Effluent Regulations (C.R.C. 1978, c. 829)
- Pulp and Paper Effluent Regulations (SOR/92-269)

The pulp and paper regulations and the metal mining regulations are highlighted below to show how regulatory expectations are evolving. The regulations for meat and poultry products facilities and potato processing facilities are identified for repeal.

The two pulp and paper effluent regulations apply to pulp and paper mills discharging effluent to receiving water where the quantity of biochemical oxygen demanding matter (measured as BOD) is higher than 5,000 kg per day, and to mill discharge to off-site treatment facilities and account for more than 20% of the total quantity of BOD received by the treatment facility.

Every three years, all mills are required to provide Environment Canada with an EEM report and supporting data based on tests conducted in accordance with requirements published by Environment Canada. These studies provide information as to whether discharges of deleterious substances in water frequented by fish have altered, disrupted or destroyed fish habitat. An EEM study also provides information to evaluate the need for further control measures by evaluating the effectiveness of existing control measures and by assessing changes in the receiving environment.

Environment Canada has been proposing amendments to the pulp and paper regulations since 1997. Consideration is being given to streamlining the regulations without altering the discharge limits, for example, revoking the regulation for the Port Alberni mill and regulating that mill by the generic regulation. Changes to the EEM document also are being proposed. More details about EEM are provided in Section 13.3.

The regulations for liquid effluents from metal mining operations are based on the application of "best practicable technology" and set limits for the discharge of deleterious substances (arsenic, copper, lead, nickel, zinc, total suspended matter, radium-226 and pH) in liquid effluents from base metal, uranium and iron ore mines. They apply to all new, expanded and reopened mines after 1977. The regulations do not apply to gold mines using a cyanide process because a demonstrated technology for the treatment of cyanide-bearing effluents did not exist when the regulations were promulgated in 1977.

The current regulations (C.R.C. 1978, c. 819) are to be repealed in December, 2002. At that time, the Metal Mining Effluent Regulations (SOR/2002-222) will come into force and apply to all mines. The new regulations will introduce more comprehensive and stringent effluent quality standards based on "best available technology that is economically achievable". Monitoring and reporting will include quarterly effluent quality monitoring, quarterly monitoring of the water quality of the receiving waters, monthly monitoring of acute toxicity and biannual monitoring of chronic toxicity. All mines will need to conduct extensive EEM studies (see Section 13.3 for details).

Several guidelines have been produced to help dischargers in several industrial sectors understand regulatory requirements. There are guidelines for liquid effluents from fish processing operations, meat and poultry processing, metal mines, petroleum refineries and potato processing plants. Many of these guidelines were produced in the 1970s.

3.2.5   Other Federal Acts

There are other federal acts that pertain to water, but are seldom relevant in the context of environmental compliance in Ontario. For example, the Canada Waters Act contains provisions for formal consultation and agreements with the provinces. The International River Improvements Act provides for the licensing of activities that may alter the flow of rivers into the United States.

There also are federal acts that address shipping such as the Canada Shipping Act and the Navigable Waters Protection Act, and acts that address waters outside Ontario such as the Arctic Waters Pollution Prevention Act and the Yukon Waters Act. These are not relevant to environmental compliance in Ontario.

3.2.6   Other Agreements and Initiatives

The Canada-Ontario Agreement (COA) is the framework by which the federal and provincial governments work co-operatively to understand, restore and protect the environmental quality of the Great Lakes Basin ecosystem. COA is a tool by which the two levels of government can meet some of their commitments to the Canada-U.S. Great Lakes Water Quality Agreement. Priority areas such as reducing nutrient loadings, creating remedial action plans and lakewide management plans as well as the virtual elimination of persistent toxic substances are reflected in several annexes of the Great Lakes Water Quality Agreement as well as in the COA (MOE, 2001a)

The COA focuses on cleaning up the following areas of concern: Thunder Bay, Nipigon Bay, Jackfish Bay, Peninsula Harbour, St. Mary's River, Spanish Harbour, Severn Sound, St. Clair River, Detroit River, Wheatley Harbour, Niagara River, Hamilton Harbour, Toronto and region, Port Hope Harbour and St. Lawrence River.

The Great Lakes/St. Lawrence Pollution Prevention Initiative was announced in March, 1991. It is dedicated to the source reduction of toxic pollutants and includes a bilateral partnership with the United States. There are three program areas: the Binational Lake Superior Program, sectoral programs, and the Great Lakes Pollution Prevention Centre.

The first sectoral program is the Automotive Pollution Prevention Program (APPP), a bilateral effort directed at the voluntary reduction in toxic substance use, generation or release. The Canadian Automotive Manufacturing Pollution Prevention Project is a component of the APPP and involves the federal Department of Environment, MOEE and the participating member companies of the Motor Vehicle Manufacturers' Association (MVMA). A similar project is underway in the United States. The goal of the project is to produce a verifiable reduction of persistent toxic substances as well as other environmental contaminants (toxic substances) used, generated or released by the participating member companies of the Canadian MVMA.

As noted in Section 3.2.1, CEPA, 1999 has a section devoted to pollution prevention and the development of P2 plans (both those undertaken voluntarily and those required by notice under CEPA, 1999). A P2 planning handbook has been prepared (PWGSC, 2001). In May, 2002, the Environment Minister proposed that P2 plans be required of facilities that use acrylonitrile (largely the sector that manufactures synthetic rubber). Earlier in 2002, Environment Canada indicated that it was developing a P2 planning document for dichloromethane. That initiative would target five industrial sectors: aircraft paint stripping; flexible polyurethane foam blowing, pharmaceuticals, industrial cleaning and adhesive formulation. The overall objective of P2 plans is to reduce releases using best available technologies that are economically achievable. More details on P2 initiatives at the federal level can be obtained at **www.ec.gc.ca/nopp**. Another resource is the Canadian Pollution Prevention Roundtable (CPPR), which was formed in 1997. Its mission is to promote the shift to pollution prevention. More information can be found at **www.c2p2online.com**.

3.2.7   Federal Water Quality Guidelines

Since the mid-1980s, the Canadian Council of Ministers of the Environment (CCME) and its predecessor the Canadian Council of Resource and Environment Ministers have developed limits (usually numerical) collectively referred to as Canadian Water Quality Guidelines (CWQGs). The CWQGs are divided into four categories:

- guidelines for the protection of aquatic life. These are intended to avoid negative effects in plants and animals that live in lakes, rivers and oceans (CCME, 2001a).
- guidelines for the protection of agricultural water uses. These are intended to avoid negative effects due to chemicals in water used for irrigation or watering livestock (CCME, 1999).
- guidelines for drinking water quality establish maximum acceptable concentrations for more than 85 physical, chemical and biological characteristics of water quality (CCME, 2001b).
- guidelines for recreational water help protect the health of people who use water for activities such as swimming, diving, boating and fishing and deal mainly with potential health hazards such as infections transmitted by micro-organisms, aesthetics and nuisance conditions (HWC, 1992).

The CCME also has developed protocols for setting guidelines for the protection of aquatic life (CCME, 1991) and for the protection of agricultural uses (CCME, 1993).

The CCME cautions that the guidelines should not be regarded as values suitable for all locations or regions across the country. Various local conditions need to be taken into account such as assimilative capacity and the sensitivity of local species and habitat. CCME guidelines can be adopted by a provincial or territorial government. Summary tables of the CCME guidelines and supporting protocols are available at **www.ec.gc.ca/ceqg-rcqe**.

3.3   Provincial Acts, Policies and Guidelines

3.3.1   Environmental Protection Act

The Environmental Protection Act (EPA) is broadly directed toward the protection and conservation of the natural environment. Much of the text does not specifically mention water quality or liquid discharges, but the broad scope of this legislation encompasses water quality and the protection of aquatic environments.

Section 14(1) of the EPA states that no person shall discharge a contaminant or cause or permit the discharge of a contaminant into the natural environment that causes or is likely to cause an adverse effect. While the EPA definition of adverse effect includes many components, some of the most appropriate in the context of water management include:

- impairment of the quality of the natural environment for any use that can be made of it
- injury or damage to property or to plant or animal life
- an adverse effect on the health of a person
- loss of enjoyment of normal use of property
- interference with the normal conduct of business

Other sections of the EPA address water quality indirectly. For example, Part IV deals with the discharge of waste upon or over ice.

Despite its initial orientation toward the total natural environment, the EPA has become a cornerstone in the control of liquid discharges because it embraces all legislation for the Municipal-Industrial Strategy for Abatement (MISA) program (see Section 3.3.2).

3.3.2   Sector-Specific Regulations (MISA Program)

Several regulations concerning effluent monitoring and limits for specific industrial sectors were developed in the 1990s under the Municipal-Industrial Strategy for Abatement (MISA) program. The regulations are:

- Effluent Monitoring – General (O. Reg. 695/88)
- Electric Power Generation Sector (O. Reg. 215/95)
- Industrial Minerals Sector (O. Reg. 561/94)
- Inorganic Chemical Sector (O. Reg. 64/95)
- Iron and Steel Manufacturing Sector (O. Reg. 214/95)

- Metal Casting Sector (O. Reg. 562/94)
- Metal Mining Sector (O. Reg. 560/94)
- Organic Chemical Manufacturing Sector (O. Reg. 63/95)
- Petroleum Sector (O. Reg. 537/93)
- Pulp and Paper Sector (O. Reg. 760/93)

All have subsequently been amended.

The goal of the MISA program is the virtual elimination of persistent toxic substances from all discharges to Ontario's waterways. Since the MISA regulations came into effect, the following reductions in pollutant loadings to Ontario waterways from 1990 to 1998 have been reported (MOE, 2002):

- Benzo(a)pyrene, mainly from the iron and steel sector, has been reduced by 74% and total polycyclic aromatic hydrocarbons (PAHs) reduced by 80%
- the pulp and paper sector reduced by 82% the adsorbable organic halides (AOX) B, an indicator of chlorinated organic compounds generated as a result of the bleaching process. In addition, chloroform was cut by 99%, toluene reduced by 71%, and dioxin and furan compounds were reduced to levels so low that they could not be detected
- the organic chemical manufacturing sector reduced compounds such as benzene, vinyl chloride, ethylbenzene and carbon tetrachloride by 78%. More than 90% of hexachlorobenzene was eliminated and mercury was reduced by 70%

O. Reg. 695/88 addresses various aspects of effluent monitoring such as analytical procedures, sampling protocols, reporting requirements and flow measurement.

In general terms, the other MISA regulations consist of the following sections:

Part I   – General
Part II   – Sampling Points
Part III   – Calculation of Loadings
Part IV   – Parameter and Lethality Limits
Part V   – Monitoring
Part VI   – Effluent Volume
Part VII   – Storm Water Control Study
Part VIII   – Records and Reports
Part IX   – Commencement and Revocation Provisions

Some key components that appear in the MISA regulations are discussed below. Sector-specific regulations may be slightly different.

1. **Assessment parameter** – A parameter for which there is no limit, but the regulation requires that it be monitored.
2. **Limited parameter** – A parameter for which a numerical limit is specified in the regulation.
3. **Sample points** – Sampling points must be established for each process effluent and cooling water effluent stream so that accurate estimates can be made of the discharge of each limited parameter and assessment parameter. All process effluent discharges must have a sampling point that is used in accordance with the regulation.
4. **Sample collection** – The publication *Protocol for the Sampling and Analysis of Industry/Municipal Wastewater* must be followed (MOEE, 1993a). Daily samples are to be collected during an operating day (in some circumstances it may be any period of 24 consecutive hours, while in others cases, samples must be collected between 7:00 A.M. and 10:00 A.M.), weekly samples are not to be collected within four days of each other (all process samples must be collected on the same day), and quarterly samples are not to be collected within 45 days after the previous weekly sample. Process effluent and cooling water do not have to be collected if there is no industrial processing going on at the plant.

   In some cases, the regulations require on-line measurement of pH. Duplicates, travelling blanks and travelling spiked blanks are required at different frequencies as part of the quality assurance/quality control program.
5. **Loading calculations** – If the actual analytical result is > 1/10 the method detection limit (MDL), then the analytical result must be employed when calculating loadings. If the actual analytical result is < 1/10 the MDL, then the value "zero" must be employed. Monthly average loadings are to be calculated using the arithmetic mean of the daily loadings.
6. **Loading limits** – The discharger is responsible for ensuring that each daily and monthly process effluent plant loading for each parameter does not exceed the values specified in the regulation. Each discharger must also control the quality of each process effluent to ensure that the pH value of any sample collected at a process effluent sampling point is within 6.0 and 9.5.
7. **Toxicity testing** - Acute lethality shall be tested using *Daphnia magna* and rainbow trout. Lethality tests shall be carried out as a single concentration test using 100% effluent. A period of 15 days is required between the collection of toxicity samples.

   In addition to the acute lethality testing, dischargers may be required to perform chronic toxicity testing using *Ceriodaphnia dubia* and fathead minnow. (See Chapter 13 for additional information about aquatic toxicity testing.)

8. **Toxicity limits** – The quality of each process effluent monitoring stream and each cooling water effluent monitoring stream shall not result in a mortality of more than 50% of the test organisms (*Daphnia magna* and rainbow trout) in 100% effluent.

   If toxicity tests indicate non-toxic conditions for 12 consecutive months, the frequency of testing can be reduced to quarterly. The frequency goes back to monthly if any of the quarterly results indicate that the discharge is toxic.

9. **Flow measurement** – Flow measurement devices must provide for an accuracy of within ± 15% for process effluents and ± 20% for cooling water effluents.

10. **Storm water** – The goal of storm water control is to reduce contaminant loading to the maximum extent practicable and to ensure that storm water discharges are not acutely lethal. Storm water is defined as rainwater runoff, snow melt, surface runoff and natural drainage from a plant site.

    MISA regulations do not set limits for storm water discharges, but require a storm water control study (SWCS). Key components of a SWCS include:

    - record-keeping and reporting to determine quantity and quality of storm water discharges
    - identifying sources of storm water contamination
    - identifying the need to control based on nature of problem(s)
    - evaluating prevention and control measures
    - identifying preferred prevention and control measures
    - implementing control measures

    Additional information on undertaking a SWCS can be obtained from Protocol for Conducting a Storm Water Control Study (MOEE, 1993b).

11. **Record keeping** – Records of all monitoring for limits (concentration and loading), toxicity, flow and quality control stipulated by the regulation must be stored in an electronic format acceptable to the MOE director. The discharger must also keep records of all problems or malfunctions related to sampling, chemical analyses, on-line analysis for pH measurement, lethality tests, flow measurement or other problems encountered that interfere with fulfilling the requirements of the regulation. The duration and course of each malfunction along with a description of any remedial action taken must also be kept.

12. **Reporting** – A discharger must notify the director, during normal business hours and as soon as the results are available, if a daily plant loading or monthly average plant loading exceeds the limit, or if the lethality limits are exceeded.

Information must be submitted quarterly in an electronic format acceptable to the MOE director along with a hard copy signed by the direct discharger no later than 45 days after the end of the quarter.

On or before the first day of June in each year, the discharger must prepare a report relating to the previous calendar year and include summaries of:

- plant loadings
- monitoring data
- concentrations that exceed limits
- incidents where sample points were by-passed

These reports must be available to any person at the plant and a copy must be sent to the MOE director.

3.3.3   Ontario Water Resources Act

The Ontario Water Resources Act (OWRA) contains a general prohibition against the discharge of any material into water. Section 30(1) requires that "every person that discharges or causes or permits the discharge of any material of any kind into or in any waters or on any shore or bank thereof or into or in any place that may impair the quality of the water of any waters is guilty of an offence."

Section 28 of the OWRA indicates that the quality of water shall be deemed to be impaired if any material (or any derivative of such material) is discharged or deposited and causes or may cause injury to any person, animal, bird or other living thing as a result of the use or consumption of any plant, fish or other living matter or thing in the water or in the soil in contact with the water.

Under Section 33(2)(a), it is an offence to place, discharge or allow to remain within an area defined as a source of public water supply, any material that may impair the quality of the water.

Several regulations have been promulgated under the OWRA. Those that pertain to drinking water are discussed in Section 3.3.4. Those that pertain to water and sewage works are described in Section 3.5. Those that pertain to obtaining approvals to take water and to discharge water or waste water are also described in Section 3.5.

In addition, there are several documents under the OWRA that describe water management polices and water quality limits. These are described in Section 3.3.5.

3.3.4   Drinking Water Regulations

In Ontario, two key regulations under the OWRA govern drinking water protection:

- Drinking Water Protection – Smaller Water Works Serving Designated Facilities (O. Reg. 505/01)
- Drinking Water Protection – Larger Water Works (O. Reg. 459/00)

Examples of facilities covered by the regulation for smaller waterworks include senior care facilities, nursing homes, retirement homes, childcare facilities, nurseries, schools, hospitals, shelters, universities, colleges and homes for special care needs where water treatment or distribution system is used for human consumption.

Key requirements in both regulations include:

1. Any service water or foreign material must be prevented from flowing into or entering into a well that is used as a water source.
2. If the system collects water from a ground water source, the owner is required to ensure that the chlorination or other disinfection equipment (that is equivalent or better than chlorination) complies with the regulation.
3. If the system collects water for a surface water source, the owner is required to ensure that the operation of the filtration and disinfection equipment meets the requirement set out in the regulation.
4. The plumbing at any school or nursing home must be flushed weekly on the first day that the facility is opened each week.
5. The owner is required to have trained personnel perform at least a weekly check of all water treatment equipment.
6. If chlorination is not used to treat water, a water sample must be analyzed weekly, whereas if chlorination is used, a sample must be analyzed every two weeks.
7. A monthly sample must also be collected and analyzed from the water source prior to treatment.
8. Every five months chemical parameters are to be analyzed as set out in the regulation.
9. An accredited laboratory must do all analysis.
10. If a water sample fails to meet applicable criteria, notice must be immediately provided to the local medical officer of health, the MOE and the operator of the designated facility. A warning notice must be posted at each designated facility served by the water treatment and distribution system.

For the purpose of the regulations, a laboratory is an accredited laboratory for a parameter if the laboratory is accredited for analysis of that parameter by the Standards Council of Canada or, in the director's opinion, is equivalent to the accreditation by the Standards Council of Canada.

3.3.5   Water Management Policies

Numerous MOE publications describe policies for managing various types of water features and aquatic environments. These do not have force of law except when incorporated into a Certificate of Approval (C of A) or order issued by the MOE to a specific facility owner or operator. While some of the key publications are named below, Table 3.1 identifies others.

For surface water and ground water, a key document is *Water Management: Policies, Guidelines, Provincial Water Quality Objectives* (MOEE, 1994a).

The underlying theme of the policies is the protection and preservation of the province's water resources. Principles behind the policies include: using an ecosystem approach that considers the cumulative effects on the environment; setting requirements for water that must take into consideration the protection of other elements such as land and air; preventing pollution is more desirable than end-of-pipe treatment; and banning or phasing out certain hazardous substances (often referred to as persistent, bioaccumulative toxic substances).

For surface water quality, the main goal is to protect, preserve and restore it to permit the greatest number of uses. Since water that protects aquatic life and recreation usually is suitable for other uses, this goal often is stated as ensuring that surface waters are of a quality that is satisfactory for aquatic life and recreation. As a result, meeting the provincial water quality objectives (PWQO — described in Section 3.3.6) is the minimum requirement.

The goal also is expressed in five policies:

1. In areas where water quality is better than PWQO, water quality shall be maintained at or above the objectives.
2. In areas where water quality does not meet the objectives, all practical measures shall be taken to upgrade water quality to meet the objectives.
3. For certain hazardous substances that have been banned by the MOE, releases shall be prevented. The banned hazardous substances include aldrin, chlordane, Kepone, DDT, Dieldrin, Endrin, Mirex, polybrominated biphenyls, PCBs, polychlorinated terphenyls and Toxaphene.
4. For hazardous substances that have not been banned, ensure that special measures are taken on a case-by-case basis to minimize releases.
5. Mixing zones (the zone in the receiving water where the effluent mixes with the water that is there naturally) should be as small as possible and should not interfere with beneficial uses. Mixing zones are not to be used as an alternative to reasonable and practical treatment. Conditions within a mixing zone must not result in irreparable damage to ecosystem integrity and human health or interfere with other water uses.

Where it is necessary or unavoidable to discharge an effluent into the environment, limits will be set and incorporated into legally enforceable documents such as a C of A. These limits are used most often for municipal and industrial effluent releases but may also be applied to cooling water, storm water or other liquid discharges. Detailed directions for setting effluent requirements are described in *Deriving Receiving-Water Based, Point-Source Effluent Requirements for Ontario Waters* (MOEE, 1994b).

Some surface waters have additional considerations. For the Great Lakes, Ontario has agreed that the more stringent of the objectives described in the Great Lakes Water Quality Agreement and the provincial objectives are used to assess and maintain water quality in the Great Lakes (see Section 3.2.6).

For ground water quality, the main policy is to protect ground water for the greatest number of beneficial uses. The emphasis is on pollution prevention since ground water remediation is usually difficult and expensive.

Various types of activities that can affect ground water require MOE approval under the EPA or OWRA. Examples include waterworks, wastewater treatment facilities, landfills, spray irrigation, sludge utilization practices, septic tank systems and mining operations. Requirements usually are established on a case-by-case basis and described in a C of A. The document entitled *Incorporation of the Reasonable Use Concept into MOEE Groundwater Management Activities* (MOEE, 1994c) addresses ground water concerns around landfills, exfiltration lagoons and large subsurface disposal systems.

There are activities that do not require specific approval under the EPA or OWRA but that can have an impact on ground water. Examples include crop fertilization, road de-icing, closed landfills, leaks and site remediation activities. The MOE will require the treatment or elimination of impacts from unregulated sources to prevent or stop adverse effects or degradation of water quality.

While the focus of water management often is on water quality, the MOE also addresses water quantity. Its main policy is that the taking and using of water should be conducted in ways that ensure the fair sharing and conservation of water in the province. The main involvement of the MOE is to issue permits to take water under the OWRA. When water is to be taken from the ground or a surface feature at a rate in excess of 50,000 L per day, a permit is required. This includes temporary activities such as testing a well. Taking water can be prohibited except if the water is taken by an individual for ordinary household purposes, livestock watering and firefighting. Where taking water interferes with existing supplies, the party taking the water is required to provide a temporary supply to those experiencing the difficulties. Details are described in *Guide for Applying for Approval of Permit to Take Water* (MOEE, 2000).

3.3.6   Water Quality Standards and Objectives

Water Management: Policies, Guidelines, Provincial Water Quality Objectives (MOEE, 1994a), described in Section 3.3.5, introduces the provincial water quality objectives (PWQO), numerical and narrative criteria for assessing the quality of water in surface water and for ground water where it discharges to the surface. The PWQO are set to provide water quality that is protective of all forms of aquatic life and life cycles during indefinite exposure to the water. The PWQO for protection of recreational water uses are based on public health and aesthetic considerations. The PWQO often are used as a starting point in deriving effluent requirements for a C of A, to assess ambient water quality conditions, to infer where uses have been impaired, to assess spills and to monitor the effectiveness of remedial actions.

The process of setting PWQO values is described in *Ontario's Water Quality Objective Development Process* (MOE, 1992). The process considers data for acute and chronic toxicity, bioaccumulation and mutagenicity, and persistence. Also considered is information concerning fate, physical/chemical properties, taste, odour, tainting of fish, impacts on wildlife, recreation (bathing and aesthetics), sediment quality and the standards of other agencies.

The objectives themselves are contained in a document entitled *Provincial Water Quality Objectives* (MOE, 1999a). This document is updated more frequently that the Water Management document. Some PWQO values are described as "interim" which indicates that the objective is based on incomplete information and are to be upgraded when sufficient information becomes available and has undergone external peer review.

PWQO values have been established for many parameters. They cover general indicators, many individual organic chemicals, many inorganic parameters, radioactive material, and microbiological parameters. Table 3.2 lists the PWQO for many parameters, but ends with a list of those that have been excluded.

Drinking water quality standards and objectives are described in the document entitled *Ontario Drinking Water Standards* (MOE, 2001b). The standards and objectives are considered to be the minimum level of quality and should not be regarded as implying that water supplies should be allowed to degrade to the prescribed levels.

There are two types of drinking water standards. The maximum acceptable concentration (MAC) is a health-related concentration above which there are known or suspected adverse health effects. Interim maximum acceptable concentration (IMAC) describes a concentration that is health based but lacks sufficient toxicological data to establish a MAC with reasonable certainty.

There are two types of objectives. An aesthetic objective (AO) is used for a substance that may impair the taste, odour or colour of water, or which may interfere with good water quality control practices. For some chemicals, both AO and MAC values have been set. The term operational guideline (OG) is used for a substance that needs to be controlled to ensure efficient and effective treatment and distribution of water. Table 3.3 and Table 3.4 list both types of standards and both types of objectives.

The MOE applies the standards and objectives in approving waterworks capable of supplying water at a rate of 50,000 L per day or waterworks that supply drinking water to more than five private residences. (See Section 3.5.4 for more information.)

The *Ontario Drinking Water Standards* (MOE, 2001b) include sections that describe the sampling and assessment of drinking water supplies, as well as the actions required when a standard is exceeded. Appended to the document is a supporting document that describes the chlorination of potable water supplies (also referred to as Procedure B13-3). These aspects of drinking water supply have been subjected to a high level of scrutiny and review since the events at Walkerton in 1999 and the two drinking water regulations — O. Reg. 505/01 and O. Reg. 459/00 — noted in Section 3.3.4.

3.4   Municipal Requirements

3.4.1   Overview

Most municipalities in Ontario have by-laws to control discharges to sewers (sanitary, combined and storm). In 1988, the MOE developed a draft *Model Sewer-Use By-law* to provide a uniform basis to regulating sanitary, combined and storm sewers across Ontario (MOE, 1988). Many municipal governments used the model by-law for that purpose.

In 1998, a new version of the model sewer use by-law was proposed (MOE, 1998). The 1998 version includes limits for more organic chemicals, reductions in the limits of some chemicals (notably cadmium, lead and mercury), a new approach for storm water with an emphasis on pollution prevention, new sampling and analysis protocols modelled after those in the MISA regulations, and new monitoring requirements for dischargers. Since municipalities do not update sewer use by-laws frequently, it remains to be seen if the 1998 version of the model by-law will prompt modifications at the municipal level.

In the following sections, components of the model sewer-use by-law are used to provide an overview of potential municipal requirements. When assessing a specific discharge it is imperative to use the most recent sewer-use by-law for that municipality.

3.4.2   Discharges to Sanitary and Combined Sewers

Municipal sewer-use by-laws usually include numerical limits for some parameters and prohibitions for selected types of material and waste.

Table 3.5 presents the allowable concentrations stipulated in the 1998 version of the MOE model sewer-use by-law for discharges to sanitary and combined sewers. The by-law expressly prohibits dilution to achieve those limits. Also included in Table 3.5 are the limits put in place by Toronto in 2000.

Most municipal by-laws also prohibit the discharge of various materials such as fuels, severely toxic material, PCBs, waste radioactive material and pesticides. PCBs can be exempted if the owner/operator has a C of A that expressly allows the discharge, or the owner/operator has approval from the municipality, or the concentration is less than 5 µg/L of PCBs. The discharge of radioactive materials is allowed if it is in accordance with a licence from the Canadian Nuclear Safety Commission.

Most by-laws also use the term "wastes" to identify materials prohibited from being discharged to sewers. These wastes are defined using definitions that appear in Regulation 347 under the EPA (see Section 4.4) and include:

- acute hazardous waste chemicals
- pathological wastes
- hazardous industrial wastes
- PCB wastes
- hazardous waste chemicals
- reactive waste
- ignitable wastes

Exemptions are allowed for the discharge of pathological waste if it has been decontaminated prior to discharge and the owner/operator has a C of A from the MOE, or permission from the municipality.

Water that does not originate from a municipality (for example, ground water from an excavation) may be discharged to the municipality sanitary sewer system if the municipality is provided with information in advance that describes the amount of water and its source. The municipality may require the discharger to enter into a water surcharge agreement (see Section 3.4.4). In some municipalities, storm water is not allowed into the sanitary sewer system regardless of any surcharge agreements. Dischargers to municipal sewer systems need to keep aware of all by-law requirements.

3.4.3   Discharges to Storm Sewers

Like discharges to sanitary and combined sewers, there are numerical limits, general prohibitions and prohibitions with exceptions for discharges to storm sewers.

The 1988 model by-law contains numerical limits for a few parameters. These included temperature (< 40° C), pH in the range of 6 to 9, suspended solids of not more than 15 mg/L, limits for a few metals, and a limit of not more than 200 fecal coliform colonies per 100 mL. In the 1998 version of the model by-law proposed by the MOE, those limits have been dropped with the exception of pH, which should not be outside the range of 6.5 to 8.5. The proposed model by-law notes that municipalities are responsible for storm water that enters municipal sewers and is subsequently released to the natural environment. It then lists various qualitative conditions that a municipality may use to assess the suitability of storm water. Table 3.6 presents the allowable limits for discharges to storm sewers declared by Toronto in 2000. The limits address numerous parameters and may signal how other municipalities respond to this responsibility.

The 1998 version generally recommends that only storm water be discharged to storm water sewers with the exceptions of "uncontaminated" water or cooling water that does not interfere with the proper operation of the storm sewer, does not result in any form of adverse impact and does not contravene any regulatory requirements or C of A.

The types of materials explicitly prohibited from discharges to storm sewers typically include: biomedical waste, combustible liquids, fuels, hauled sewage, ignitable waste, PCB waste, pesticides, reactive wastes, waste radioactive substances, leachate from waste disposal sites and any discharge that contains contaminants from raw materials, intermediate or final products, or waste water from an industrial operation.

The 1998 version of the by-law proposed by the MOE notes that dischargers (principally industrial and commercial facilities) may be required by municipalities to complete storm water studies and subsequently take steps to manage storm water. Storm water studies are similar to the Storm Water Control Study (SWCS) described in the MISA regulations (see Section 3.3.2). In the past, these also were referred to as "best management practice" (BMP) plans. A storm water study should describe the anticipated quantity and quality of the storm water, the types and locations of facilities needed to control or treat storm water, and the implementation of pollution prevention techniques and measures.

The *Stormwater Pollution Prevention Handbook* (2001) has been produced by a collaborative effort by several organizations including the MOE, the Toronto Region Conservation Authority and the Municipal Engineers Association. Copies can be obtained at **www.ene.gov.on.ca/envision/water/stormwaterpph.htm.**

3.4.4   Discharge Reports, Plans and Agreements

Reports

Prior to any discharge to a sewer, industrial dischargers must provide a discharger information report to the municipality. Municipalities may require more information from larger facilities and can exempt existing dischargers.

Generic versions of the reports are included in the 1998 version of the model sewer-use by-law. A discharger information report is similar to the waste survey report used in the past. The required information includes a description of on-site processes; the types of chemicals stored, used or produced; descriptions of the types of discharges (process wastewater, cooling water, uncontaminated water, other) and the points of discharge; known characteristics of the quality of the discharges; waste generator information; information about any agreements with the municipality; information about any pre-treatment works and sewer connections; and copies of related environmental management plans.

The municipality reviews completed reports in order to assess the potential effects of the discharges on sewage treatment processes. Depending on the quantities and characteristics of its discharges, a discharger may be required to enter into any one of several types of agreements with the municipality.

Agreements

If a discharge to a sewer exceeds by-law limits and the discharger believes that this will continue to be the case, the discharger may be able to enter into an extra strength surcharge agreement (ESSA) with the municipality whereby the municipality is paid to treat the discharge. ESSAs (formerly called overstrength agreements) can only apply to the following parameters and only if the excess capacity exists at the municipal wastewater treatment plant:

- biochemical oxygen demand (BOD)
- total phenolics
- total suspended solids (TSS)
- total phosphorus
- total Kjeldahl nitrogen (TKN)
- solvent extractable matter

The payment schedule is based on the excess amount of material being deposited to the sewer. For example, a municipality might impose a surcharge of 28 cents per kg of excess BOD in a discharge. Payment typically is made quarterly and is based on the actual flow rate multiplied by a concentration specified in the ESSA. Rebates may be possible provided that there is sufficient data to support such a claim. Most ESSAs require the installation of a flow-measuring device.

For parameters that cannot be covered by an ESSA, the municipality may enter into a compliance agreement with the discharger. A compliance agreement allows for a period of grace while action is being taken to reduce contaminant levels.

A compliance agreement specifies parameter limits that are not to be exceeded while the program is in effect. A compliance program also can specify activities to be undertaken, which may include: retaining an engineer, source characterization, treatability studies, start-up of a treatment system or alteration in the process, and reporting to the municipality.

Plans

In addition to the SWCS, there is growing support among regulatory agencies for industrial and commercial dischargers to prepare and submit Pollution Prevention (P2) plans if their discharges contain pollutants of concern (usually those listed as "subject pollutants" in a by-law), or if the facility falls in certain sectors. Such plans are directed at achieving the maximum feasible reduction — and preferably elimination — of discharges of subject pollutants.

The City of Toronto has been requiring P2 plans since 2001 (City of Toronto, 2000) (for additional information, see the Toronto Municipal Code, s. 681). The city has outlined what the plans and plan summaries should address. Plan summaries are submitted to the city for approval. If a plan summary is not approved, the discharger has 90 days to amend and resubmit it. An updated plan summary is to be submitted every two years and an updated plan every six years.

The city's goal is to stop incinerating sewage sludge. Discharge limits are designed to ensure the sludge will be suitable for constructive uses such as land application. Additional information on the Toronto Municipal Code and by-law 457-2000 can be obtained at **www.city.toronto.on.ca.**

3.5   Obtaining Approvals for Discharges

3.5.1   Direct Discharges

A direct discharge is a discharge directly to a receiving water. Most direct discharges are effluents from privately-owned or municipal treatment plants. Owner/operators of such plants must obtain a C of A from the MOE. The MOE provides guidelines and procedures for applying for an approval of municipal and private water and sewage works. Examples of these guidelines are listed in Table 3.1. The listed documents include protocols for updating Cs of A for both water and sewage works issued in 2002.

Requirements

Prior to the establishment, extension or change in a water or sewage works, a C of A must be obtained. For example, section 53 of the OWRA requires that:

> (1) No person shall establish, alter, extend or replace new or existing sewage works except under and in accordance with an approval granted by the Director.

. . . . .

(5) No person shall use or operate sewage works for which an approval is required under subsection (1) unless the required approval has been granted and complied with.

It is important to note the following definitions, as per s. 1 of the OWRA, when reviewing the above requirements:

"sewage" includes drainage, storm water, commercial wastes and industrial wastes and such other matter or substance as is specified by the regulations;

"sewage works" means any works for the collection, transmission, treatment and disposal of sewage or any part of such works, but does not include plumbing to which the *Building Code Act, 1992* applies

Section 53(6) of the OWRA provides that ss. 53(1)and (5) of the OWRA do not apply:

(a) to a sewage works from which sewage is not to drain or be discharged directly or indirectly into a ditch, drain or storm sewer or a well, lake, river, pond, spring, stream, reservoir or other water or watercourse;

(b) to a privately-owned sewage works designed for the partial treatment of sewage that is to drain or be discharged into a sanitary sewer;

(c) to a privately-owned sewage works serving only five or fewer private residences;

(d) to a sewage works the main purpose of which is to drain agricultural lands;

(e) to a drainage works under the Drainage Act, the Cemeteries Act, the Public Transportation and Highway Improvement Act or the Railways Act;

(f) to such sewage works as may be exempted by regulations made under this Act, but s. 53 does apply to a sewage works for the distribution of sewage on the surface of the ground for the purpose of disposing of the sewage.

Section 53(6) essentially excludes sewage systems having a subsurface disposal and all associated collection works from the requirements of approval under s. 53 of the OWRA, however, such works would require approval under Part VIII of the EPA. Similarly, other works that are exempted from the OWRA may require approval under other legislation.

Additional exclusions related to actual alteration of service connection or relining/replacement of watermain or sewer are described in O. Reg. 525/98.

Approval Process

The following are the key steps in obtaining a C of A (MOE, 1999b):

1. Pre-application consultation with MOE provides an opportunity to define objectives for the project, discuss treatment/disposal options and expected effluent quality and environmental impacts, discuss approval requirements, identify public concerns and potential requirements under other statutes, for example, Niagara Escarpment Planning and Development Act, and the Environmental Assessment Act. Pre-application consultation is required for all projects involving construction of sewage/wastewater treatment and disposal facilities, expansion or re-rating of existing facilities, major modifications/upgrades to existing facilities, introduction of an innovative technology, and projects involving prescribed instruments under the Environmental Bill of Rights (EBR) Act.

2. It is important that the applicant carefully review the EBR requirements, as inclusion in the EBR process will require a minimum 30-day posting for public input.

3. If a surface water impact assessment is required, the assessment must be completed by the proponent and accepted by the technical support section of the respective regional office of the MOE before the formal application is submitted to the Environmental Assessment and Approvals Branch (EAAB).

4. Two copies of application forms plus supporting documentation must be submitted to the MOE. One copy must be sent to the director of the EAAB with corresponding fee (where applicable), and the second copy must be sent to the MOE district office. If the application involves sewage works with subsurface disposal of effluent, where the project requires submission of an environmental impact assessment report, two copies must be submitted to the EAAB.

5. The MOE will review the proposal to assess whether the works comply with relevant Ministry guidelines and policies.

The following steps are performed as part of the review of an application (MOE, 1999b):

1. Upon receipt of the submission, the application is entered into a computer database by administration. The application is then forwarded to the appropriate unit and assigned to a review engineer.

2. An application processor will review the file and ensure basic information and fee, if applicable, are present and will generate a letter of acknowledgement to the proponent. This letter will stipulate how long the proponent has to provide any missing fees or information. The letter may also indicate the estimated turn-around time for the application.

3. The application processor will also place the EBR proposal for the application on the EBR registry. The proposal will be available on the registry for at least 30 days for a public comment period. During this period, the public is given the opportunity to review the application and submit comments on the proposal to EAAB.

4. The review engineer conducts a short initial screening and review to determine whether there are any gross errors or omissions of information and whether the application requires any supplementary review (for example, comments on the submitted environmental impact analysis from the technical support section of the appropriate regional office). Incomplete applications are returned to the proponent. Examples of gross errors that would result in the return of an application are a submission where there was no pre-application consultation, or where the technical support staff advises the review engineer that they are unable to assess the environmental impact analysis submitted with the application without further data or other information readily available from the proponent.

5. The review engineer will consider any public comments received during the public comment period for the proposal.

6. The review engineer performs a detailed review to assess the completeness and adequacy of the submitted detailed design documentation and other supporting information; the compliance of the proposal with other acts, regulations, policies, objectives and environmental guidelines; the conformance of the engineering design to the principles of sound engineering; and the adequacy of controls and contingencies provided to facilitate the proper operation of the works.

7. If required, requests for additional information, design changes or submission of additional fees are communicated in writing to the proponent with a deadline for response. Failure to respond will result in the assumption that the proponent no longer wishes to proceed with the application and the application will be returned less the non-refundable portion of the fee.

8. A public hearing under ss. 54 or 55 of the OWRA may be required prior to the issuance of an approval under the Act. Pursuant to s. 7 of the OWRA, when the Environmental Assessment Board holds a public hearing, it shall serve notice of its decision resulting from the hearing, and the director shall implement this decision. Should a public hearing be required, the fee submitted with the application will be refunded. In these circumstances the fee will be required only when the certificate is ready to be issued. An invoice will be sent to the proponent.

9. Upon recommendation of the review engineer, when satisfied, the director will grant approval by signing a C of A. If a C of A already exists for the proposal, it may be amended to include the new works.

According to s. 53(4) of the OWRA, if it is in public interest to do so, the director may:

(a) refuse to grant the approval;

(b) grant the approval on such terms and conditions as the Director considers necessary;

(c) impose new terms and conditions to the approval;

(d) alter the terms and conditions of the approval; or

(e) revoke or suspend the approval.

Should the director decide to do any of the above, s. 100 of the OWRA requires that the director provide written notice of intentions with reasons. This notice may be appealed to the Environmental Appeal Board provided that the appeal is filed within 15 days of receipt of the notice.

If approved, the MOE will either issue a new C of A, an amended C of A, or a notice of amendment to C of A. A notice of amendment to C of A usually is issued to approve modifications to existing, previously approved works, and/or to impose new or modify existing terms and conditions on an existing C of A.

Upon the director's approval, the applicant may then proceed to construct and operate the works. A copy of the C of A should be attached to the application (it is usually referenced) and a copy of the conditions should be attached to the piece of equipment for which it applies.

Fees

The MOE charges fees for Cs of A for commercial, industrial and private sector undertakings. The fees are presented in O. Reg. 364/98 (Fees – Approvals). Fees typically consists of several components. For example, the fee for a new sewage works can include an administrative processing fee ($200), a fixed cost for general technical review (a function of the type of equipment or processes in the application and the capacity of the facility), a hydrogeological review ($3,000 if required as determined by the MOE), a review of the criteria to be met by the effluent or discharge ($1,400 to $6,000 if required as determined by the MOE), and the cost of a hearing ($18,000 if necessary). Total fees can range from $1,000 to $37,000. A hearing is mandatory if the works cross municipal boundaries.

Similarly, the costs for amendments to existing approvals are based on the same or similar components.

The MOE takes a firm stand when the construction of facilities begins before a C of A is issued. There are cases of construction being stopped by the owner so as to not jeopardize a pending C of A, and of the MOE refusing to issue a C of A for completed facilities. The supplying of utility services to a site, such as those covered by a municipal building permit, do not seem to be of a concern to the MOE, however, the regional office of the MOE should be contacted.

Timing

Approval often is received with a few weeks, but may take up to three months if applications are complex or voluminous. The expected processing time goals for applications include:

1. Applications that arrive with complete documentation and with environmental impact information, effluent criteria, hydrological criteria or raw water quality analysis that have been accepted/endorsed by the regional office of the MOE during presubmission consultation are likely to be processed within four weeks (and often quicker).

2. The review time is as long as necessary for applications that arrive with complete documentation, but with environmental impact information, effluent criteria, hydrological criteria or raw water quality analysis that have not been accepted/endorsed by the regional office.

3. The review time will be as along as necessary for applications for which a public consultation/hearing is required.

4. Applications that arrive without environmental impact information, effluent criteria, hydrological criteria or raw water quality analysis where required will be returned to the applicant.

5. All other incomplete applications will be returned to the applicant. A complete application of approval consists of a completed and signed application form and all relevant supporting information as specified in the application form.

It is in a discharger's best interest to make sure that an application is submitted at the earliest possible stage of a project, and that the application provides sufficient information for its evaluation.

3.5.2   Other Possible Permits for Direct Discharges

If a proponent is going to construct or modify an outlet to a watercourse and the outlet is within a flood line or fill line that is regulated by a Conservation Authority, a permit is required by regulations under the Conservation Act. The name of the permit varies among Authorities but often is referred to as a "fill permit". In Ontario, most water courses that drain more than 100 hectares fall under the jurisdiction of a local Conservation Authority.

Approval from the Ontario Ministry of Natural Resources (MNR) may also be required. The approval is called a Work Permit, or Part F – Works

Within a Waterbody and is administered under the Lakes and Rivers Improvement Act and the Public Lands Act. Since the early 1990s, the MNR has generally reserved this requirement to relatively larger projects such as creek diversions, but it can be applied to projects such as the construction of a water treatment plant discharge or a storm water discharge.

Changes to treatment plant process, and the upgrading of storm sewer outfalls also may need to meet the requirements of the Ontario Environmental Assessment process where the discharge is to a public water body and the discharge needs to be treated prior to being released.

3.5.3　Indirect Discharges

Indirect discharges are those to sewer systems connected to treatment plants typically operated by or for the municipality. The Ontario Clean Water Agency (OCWA) is a provincial Crown corporation that operates more than 400 water and wastewater facilities for municipalities.

Permission from the operator of the treatment plant must be obtained prior to establishing new sewer connections. Most municipalities require dis-chargers to complete and submit discharger information reports (as noted in Section 3.4.4) They also require that the report be updated and resubmitted whenever changes are made to the quality or quantity of a discharge.

3.5.4　Permit to Take Water

Section 34 of the OWRA requires that "no person shall take more than a total of 50,000 L of water in a day" from ground and/or surface waters without a permit issued by the director.

An application to the Regional Technical Assessment Section is required to receive a permit to take water. The types of information that must be included with the application include a design brief, a hydrology reprint, a construction reprint and water quality analysis.

If a proposal for a waterworks, subject to the requirements of s. 42 of the Act, involves a new water intake/well or an existing intake/well with a change in quantity or rate of withdrawal of water, the proponent must obtain a permit to take water from the appropriate regional office of the MOE before submitting the application for approval of water works under s. 52 of the OWRA.

3.6　Monitoring and Operation

3.6.1　Monitoring Requirements

As described in previous sections of this chapter, monitoring requirements may come from various sources: federal and provincial regulations, an MOE

C of A, control orders, clean-up orders, an ESSA or a compliance agreement with a municipality.

In addition to the monitoring required by regulatory agencies, a facility may perform in-house monitoring for various reasons:

- to assess compliance
- to help operate a wastewater treatment facility effectively
- to obtain a rebate for an overstrength agreement
- to assess the sources and fate of elevated contaminants levels
- as part of a toxicity reduction evaluation (see Chapter 13)
- to assess the effects of a specific event or incident
- to determine net loadings
- to assess "background" conditions prior to constructing a facility

In some cases, the procedures used for sampling and analysis may be different for the non-regulated monitoring compared to those of regulated monitoring programs. For example, higher method detection limits (MDL) may be suitable for an in-house program to track down the source of a substance in a wastewater stream compared to measurements in final effluent.

3.6.2   Chemicals of Interest

The chemicals present in an effluent or discharge stream are influenced by various factors that include the use of the water (for example, non-contact versus contact cooling water), the source of the water (for example, river water versus lake water), whether chemical additives such as chlorine or algaecides are used, pre- and final treatment, and whether sources include ground water inflow or storm runoff.

The following parameters, as identified by the MOE analytical test group (ATG) numbers, may need to be analyzed, especially when assessing the operations of a wastewater treatment system or considering the possible re-use of an effluent stream:

General alkalinity and acidity	ATG #3	hydrogen ion (pH)
biochemical oxygen demand (BOD)	ATG #4a	nitrogen (ammonia
chlorine and chloride		plus ammonium total
cyanide (free)		Kjeldahl Nitrogen)
dissolved oxygen	ATG #4b	nitrogen (nitrate +
fluoride		nitrite)
iron	ATG #5a	dissolved organic car-
sulphate		bon (DOC)
temperature	ATG #5b	total organic carbon
ATG #1 chemical oxygen		(TOC)
demand (COD)	ATG #6	total phosphorus
ATG #2 total cyanide	ATG #7	specific conductance

ATG #8	total suspended solids (TSS), volatile suspended solids (VSS)		carbons and several phthalates)
ATG #9	total metals (aluminum, beryllium, cadmium, chromium, cobalt, copper, lead, molybdenum, nickel, silver, thallium, vanadium, and zinc)	ATG #20	extractables, acid – phenolics (organics, *e.g.* phenol and cresol)
		ATG #21	extractables, phenoxy acid herbicides
		ATG #22	extractables, organochlorine pesticides
ATG #10	hydrides (antimony, arsenic and selenium)	ATG #23	extractables, neutral-chlorinated
ATG #11	chromium (hexavalent)		
ATG #12	mercury	ATG #24	chlorinated dibenzo-p-dioxins and dibenzofurans
ATG #13	total alkyl lead		
ATG #14	total phenolics (4AAP)		
ATG #15	sulphide	ATG #25	solvent extractables (mineral, animal and plant)
ATG #16	volatiles, halogenated (volatile organics, *e.g.*, methylene chloride and trichloroethylene)		
		ATG #26	fatty and resin acids
		ATG #27	PCBs
ATG #17	volatiles, non-halogenated (benzene, ethylbenzene, styrene, toluene, o-xylene, m-xylene and p-xylene).	ATG #28a	open characterization (volatiles)
		ATG #28b	open characterization (extractables)
ATG #18	volatiles, water soluble (acrolein and acrylonitrile)	ATG #29	open characterization – elemental (approximately 70 elements)
ATG #19	extractables, base neutral (organics, *e.g.* polycyclic aromatic hydro-		

Biochemical oxygen demand (BOD) is the amount of molecular oxygen required to stabilize the decomposable matter present in water by aerobic biochemical action. Because the complete stabilization of a given waste may take too long, a standard laboratory BOD test has been developed that incubates the material for a period of five days at 20°C. The results of a BOD_5 test are often used to assess the efficiency of both municipal and industrial biological treatment plants and the appropriateness of the wastewater.

Chemical oxygen demand (COD) measures the non-biodegradable as well as the ultimate biodegradable organic compounds in terms of oxygen consumption to stabilize the decomposition.

3.6.3   Surrogate Parameters

There are many situations where it is sufficient or appropriate to use surrogate parameters to assess effluent or water quality. Most surrogate parameters provide information that is similar to the parameters or conditions that are of actual concern. For example, an effluent might be analyzed for total phenols when, in fact, it is a specific phenolic compound that is of interest.

Most surrogates are used because the sample requirements are simpler, or the samples require less time or expense to analyze than the compound of interest. Some surrogates lack the sensitivity of specific compounds, but this may not be a limitation for assessing certain situations.

Surrogates should be used only after a correlation between it and the parameter of concern has been developed. For example, some effluents have shown a strong correlation between absorbed metals and TSS. In such cases, TSS can be used to monitor a discharge stream on a frequent basis and specific metals analyzed only if a pre-defined TSS concentration is exceeded.

Another example is the COD test, which is more reproducible and less time-consuming than a BOD test, however, the COD/BOD correlation is such that it is normally only a qualitative value. A change in the ratio of biodegradable to non-biodegradable organic compounds affects the correlation.

The analysis of organic parameters is relatively expensive when compared to conventional parameters and many of the surrogate parameters deal with organic compounds. For example, solvent extractables, dissolved organic carbon, total organic carbon, total halogens or total phenols can be monitored on a regular basis with the provision for analyzing specific compounds if a threshold concentration is exceeded.

For some groups of organic compounds, specific members can be used as indicators of the possible presence of other group members or may become recognized as the compounds of greatest potential concern and therefore the prime objective of frequent monitoring. For example, benzene may be the only non-halogenated volatile compound of concern in an effluent. Similarly, benzo[a]pyrene can be used for other polyaromatic hydrocarbons. Such selections are best based on substantial supporting data.

3.6.4   Sample Collection and Preservation Methods

Prior to the collection of samples, available guidelines such as the *Protocol for Sampling and Analysis of Industrial/Municipal Waste Water* (MOEE, 1993a) and industrial associations such as the Water Pollution Control Federation and the American Society of Civil Engineers should be reviewed.

The most common form of sampling, prior to the introduction of the MISA monitoring program, was grab sampling. It is still widely used when only a few samples are required or when sampling the following ATGs: 15

– sulphide; 16 – halogenated volatiles; 17 – non-halogenated volatiles; 18 – water soluble volatiles; and 28a – open characterization for volatiles.

One weakness of grab samples is that the samples may only be indicative of effluent quality at the time of sample collection. To improve the representativeness of effluent sampling efforts, automatic samplers or on-line analyzers can be used.

Automatic samplers can operate in one of two modes: collecting and combining equal volume subsamples at equal time intervals, or collecting and combining samples that are flow proportional. The benefits of these types of devices are that the manpower commitment is reduced and the data represents a truer mean of the concentration of the discharge stream as more subsamples can be collected.

On-line analyzers have been available for many years but capabilities have only recently been extended beyond conventional parameters such as pH and conductivity. On-line analyzers are now available that continuously analyze effluents for organic compounds.

Depending upon the chemicals to be analyzed in a sample, preservatives should be added to either the sample container or laboratory container to ensure a concentration that is representative of the discharge quality. The chemicals that are used to preserve samples sometimes are categorized as either pre-preservatives or preservatives. Pre-preservatives are added to the sample container before the sample itself. For example, a pre-preservative should be used if an automatic sampler is collecting samples to be analyzed for ATG 2 – cyanide, or ATG 14 – phenols. Preservatives are added after the sample has been collected in the sample container. Table 3.7 lists the preservatives for various ATGs.

3.6.5   Quality Assurance and Quality Control

It is essential that an effective quality assurance/quality control (QA/QC) program be implemented with all monitoring programs. It assures the controllability, accountability and retractability of the work being performed.

A good QA/QC program involves close supervision and surveillance of all field operations, thorough documentation and review of sampling procedures, and the assurance that appropriate laboratory analysis techniques are employed.

The QC program should ensure that data are generated within known limits of accuracy and precision. This involves the application of method detection limits and travelling blank samples, and the collection of replicate samples and duplicate samples.

Examples of these requirements in the MISA regulations include:

- an MDL must be calculated for each parameter
- "standard reference materials" must be employed to ensure that the laboratory standard solutions be validated

- reasonable control limits must be developed for the analysis of method blank samples
- during the course of an analytical run for specified ATGs, a replicate sample, a method blank sample and a method blank sample spiked with a standard solution must be included
- travelling blanks and travelling spiked blanks must be used
- replicate and duplicate samples must be provided

A replicate sample refers to one of at least two samples removed from a single sample container in a manner that minimizes the difference between the samples. A duplicate sample is similar, but means one of two samples collected at a sampling point.

3.6.6   Flow Measurement/Estimation Methods

Flow systems are assigned to two basic categories: flow in closed channels and flow in open channels. Closed channel flow is defined as flow in completely filled pressure conduits (pipes). Pressure conduits are usually used for fresh water lines or for industrial processes, and flow through these is often measured by some type of device inserted into the line. Measuring devices for closed channel flow include the venturi meter, flow nozzle, orifice meter, magnetic flow meter and pitot tube flow meter.

Open channel flow is defined as flow in any channel in which the liquid flows with a free surface. Examples are runoff ditches, canals, flumes and other uncovered conduits. Certain closed channels, such as sewers and tunnels, when flowing partially full and not under pressure are classified as open channels. There are numerous methods of determining the rate of flow in an open channel.

1. **Timed gravimetric** – Collection of the entire contents of the flow during a fixed length of time.
2. **Dilution** – The flow rate is measured by determining the degree of dilution of an added tracer solution in the flowing water.
3. **Velocity** – The flow rate is calculated by determining the mean flow velocity across a cross-section and multiplying this rate by the width of the channel at this point and the measured liquid level. The latter is measured by a second device. The sophistication of this type of system and accuracy has improved considerably in recent years.
4. **Hydraulic structure** – Some type of hydraulic structure, such as a weir, is introduced into the flow stream. The function of the hydraulic structure (primary device) is to produce a flow that is characterized by known relationship (usually nonlinear) between a liquid level measurement (head) at some location and the flow rate of the stream.
5. **Slope-hydraulic radius area** – Measurements of water surface slope, cross-sectional area and wetted perimeter over a length of

uniform section channel are used to determine the flow rate, utilizing a resistance equation such as the Manning formula. The flow channel itself serves as the primary device.

The MISA regulations specify the accuracy by which various types of flows must be measured or estimated. In general terms, the following apply:

- **Combined and process subcategory effluent** – Combined effluent streams should be continuously monitored with an overall flow device accuracy of ± 15%.
- **Cooling water** – Should be measured to an accuracy of ± 20% of the actual flow.
- **Other types of effluents** – Flow device accuracy for other types of discharges (for example, storm water and waste disposal site effluents) is usually ± 20%. Most industrial sectors do not require flow measurement for storm water discharges or waste disposal site effluent. Instead, the duration and volume of these discharges may be estimated or measured.

The accuracy of the primary devices may be determined by either calibration or certification reports. A certification report must certify that the primary flow device was installed according to international standards.

Flow measurement is a key element in the determination of the mass loading of a particular chemical. As such, it has become an integral part of the MISA program and municipal agreements.

3.7   Reporting Requirements

Many of the events described in Section 3.6 require the timely reporting of data. The actual reporting requirements are a function of the particular monitoring program (for example, a MISA regulation, a C of A, etc.).

It is in a discharger's interest, regardless of reporting requirements, to create a computerized environmental database. The database should assist in the generation of reports, perform error checks and allow for easier assessment of past trends in contaminant loadings and overall compliance.

3.8   Summary

Key legislation that governs direct discharges in Ontario includes the provincial EPA, OWRA, and sector specific regulations under the MISA program, and the federal Fisheries Act, CEPA, 1999 and corresponding regulations. This legislation makes the owners and/or operators of facilities that discharge effluents responsible for rigorous monitoring, effluent control and reporting.

Both the CEPA, 1999 and the MISA regulations are focussed on the virtual elimination of persistent toxic substances that remain in the environment for extended periods of time before breaking down and of bioaccumulative toxic substances that accumulate within living organisms.

Indirect dischargers must meet the requirements of the operators of wastewater treatment facilities — typically municipalities. Most municipalities have by-laws, which stipulate allowable concentrations and prohibit specified materials and waste from being discharged. Recently enacted or amended by-laws suggest that limits are becoming more stringent, address more chemicals, demand extensive monitoring and require that industrial facilities prepare various reports that have pollution prevention as a dominant theme.

References

Canadian Council of Ministers of the Environment (CCME). 1991. *A Protocol for the Derivation of Water Quality Guidelines for the Protection of Aquatic Life.*

Canadian Council of Ministers of the Environment (CCME). 1993. *Protocols for Deriving Water Quality Guidelines for the Protection of Agricultural Uses.*

Canadian Council of Ministers of the Environment (CCME). 1999. *Canadian Water Quality Guidelines for the Protection of Agricultural Uses – Summary Table.*

Canadian Council of Ministers of the Environment (CCME). 2001a. *Canadian Water Quality Guidelines for the Protection of Aquatic Life – Summary Table.*

Canadian Council of Ministers of the Environment (CCME). 2001b. *Summary of Guidelines for Canadian Drinking Water Quality.*

City of Toronto. 2000. By-law 457-2000.

Environment Canada. 1997. *Aquatic Environmental Monitoring Effects Monitoring Requirements.* EEM/1997/1.

Health and Welfare Canada (HWC). 1992. *Guidelines for Canadian Recreational Water Quality.* ISBN 0-660-14239-2.

Ontario Ministry of the Environment (MOE). 1988. *Model Sewer-Use By-Law.* ISBN O-7729-4419-9.

Ontario Ministry of the Environment (MOE). 1992. *Ontario's Water Quality Objective Development Process.* Aquatic Criteria Development Committee.

Ontario Ministry of the Environment (MOE). 1998. *The Proposed 1998 Model Sewer-Use By-Law.*

Ontario Ministry of the Environment (MOE). February 1999a. *Provincial Water Quality Objectives.*

Ontario Ministry of Environment (MOE). November 1999b. *Guide for Applying for Approval of Industrial Sewage Works.*

Ontario Ministry of Environment (MOE). September 2001a. *EBR Registry Number PA01E0023 – The Canada — Ontario Agreement Respecting The Great Lakes Basin Ecosystem.*

Ontario Ministry of the Environment (MOE). January 2001b. *Ontario Drinking Water Standards.*

Ontario Ministry of Environment (MOE). March 2002. "Are Things Getting Better? or Worse?" Article posted at **www.ene.gov.on.ca/envision/faq/index.htm #Arethings.**

Ontario Ministry of Environment and Energy (MOEE). July 1993a. *Protocol for the Sampling and Analyses of Industry/Municipal Wastewater.*

Ontario Ministry of Environment and Energy (MOEE). January 1993b. *Protocol for Conducting a Storm Water Control Study.*

Ontario Ministry of the Environment and Energy (MOEE). July 1994a. *Water Management: Policies, Guidelines, Provincial Water Quality Objectives of the Ministry of the Environment and Energy.*

Ontario Ministry of the Environment and Energy (MOEE). July 1994b. *Deriving Receiving-Water Based, Point-Source Effluent Requirements for Ontario Waters.* Procedure B-1-5.

Ontario Ministry of the Environment and Energy (MOEE). 1994c. *Incorporation of the Reasonable Use Concept into MOEE Groundwater Management Activities.* Guideline B-7.

Ontario Ministry of the Environment and Energy (MOEE). June 2000. *Guide to Applying for Approval of Permit to Take Water.*

Public Works and Government Services Canada (PWGSC). 2001. *P2 Planning Handbook.*

Table 3.1

Selected MOE Publications Related to Water Quality

Many of the following documents (and others) can be obtained at the MOE Web site: **www.ene.gov.on.ca**.

Protection and Management of Aquatic Sediment Quality in Ontario. Guideline B-1-3.

Fill Quality Guidelines for Lakefilling in Ontario. Guideline B-1-4.

Snow Disposal and De-Icing Operations in Ontario. Guideline B-4.

Evaluating Construction Activities Impacting Water Resources. Part III – Handbook for Dredging and Dredged Material Disposal in Ontario.

Guideline for Evaluating Construction Activities Impacting on Water Resources. Guideline B-6.

Determination of Contaminant Limits and Attenuation Zones. Guideline B-7-1.

Resolution of Groundwater Quality Interference Problems. Guideline B-9-1 (revised 1993).

Private Wells: Water Supply Assessment. Guideline D-5-5.

Resolution of Well Water Quality Problems Resulting from Winter Road Maintenance. Guideline B-3.

The Drinking Water Protection Regulation for Smaller Waterworks: Kit for Waterworks Owners.

Drinking Water Treatment: A Guide for Owners of Private Communal Works and Other Small Water Supply Systems.

Guide for Applying for Approval of Industrial Sewage Works – November 1999.

Guide for Applying for Approval of Municipal and Private Water and Sewage Works – August 2000.

Guide: Application Costs for Waterworks – August 1998.

Guide: Application Costs for Sewage Works – August 1998.

Protocol for Updating Certificates of Approval for Sewage Works, April 2002.

Protocol for Updating Certificates of Approval for Waterworks, April 2002.

Levels of Treatment for Municipal and Private Sewage Treatment Works Discharging to Surface Waters. Guideline F-5.

Sampling and Analysis Requirements for Municipal and Private Sewage Treatment Works.

Planning for Sewage and Water Services. Guideline D-5.

Compatibility between Sewage Treatment Facilities and Sensitive Land Uses. Guideline D-2.

Notification of Laboratory Services Provided to Waterworks. September 2002.

Table 3.2

Provincial Water Quality Objectives

Narrative Objectives – All waters shall be free from contaminating levels of substances and materials attributable to human activities which in themselves or in combination with other factors can:

- settle to form objectionable deposits;
- float as debris, scum or oil or matter forming a nuisance;
- produce objectionable colour, odour, taste or turbidity;
- injure, are toxic to, or produce adverse physiological or behavioural responses in humans, animals or plants; or
- enhance the production of undesirable aquatic life or result in the dominance of nuisance species.

Parameter	Objective (in µg/L)
Aldrin + Dieldrin	0.001
alkalinity	no decrease > 25%
ammonia (unionized)	20[a]
anthracene	0.0008 interim
antimony	20 interim
arsenic	100
arsenic (revised)	5 interim[b]
bacteria	[c]
benzene	100 interim
biphenyl	0.2 interim
boron	200 interim
cadmium	0.2
cadmium (revised)	0.1 to 0.5 interim[b][d]
Chlordane	0.06
chlorine	2
chlorobenzene (monochlorobenzene)	15
chromium (hexavalent)	1
chromium (trivalent)	8.9
chrysene	0.0001 interim
cobalt	0.9
copper	5
copper (revised)	1 to 5[b][d]
cresol (three isomers)	1 interim
cyanide (free)	5
DDT + metabolites	0.003
Diazinon	0.08
1,1-dichloroethane	200 interim
1,2-dichloroethane	100 interim
1,1-dichloroethylene	40 interim
1,2-dichloroethylene (both isomers)	200 interim
dichlorophenols (applies to six isomers)	0.2

Parameter	Objective (in µg/L)
1,4-dioxane	20 interim
dissolved gases	< 110% of saturation
dissolved oxygen	4 to 8 mg/L[e]
Endrin	0.002
ethylbenzene	8 interim
ethylene dibromide	5 interim
ethylene glycol	2,000 interim
Heptachlor + Heptachlor epoxide	0.001 interim
hexachlorobenzene	0.0065
hydrogen sulphide	2
iron	300
lead	5 to 25[b]
lead (revised)	1 to 5 interim[b][d]
Lindane	0.01
Malathion	0.1
mercury (filtered)	0.2
Methoxychlor	0.04
methyl ethyl ketone	400 interim
methylene chloride	100 interim
Mirex	0.001
molybdenum	40 interim
monochlorophenols (applies to all 3 isomers)	7
MTBE	200 interim
naphthalene	7 interim
nickel	25
nitrobenzene	0.02 interim
Parathion	0.008
pentachlorophenol	0.5
pH	6.5 to 8.5
phenanthrene	0.03 interim
phenol (monohydroxybenzene)	5 interim
phenols, total reactive	1
polychlorinated biphenyl (PCBs, total)	0.001
selenium	100
silver	0.1
Simazine	10
styrene	4 interim
tetrachloroethylene	50 interim
tetrachlorophenols (applies to three isomers)	1 interim
tetraethyl lead	0.0007 interim
tetramethyl lead	0.006 interim
thallium	0.3 interim
toluene	0.8 interim
Toxaphene	0.008
tributyltin	0.000005 interim
1,1,1-trichloroethane	10 interim
1,1,2-trichloroethane	800 interim

Parameter	Objective (in µg/L)
trichloroethylene	20 interim
turbidity	(f)
vanadium	6 interim
vinyl chloride	600 interim
m-xylene	2 interim
o-xylene	40 interim
p-xylene	30 interim
zinc	30
zinc (revised)	20 interim

(a) A table in MOE, 1999a shows how to determine the percentage of ammonia present in un-ionized form.

(b) Where both a PWQO and an interim PWQO exist (as is the case for arsenic, cadmium, copper, lead and zinc) the PWQO can be applied in most situations and the interim PWQO used where a greater level of aquatic protection is appropriate.

(c) For bacteria, the PWQO is 100 *Escherichia coli* per 100 mL. There also is a narrative in MOE, 1999a that should be consulted.

(d) For some metals, PWQO values are a function of hardness (as mg/L of $CaCO_3$).

	Hardness (mg/L)	PWQO (mg/L)
cadmium	0 - 100	0.1 interim
	> 100	0.5 interim
copper	0 - 20	1 interim
	> 20	5 interim
lead	< 20	5
	20 to 40	10
	40 to 80	20
	> 80	25
	<30	1 interim
	30 to 80	3 interim
	> 80	5 interim

(e) For dissolved oxygen, MOE, 1999a provides a table of PWQO values that are a function of temperature and whether the water supports cold water biota or warm water biota.

(f) For turbidity, suspended matter should not be added to surface water in concentrations that will change the natural Secchi disc reading by more than 10%.

Objectives have been set for the following parameters, but they are *not listed* above:

abietic acid	acetamide	acetanilide
acrolein	aesthetics	aluminum
4-aminoazobenzene	aminoethyl piperazine	aniline
Aroclor (numerous isomers)	benzaldehyde	benz[a]anthracene
benzidene	benzothiazole	benzo[k]fluoranthene
benzyl alcohol	beryllium	biphenol A
bis(2-chloroethyl) ether	bromodichloromethane	bromoform
bromomethane (methyl bromide)	4-bromophenyl phenyl ether	butanal
butyl benzyl phthalate	camphene	Carbaryl
chlorodibromomethane	chloromethane	1-chloronaphthalene
2- chloronaphthalene	4-chlorophenyl phenyl ether	Chlorpyrifos
4-chloro-3-methyl phenol	cineole	cyclohexanamine
2,4-D (BEE)	Dalapon	dehydroabietic acid

dibenzofuran
dibutylphthalate
1,3-dichlorobenzene
3,3'-dichlorobenzidine
1,2-dichloropropane
N,N-diethyl-m-toluamide (DEET)
dimethylbenzylamine
1,3-dimethylnaphthalene
2,6-dimethylphenol
o-dinotrobenzene
2,6-dinitrotoluene
diphenylamine
Diuron
di-n-butyltin
Endosulphan
ethylene thiourea
fluoranthene
guaiacol
hexachlorocyclopentadiene
iodine
levopimaric acid
1-methylnaphthalene
Metachlor
neoabietic acid
2-nitrophenol
NDMA
oil & grease
pentachlorobenzene
phosphorus, total
polychlorinated naphthalenes
Pyrethrum
resin acids, DHA and total
swimming and bathing
1,2,3,4-tetrachlorobenzene
1,2,4,5-tetrachlorobenzene
1,1,2,2-tetrachloroethane
1,2,3-trichlorobenzene
1,3,5-trichlorobenzene
triethyltin
uranium

dibenz[a,h]anthracene
Dicamba
1,4-dichlorobenzene
1,2-dichlorobut-3-ene
trans-2-dichloropropylene
dimethyl disulphide
N,N-dimethylformamide
2,6-dimethylnaphthalene
3,4-dimethylphenol
p-dinitrobenzene
4,6-dinitro-o-cresol
1,2-diphenylhydrazine
divinyl benzene
2,6-di-t-butyl-4-methylphenol
ethanolamine
Eugenol
formaldehyde
Guthion
hexachloroethane
isopimaric acid
limonene
2-methylnaphthalene
monomethylamine
nitrobenzene
3-nitrophenol
N-nitrosomorpholine
oleic acid
perylene
phthalates, other
propyl diphenyl
quinoline
sandaracopimaric acid
temperature
1,2,3,5-tetrachlorobenzene
1,1,1,2-tetrachloroethane
tetrachloroguaiacol
1,2,4-trichlorobenzene
3,4,5-trichloroguaiacol
triphenyltin
water clarity

dibutylamine
1,2-dichlorobenzene

4,5-dichloroguaiacol
diethylhexylphthalate
dimethylamine

2,4-dimethylphenol
m-dinitrobenzene
2,4-dinitrotoluene
diphenyl ether
Diquat
di-n-buylamine

ethylene diamine
Fenthion
furfuryl alcohol
hexachlorobutadiene
2-hydroxybinphenyl
isopropyl alcohol
methanol
4-methyl-2-pentanol
morpholine
1-nitronaphthalene
4-nitrophenol
nonyl phenol
palustric acid
phenylxlylethane
pimaric acid
1,2-propylene glycol
radionuclides, several
strontium

tolytriazole

triethyl lead
tungsten
zirconium

Reference: MOE, 1999a.

Table 3.3
Drinking Water Quality Standards and Objectives

Parameter	MAC (mg/L)	IMAC (mg/L)	AO (mg/L)
Alachor		0.005	
Aldicarb	0.009		
Aldrin + Dieldrin	0.0007		
Arsenic		0.025	
Atrazine + metabolites		0.005	
Azinphos-methyl	0.02		
Barium	1.0		
Bendiocarb	0.04		
Benzene	0.005		
Benzo(a)pyrene	0.00001		
Boron		5.0	
Bromoxynil		0.005	
Cadmium	0.005		
Carbaryl	0.09		
Carbofuran	0.09		
Carbon Tetrachloride	0.005		
Chloramines	3.0		
Chlordane	0.007		
Chlorpyrifos	0.09		
Chromium	0.05		
Cyanazine		0.01	
Cyanide	0.2		
Diazinon	0.02		
Dicamba	0.12		
1,2-Dichlorobenzene	0.2		0.003
1,4-Dichlorobenzene	0.005		0.001
DDT + metabolites	0.03		
1,2-Dichloroethane		0.005	
1,1-Dichloroethylene	0.014		
Dichloromethane	0.05		
2,4-Dichlorophenol	0.9		0.0003
2,4-Dichlorophenoxy acetic acid (2,4-D)	0.1	0.1	
Diclofop-methyl	0.009		
Dimethoate		0.02	
Dinoseb	0.01		
Dioxin and Furan		0.000000015[a]	
Diquat	0.07		
Diuron	0.15		
Fluoride	1.5[b]		
Glyphosate		0.28	
Heptachlor + H. Epoxide	0.003		

Parameter	MAC (mg/L)	IMAC (mg/L)	AO (mg/L)
Lead	0.01[c]		
Lindane	0.004		
Malathion	0.19		
Mercury	0.001		
Methoxychlor	0.9		
Metolachlor		0.05	
Metribuzin	0.08		
Monochlorobenzene	0.08		0.03
Nitrate (N)	10.0[d]		
Nitrite (N)	1.0[d]		
Nitrate + Nitrite (N)	10.0[d]		
Nitrilotriacetic Acid	0.4		
NDMA		0.000009	
Paraquat		0.01	
Parathion	0.05		
Pentachlorophenol	0.06		0.03
Phorate		0.002	
Picloram		0.19	
Polychlorinated Biphenyls		0.003	
Prometryne		0.001	
Selenium	0.01		
Simazine		0.01	
Temephos		0.28	
Terbufos		0.001	
Tetrachloroethylene	0.03		
2,3,4,6-Tetrachlorophenol	0.10		0.001
Triallate	0.23		
Trichloroethylene	0.05		
2,4,6-Trichlorophenol	0.005		0.002
2,4,5-T	0.28		0.02
Trifluralin		0.045	
Trihalomethanes	0.1[e]		
Turbidity	1.0[f]		5.0 [f]
Uranium	0.10		
Vinyl Chloride	0.002		

Notes:

MAC	— maximum acceptable concentration
IMAC	— interim maximum acceptable concentration
AO	— aesthetic objective
OG	— operational guideline
NTU	— nephelometric turbidity unit
mg/L	— milligrams per litre
pg/L	— picograms per litre

(a) Total toxic equivalents when compared with 2,3,7,8-TCDD (tetrachlorodibenzo-p-dioxin).
(b) When fluoride is added to drinking water, the concentration should be adjusted to 1 ±0.2 mg/L. Where supplies naturally exceed 1.5 mg/L but less than 2.4 mg/L, local boards of health should raise awareness to control excessive exposure. Levels above the MAC should be reported to the local medical officer of health.
(c) This objective applies to water at the point of consumption. Since lead is a component in some plumbing systems, first flush water may contain higher concentrations of lead than water that has been flushed for five minutes.
(d) Where nitrate and nitrite are present, the total of the two should not exceed 10 mg/L.
(e) This standard is expressed as a running average of quarterly samples measured at a point that reflects maximum residence time in the distribution system.
(f) A MAC for turbidity of 1 NTU in drinking water leaving the treatment plant was established to ensure the efficiency of the disinfection process. Distribution system protection processes can increase turbidity in the distribution system. To ensure that the aesthetic quality is not degraded, an aesthetic objective for turbidity at the free flowing outlet of the ultimate consumer has been set at 5 NTU.

There also are standards for microbiological parameters including total coliform and general bacterial population, and MAC for numerous radionuclides.

Reference: MOE, 2001b.

Table 3.4

Non-Health Related Drinking Water Quality Objectives

Parameter	AO (mg/L)	OG (mg/L)
Alkalinity		350[a]
Aluminum		0.1
Chloride	250	
Colour (in TCU)	5	
Copper	1.0	
Dissolved Organic Carbon	5.0	
Ethylbenzene	0.0024	
Hardness		80 to 100[a]
Iron	0.3	
Manganese	0.05	
Methane	3.0 L/m³	
Odour	inoffensive	
Organic nitrogen		0.15
pH		6.5 to 8.5 (no units)
Sodium	200[b]	
Sulphate	500[c]	
Sulphide	0.05	
Taste	inoffensive	
Temperature	15°C	
Toluene	0.024	
Total Dissolved Solids	500	
Xylenes	0.3	
Zinc	5.0	

(a) Expressed as $CaCO_3$

(b) The local medical officer of health should be notified when sodium concentrations exceed 20 mg/L so that this information may be communicated to physicians and their patients.

(c) When sulphate concentrations exceed the AO, water may have a laxative effect on some people.

Reference: MOE, 2001b.

Table 3.5

Limits on Discharges to Sanitary and Combined Sewers

Parameter	MOE Model By-law	Toronto
General Parameters		
Temperature	< 60°C	same
pH	6.0 to 10.5	6 to 11.5
Total Suspended Solids	350 mg/L	same
Biological Oxygen Demand	300 mg/L	same
Phenolic Compounds	1 mg/L	same
Total Kjeldahl Nitrogen	100 mg/L	same
Solvent Extractables (mineral)	15 mg/L	same
Solvent Extractables (animal/veg.)	150 mg/L	same
Total Cyanide	2 mg/L	same
Fluoride	10 mg/L	same
Phosphorus	10 mg/L	same
Metals		
Aluminum	NA	50 mg/L
Antimony	5 mg/L	same
Arsenic	1 mg/L	same
Cadmium	0.7 mg/L	same
Chromium (total)	5 mg/L	2 mg/L
Chromium (hexavalent)	NA	2 mg/L
Cobalt	5 mg/L	same
Copper	3 mg/L	2 mg/L
Lead	2 mg/L	1 mg/L
Manganese	NA	5 mg/L
Mercury	0.05 mg/L	0.01 mg/L
Molybdenum	5 mg/L	same
Nickel	3 mg/L	2 mg/L
Selenium	5 mg/L	1 mg/L
Silver	5 mg/L	same
Tin	NA	5 mg/L
Titanium	NA	5 mg/L
Zinc	3 mg/L	2 mg/L
Organics		
Benzene	0.01 mg/L	same
Bis(2-ethylhexyl) phthalate	NA	0.012 mg/L
Chloroform	0.04 mg/L	same
Cis-1,2-dichloroethylene	NA	4 mg/L
1,2-dichlorobenzene	NA	0.05 mg/L
1-4-dichlorobenzene	NA	0.08 mg/L
3,3-dichlorobenzidine	NA	0.002 mg/L
1,4-dichloromethane	0.47 mg/L	NA
trans-1,3-dichloropropylene	NA	0.14 mg/L

Organics (cont'd)

Di-n-butylphthalate	NA	0.08 mg/L
Ethylbenzene	0.16 mg/L	same
Hexachlorobenzene	NA	0.0001 mg/L
Hexachlorocyclohexane	NA	0.1 mg/L
Methylenechloride	0.21 mg/L	2 mg/L
Nonylphenols	NA	0.001 mg/L
Nonylphenolethoxylates	NA	0.01 mg/L
PAHs (total)	NA	0.005 mg/L
PCBs	NA	0.001 mg/L
Pentachlorophenol	NA	0.005 mg/L
1,1,2,2-tetrachloroethane	0.04 mg/L	1.4 mg/L
Toluene	0.27 mg/L	0.016 mg/L
Tetrachloroethylene	0.05 mg/L	1 mg/L
Trichloroethylene	0.07 mg/L	0.4 mg/L
Xylene (total)	0.52 mg/L (o-xylene)	1.4 mg/L

Pesticides

Aldrin/dieldrin	NA	0.0002 mg/L
Chlordane	NA	0.1 mg/L
DDT	NA	0.0001 mg/L
Mirex	NA	0.1 mg/L

NOTES

NA - Not Available

The 1998 MOE model by-law and the City of Toronto by-law also contain the condition that the discharge should consist of only one layer.

References: MOE, 1998 and City of Toronto, 2000.

Table 3.6
City of Toronto Limits on Discharges to Storm Sewers

Parameter	**Limit**
Temperature	< 40°C
pH	6.0 to 9.5
Total Suspended Solids	15 mg/L
Biological Oxygen Demand	15 mg/L
E. Coliform	200 per 100 mL
Phenolics (total)	0.008 mg/L
Phosphorus	0.4 mg/L

Metals	
Arsenic	0.02 mg/L
Cadmium	0.008 mg/L
Chromium (total)	0.08 mg/L
Chromium (hexavalent)	0.04 mg/L
Copper	0.04 mg/L
Lead	0.12 mg/L
Manganese	0.05 mg/L
Mercury	0.0004 mg/L
Nickel	0.08 mg/L
Selenium	0.02 mg/L
Silver	0.12 mg/L
Zinc	0.04 mg/L

Organics	
Benzene	0.002 mg/L
Chloroform	0.002 mg
Bis(2-ethylhexyl) phthalate	0.0088 mg/L
Cis-1,2-dichloroethylene	0.0056 mg/L
Cyanide	0.02 mg/L
1,2-dichlorobenzene	0.0056 mg/L
1,4-dichlorobenzene	0.0068 mg/L
3,3'-dichlorobenzidine	0.0008 mg/L
trans-1,3-dichloropropylene	0.0056 mg/L
Di-n-butylphthalate	0.015 mg/L
Ethylbenzene	0.002 mg/L
Hexachlorobenzene	0.00004 mg/L
Hexachlorocyclohexane	0.04 mg/L
Methylenechloride	0.0052 mg/L
Nonylphenols	0.001 mg/L
Nonylphenolethoxylates	0.01 mg/L
PAHs (total)	0.002 mg/L
PCBs	0.0004 mg/L
Pentachlorophenol	0.002 mg/L
1,1,2,2-tetrachloroethane	0.017 mg/L

Organics (cont'd)

Tetrachloroethylene	0.0044 mg/L
Toluene	0.002 mg/L
Trichloroethylene	0.0076 mg/L
Xylene (total)	0.0044 mg/L

Pesticides

Aldrin/dieldrin	0.00008 mg/L
Chlordane	0.04 mg/L
DDT	0.00004 mg/L
Mirex	0.04 mg/L

Reference: City of Toronto, 2000.

Table 3.7
Sample Preservatives

ATG	Preservatives
1*, 4a, 5a, 5b	Add sulphuric acid after sampling to lower pH to between 2 and 6 but not < 1.5
2	Add sodium hydroxide (cyanide free) to raise pH to 12
3, 4b, 7, 8, 10**, 11, 13, 19, 20, 23, 24, 25, 26, 27, 28a, 28b, chloride, sulphate, and fluoride	None
9, 10**, 29	Add nitric acid (containing < 1mg/L of all iron analytes) to lower pH to <2.
12	Add 1 to 2 mL of nitric acid per 250 mL sample followed by at least 0.5 mL of potassium dichromate solution to produce definite yellow colour.
14	Add sulphuric acid solution prior to sampling to lower pH to 2 but not <1.5: diluted acid may be used. Or, prior to sampling, add 1 mL of solution containing 3N phosphoric acid and 0.5 g/L copper sulphate pentahydrate per 250 mL sample.
15	Add 0.5 mL 2N zinc acetate solution per 250 mL sample followed dropwise by 5% sodium carbonate or 5% sodium hydroxide to pH 10.
16, 17 and 18	Only for samples containing residual chlorine. Prior to sampling add 80 mg sodium thiosulphate per 1 L. Store in the dark.

* No preservative may be added but the maximum storage time is reduced from 28 days to four days.

** Can be either none or same as for metals if to be analyzed from same sample.

4

Waste Management and Transportation

4.1   Overview

The generating, handling and disposing of waste materials are regulated by an extensive combination of federal and provincial acts, regulations and guidelines.

This framework defines waste and several classes or types of waste, and describes the responsibilities of those who generate, transport and manage waste. It also sets out the requirements for obtaining approvals to operate waste management systems and describes a comprehensive manifest system for tracking wastes from their point of generation to their recycling or disposal.

The framework is modified frequently, but the basic elements and underlying philosophy have not changed for several years. Nor are regulatory agencies signalling major changes in the near future.

For the purpose of transportation, some types of wastes are considered to be dangerous goods. The transportation of such materials on roads within the province falls under the requirements of Ontario's Dangerous Goods Transportation Act. It is strongly patterned after its federal counterpart, the Transportation of Dangerous Goods Act, 1992 (TDGA, 1992), which pertains to marine, air, rail and inter-provincial and international road transportation.

4.2   Federal Acts and Regulations

The major piece of federal legislation that influences waste management is the Canadian Environmental Protection Act, 1999 (CEPA, 1999), which has provisions to regulate all aspects of the life cycle of toxic substances including their importation, transport, distribution and ultimate disposal as waste.

Key aspects of CEPA, 1999 are summarized in Table 1.2. Those pertaining to waste management include:

1.   The federal government (through Environment Canada) is authorized to regulate waste handling and disposal practices of federal departments, boards, agencies and Crown corporations.

2. The federal government (through Transport Canada and Environment Canada) is authorized to regulate the export and import of hazardous wastes entering into, leaving or passing through Canada.
3. There are provisions to issue permits to control dumping at sea from ships, barges, aircraft and man-made structures (excluding normal discharges from off-shore facilities involved in the exploration for, exploitation and processing of sea-bed mineral resources).
4. There are provisions to create guidelines and codes for environmentally sound practices as well as objectives setting desirable levels of environmental quality.

Several regulations addressing waste management have been promulgated under CEPA, 1999:

- Regulations Respecting Applications for Permits for Disposal at Sea (SOR/2001-276)
- Disposal at Sea Regulations (SOR/2001-275)
- Export and Import of Hazardous Wastes Regulations (SOR/92-637)
- PCB Waste Export Regulations, 1996 (SOR/97-109)

The two regulations that address disposal at sea are not relevant to Ontario. The export and import regulations are discussed in Section 4.10 and the regulations pertaining to PCB waste are discussed in Chapter 7.

The transport of wastes that also are considered to be dangerous goods can trigger the requirements of TDGA, 1992 and associated regulations. These requirements are discussed in Section 4.9.

4.3   Provincial Acts and Regulations

4.3.1   Environmental Protection Act

Part V of the EPA is entitled "Waste Management". It provides definitions of several key terms and sets out the administrative processes used to control (in a regulatory sense) the management and disposal of waste in Ontario.

Section 25 defines waste as "ashes, garbage, refuse, domestic waste, industrial waste, or municipal waste and such other materials as are designated in the regulations". This is examined further in Section 4.4. Other important terms defined in s. 25 include those for owners and operators of waste management systems and disposal sites.

Section 27 describes the need for a Certificate of Approval (C of A) to operate, establish, alter, enlarge or extend a waste management system or disposal site. Subsequent sections describe the process for obtaining a C of A, the role of public hearings when the Ministry considers issuing a C of A, and the terms and conditions that can be included in a C of A.

Section 42 contains an important provision: generators are responsible for the wastes they generate until the wastes are delivered to an approved disposal site. At that point, the owners or operators of the disposal site become responsible for the wastes.

Section 46 prohibits the use of waste disposal sites for 25 years unless approved by the Minister.

4.3.2   EPA Regulations

Several regulations have been promulgated under the EPA that address waste management:

- Deep Well Disposal (Regulation 341)
- Designation of Waste (Regulation 342)
- General – Waste Management (Regulation 347)
- Industrial, Commercial and Institutional Source Separation Programs (O. Reg. 103/94)
- Landfilling Sites (O. Reg. 232/98)
- Packaging Audits and Packaging Reduction Work Plans (O. Reg. 104/94)
- Recycling and Composting of Municipal Waste (O. Reg. 101/94)
- Waste Audits and Waste Reduction Work Plans (O. Reg. 102/94)
- Waste Disposal Sites and Waste Management Systems Subject to Approval under the Environmental Assessment Act (O. Reg. 206/97)
- Waste Management – PCBs (Regulation 362)

Regulation 347 contains details in support of the concepts described in Part V (ss. 25-55) of the EPA. For example, it defines various types of waste. (See Section 4.4.1 for discussion.) It also describes various types of waste management activities or facilities including: dumps, landfilling, composting and transfer stations. It sets standards for disposal sites, incineration sites and dumps.

Among the sections in Regulation 347 that are relevant to this chapter, topics include:

- managing some specific types of waste including asbestos waste (s. 17) and refrigerant waste (in ss. 30–42).
- describing the process for registering as a waste generator (s. 18). This is examined in detail in Section 4.5.
- describing the six-part manifest to be used when transporting waste (ss. 19–26). Several aspects of the manifest system are examined in Section 4.7.
- setting out conditions for operating waste depots (ss. 43–60) and pesticide container depots (ss. 61–73).

Regulation 362 defines various types of materials including PCB waste, PCB liquid, PCB equipment and PCB materials and describes the provincial requirements for managing these materials.

From the late 1970s — when PCBs started to be closely regulated in Canada — to the late 1990s, there were few if any options for disposal of PCBs. As a result, a special regulatory regime was created for the management of PCB waste and out-of-service equipment that contained PCBs. Managing PCB waste is described further in Section 4.4.3. Other aspects of PCBs are addressed in Chapter 7.

The four regulations directed at waste reduction are addressed in Section 4.11.2.

4.3.3   Ontario Waste Diversion Act, 2002

The Ontario Waste Diversion Act, 2002 establishes Waste Diversion Ontario, a permanent non-government organization with the objectives of developing, implementing and funding waste diversion programs. The Act also allows the MOE to designate waste materials in regulations and to require a diversion program to be implemented for that waste. Waste diversion programs established under the Act must be approved by the MOE. Public consultation is required. See Section 4.11 for more details.

4.3.4   Nutrient Management Act, 2002

The Nutrient Management Act, 2002 will establish provincial requirements for managing materials containing nutrients such as biosolids generated by municipal sewage treatment plants, pulp and paper sludge, commercial fertilizers, and animal wastes and other materials generated by agricultural operations. As noted in Section 4.4.2, these materials are exempt from the requirements of the EPA and Regulation 347.

This Act is expected to come into force in 2003. It will be administered by the Office of the Ministry of Agriculture, Food and Rural Affairs. More details can be obtained at **www.gov.on.ca/OMAFRA**.

4.4   Definitions and Classifications

4.4.1   Defining Waste

Despite the framework of statutes and regulations pertaining to waste management, the simple question "what is waste?" continues to be a source of difficulty for both regulators and the regulated community.

Discussions often begin with s. 25 of the EPA, which provides the basic but somewhat vague definition of "ashes, garbage, refuse, domestic waste, industrial waste, or municipal waste and such other materials as are designated in the regulations".

Regulation 347 provides details for two of these terms. Industrial waste includes wastes from industrial or commercial activities such as manufacturing, processing and warehousing; research activities; clinics that provide medical diagnosis or treatment; and schools, hospitals and laboratories. Municipal waste is defined by what it is not. It includes any waste except subject waste (described below), gaseous waste and solid fuel. The waste does not need to be owned, controlled or managed by a municipality.

When faced with the question of what is waste, the courts often have defined waste as materials thrown aside or discarded as worthless, however, the MOE has determined that some materials with value need to be considered as waste and are therefore subject to the requirements of Part V of the EPA and Regulation 347. These include (MOE, 2001):

- materials that are sold for heating value or that are otherwise being re-used, recycled or reclaimed
- all products and by-products from waste transfer, bulking, treatment or processing facilities
- oil recovered from oily water treatment facilities as well as blended or bulked waste solvents destined for disposal or recycle, and
- commercial chemical products or by-products, including those that are off-specification or that have exceeded their expiry date

Sections of Regulation 347 have been revised on numerous occasions to separate the issue of value from the definition of waste. Specifically identified as waste are materials as diverse as: dust suppressant, inert fill, rock fill, mine tailings, processed organic waste, waste-derived fuel, used tires that have not been refurbished, wood waste, refrigerant waste, hauled sewage and building materials such as brick, concrete, steel, wood and dry wall. This applies to material as soon as it leaves a construction or demolition site, unless the material is going elsewhere for use or sale. Revisions in 1998 designate as waste the residue from an industrial, manufacturing or commercial process or operation if the residue leaves the site where the process or operation is carried out.

4.4.2   Exempt Wastes

Some types of waste materials are exempted from the requirements of Part V and Regulation 347 for one or more of the following reasons:

- the wastes are regulated by other statutes (for example, agricultural wastes, hauled sewage, oil and gas well wastes [also known as brines] and condemned and dead animals)
- the wastes are generally perceived to be innocuous (for example, inert fill, rock fill and mill tailings)
- the wastes have been specifically identified as exempted (for exam-

ple, pickle liquor going to a sewage treatment plant or waste water treatment plant and photographic waste to be processed to recover silver)
• the wastes are recyclable

To qualify as recyclable, the waste must meet the provisions of Regulation 347. Section 1 of the regulation defines recyclable material as waste transferred by a generator directly to a site where:

• it will be wholly utilized in an ongoing agricultural, commercial, manufacturing or industrial process
• it will be processed principally for functions other than waste management and in ways that do not involve combustion or land application of waste and the transporter carries an agreement from the owner or operator of the destination to accept the material, and
• it will be promptly packaged for retail sale or offered for retail sale to meet a realistic market demand

Not eligible for this exemption are used, shredded or chipped tires even when they meet the provisions noted above. Subject wastes (see Section 4.4.3) are considered to be recyclable materials only if consumed completely at the recycling operation.

The question as to whether or not a process is being used for recycling or for waste management can be determined by reviewing its viability if the waste was not available. Processes or operations that are not viable without the incoming waste are deemed to be in the business of waste management.

According to the definitions in Regulation 347, by-products or intermediates from traditional refining operations, such as mineral or metal recovery, are recyclable materials. For example, sludges from an electrolytic recovery process for metals subsequently processed to remove precious metals, such as silver or gold, are not considered to be wastes.

4.4.3   Classifying Wastes

While many materials are waste, not all wastes require the same degree of regulatory control or the same degree of attention in Part V and the associated regulations, notably Regulation 347. Excluding the wastes that are exempt (as noted in Section 4.4.2), other waste materials in Ontario can be divided into two broad categories.

There are the wastes that are subject to the full gamut of regulatory controls. These are referred to as subject waste (and often divided into hazardous waste and liquid industrial waste). Generators must be registered with the MOE. Approvals are required to transport, handle and dispose subject wastes. Stiff penalties can be imposed for not complying with the regulatory requirements.

The second category of wastes includes those that do not pose extraordinary concern and therefore do not warrant the same degree of care given to subject waste. These wastes tend to be generated widely by individuals and organizations. They include common municipal waste or household garbage. Generators do not have regulatory requirements but approvals are required to operate the systems that handle and dispose these wastes. Wastes that are not classified as subject waste are:

- waste generated by individual households (but not hotels or motels)
- waste generated by nursing homes, special care facilities and the professional offices of doctors and dentists
- waste generated at retail motor vehicle service stations that have written agreements with approved waste collection and management companies
- hauled sewage
- sewage sludge
- discharges from approved waste water treatment plants
- sludges from approved waterworks
- wastes collected at selected depots for materials such as antifreeze, waste oil filters and waste oil
- incinerator ash (but not fly ash)
- small quantities of wastes as defined in Regulation 347 and described below

Subject waste is the focus of much of Regulation 347. All subject wastes are assigned "waste numbers", a slightly misleading term, since they consist of three digits (that signify the waste class) and a single letter (which signifies the waste characteristic). Once determined, the waste number should appear in all documents that pertain to the registration, transportation and disposal of the waste. The three digit waste classes are listed in Table 4.1. The waste characteristic can be determined by following the sequence of questions shown in Figure 4.1 and described briefly below. A detailed guide to this process is available (MOE, 2001).

There are eleven types of subject waste. The first 10 described below are all specific types of hazardous waste. The eleventh type is liquid industrial waste. To determine which characteristic is appropriate for a waste, information about the waste is assessed using a sequence of questions. The first question that correctly matches a waste determines the characteristic for the waste. Some wastes may match more than one characteristic. In that case, the first match determines the primary characteristic, and the second match determines the secondary characteristic. The primary characteristic is used in the waste number for the waste.

Severely Toxic (S)

A waste that contains any of the substances listed in Schedule 3 of Regulation 347 at a concentration greater than 1 ppm is characterized as being severely toxic (S). Examples include pesticides such as 2,4,5-T and pentachlorophenol. There is no small quantity exemption (SQE) for such wastes. Empty containers and inner liners are considered to be hazardous waste. For all subject wastes, an "empty container" is defined as one that contains less than 2.54 cm (1 inch) of material at the bottom. (this also is referred to as the "heel" rule).

Pathological (P)

A waste that contains any part of the human body, including tissues and bodily fluids, but excluding fluids, extracted teeth, hair, nail clippings and the like which are non-infectious; any part of the carcass of an animal infected with a communicable disease or suspected by a licensed veterinarian to be infected; or any non-anatomical waste infected with communicable disease are characterized as pathological (P). There is no SQE for such wastes. Empty containers and liners are also hazardous unless they have been incinerated, autoclaved or otherwise sterilized to make them non-infectious. As of late 2002, the MOE was proposing to replace pathological waste with "biomedical waste" to distinguish it from municipal solid waste and so reduce the amount of municipal waste that is classified and treated as pathological waste. In addition to the change in definition, the methods for processing, treating and disposing of this waste would be specified.

PCB Waste (D)

A waste that is defined to be a PCB waste according to Regulation 362 is characterized as a PCB waste (D). Generally, PCB wastes contain PCBs at concentrations greater than 50 ppm by weight. There is no SQE for PCB waste (but some materials or equipment that contain PCBs are exempt from the definition of PCB waste as noted in Chapter 7).

Acute Hazardous Waste (A) and Hazardous Waste Chemical (B)

Any commercial products or manufacturing intermediates that are off-specification or otherwise unacceptable for use and that contain any of the substances listed in Schedules 2(A) or 2(B) of Regulation 347 are characterized as being either acute hazardous waste chemical (A) or hazardous waste chemical (B). This includes materials such as pharmaceutical or pesticide waste products that contain active ingredients from Schedules 2(A) or 2(B). Active ingredients are constituents that have been included in a formulated product for an intended effect.

The SQE for waste listed in Schedule 2(A) is 1 kg of waste per month or otherwise accumulated in an amount of less than 1 kg. Containers with a capacity greater than 20 L, and that contained the Schedule 2(A) material, are considered to be hazardous unless triple-rinsed using an appropriate solvent. Inner liners weighing more than 10 kg that contained Schedule 2(A) products are also considered hazardous unless they have been triple-rinsed using an appropriate solvent.

The SQE for waste listed in Schedule 2(B) is 5 kg of waste per month or otherwise accumulated in an amount of less than 5 kg. Empty containers and inner liners are not considered hazardous.

Hazardous Industrial Waste (H)

Any industrial waste stream that is generated from a process listed in Schedule 1 of Regulation 347 is characterized as being hazardous industrial waste (H). There is a SQE of 5 kg per month or otherwise accumulated in an amount of less than 5 kg. Empty containers and inner liners are not hazardous.

Ignitable Waste (I)

A waste that meets the Regulation 347 definition for ignitability is characterized as being ignitable waste (I). The following criteria are used to determine if a waste should be categorized as ignitable:

- it is a liquid with a flash point less than 61°C (for example, ethanol or gasoline)
- it is a solid capable, under standard temperature and pressure, of causing fire due to friction, absorption of moisture or spontaneous chemical changes and, when ignited, burns so vigorously and persistently that is creates a hazard (for example, sodium metal)
- it is an ignitable compressed gas having a critical temperature less than 50°C or an absolute vapour pressure greater than 294 kPa at 50°C, or exerts an absolute pressure, in the cylinder, packaging tube or tank in which it is contained, greater than 275 ± 1 kPa at 21.1°C or 717 ± 2 kPa at 54.4°C, and is ignitable at normal atmospheric pressure when in a mixture of 13% or less by volume with air, or have a flammability range of at least 12 (for example, methane, butane and propane)
- it is an oxidizing substance which readily yields oxygen to stimulate, or contribute to, the combustion of other materials (for example, chlorates, permanganate and nitrates)

For ignitable wastes there is an SQE of 5 kg per month or otherwise accumulated in an amount of less than 5 kg. Empty containers and inner liners are not hazardous.

Corrosive Waste (C)

A waste that meets the Regulation 347 definition for corrosivity is characterized as being corrosive waste (C). The following criteria are used to determine if a waste should be categorized as corrosive:

- it is aqueous and has a pH less than or equal to 2.0 or greater than 12.5
- it is a liquid and corrodes steel (SAE 1020) at a rate greater than 6.35 millimetres per year at a test temperature of 55°C using the National Association of Corrosion Engineers test method TM-01-69

Corrosive waste has an SQE of 5 kg per month or otherwise accumulated in an amount of less than 5 kg. Empty containers and inner liners are not hazardous.

Reactive Waste (R)

A waste that meets the Regulation 347 definition for reactivity is characterized as being reactive waste (R). The following criteria are used to determine if a waste should be categorized as reactive:

- it is normally unstable and readily undergoes violent change without detonating
- it reacts violently with water
- it forms potentially explosive mixtures with water
- when mixed with water it generates toxic gases, vapours or fumes in a quantity sufficient to present danger to human health or the environment
- it is a cyanide- or sulphide-bearing waste which, when exposed to pH conditions between 2.0 and 12.5, can generate toxic gases, vapours or fumes in an quantity sufficient to present danger to human health or the environment
- it is capable of detonation or explosive reaction if it is subjected to a strong initiating source or if heated under confinement
- it is readily capable of detonation or explosive decomposition or reaction at standard temperature and pressure
- it is a Class 1 explosive as defined in Schedule II, List I of the TDGA, 1992 regulations.

Reactive waste has an SQE of 5 kg per month or otherwise accumulated in an amount of less than 5 kg. Empty containers and inner liners are not hazardous.

Leachate Toxic (T)

Any waste that produces a leachate that contains any of the substances at a concentration greater than the concentrations listed in Schedule 4 of Regulation 347 (see Table 4.2) is characterized as being leachate toxic (T). The Toxicity Characteristic Leachate Procedure (TCLP) is described in Regulation 347. There is a SQE of 5 kg per month or otherwise accumulated in an amount of less than 5 kg. Empty containers and liners are not hazardous.

Until 2001, there also was a classification of registerable waste (N) that applied to a material if it produced a leachate that contained any of the substances at concentration between 10 and 100 times the concentrations listed in Schedule 4 of Regulation 347. That classification was eliminated in 2001 when the leachate procedure in Regulation 347 was amended by O. Reg. 558/00.

Liquid Industrial Waste (L)

If a waste meets none of the criteria for the 10 types of hazardous waste, it may be a liquid industrial waste (L). The slump test described in Schedule 5 of Regulation 347 is used to determine if a material is a liquid. This type of waste is broadly defined to include any liquid waste or sludge from industrial, commercial, manufacturing, research or experimental activities. There is an SQE of 25L/month or otherwise accumulated in an amount of less than 25L. The following wastes are exempted from this waste type:

- waste from the operation of a waterworks subject to the Ontario Water Resources Act
- waste directly discharged from a waste generation facility into a sewage system subject to the Ontario Water Resources Act
- waste from food packing, processing or preparation operations including wineries and cheese-making facilities, as well as the grease from restaurants
- drilling fluids and produced waters associated with oil and gas exploration, development and production
- processed organic waste
- asbestos waste

4.4.4   MOE Advice on Common Situations

The guidance manual for waste generators (MOE, 2001) provides the following advice in more detail for several common situations.

Hazardous wastes discharged to on-site private wastewater treatment facilities must be registered. Sludges and skimmings produced at treatment facilities must be registered. Wastewater discharges to a watercourse do not need to be registered.

For companies that have approved multi-step wastewater treatment facilities, registration is required for all of the hazardous waste discharges to the final treatment process. Wastes discharged to intermediate processes do not need to be registered.

Hazardous wastes discharged to municipal sanitary sewers must be registered. Educational or research laboratories are required to estimate the types and quantities of hazardous waste to be discharged in this manner and to register them.

All subject wastes hauled to an approved, off-site water pollution control plant are registerable.

All subject wastes disposed of by means of an on-site waste management facility, such as a landfill or a combustion facility, must be registered.

Wastes sent to another on-site process such as bulking or blending operations, do not need to be registered. If those wastes subsequently are returned to an on-site operation, they are not registerable, but if hauled off-site, they are registerable.

4.4.5   Difficult-to-Classify Materials

Although the waste classification system described in Sections 4.4.2 and 4.4.3 works well for most wastes, there are some materials for which the existing definitions do not provide a clear indication as to whether or not they are wastes or how they should be managed. An example is contaminated soil. It is well-known that there are many properties where soils might be classified as leachate toxic waste. If these materials were deemed to be wastes while in the ground, it could follow that the properties would need to be classified as waste disposal sites (according to the definition of "waste disposal site" provided in s. 25 of the EPA). It also has been argued that such soils become waste only upon being excavated or otherwise worked. Under such a scenario, the re-use of treated soil as fill could be construed as creating a waste disposal site (since the soil, even though it had been treated, would still be waste). These uncertainties have hindered clean-up efforts at some sites in Ontario (see Chapter 5).

Similar difficulties can arise when dealing with mildly contaminated soil. The presence of contaminants could be interpreted to mean that the soil that does not strictly conform to the definition of "inert fill," or that it should not be considered to be an "innocuous waste" (refer to Section 4.4.2).

In 1998, definitions for inert fill were proposed (MOE, 1998). Four classes of inert fill were proposed, based on maximum measured concentrations of chemicals. In effect, this was an attempt to clarify how to manage these materials. As of late 2002, no confirmation had been announced as to the adoption of the proposed definitions.

4.5   Responsibilities of Waste Generators

Section 1 of the EPA defines the term "person responsible" as the owner or the person having charge, management or control of a source of contaminant. Part V of the EPA expands this concept and identifies three types of responsible persons: waste generators, owners of waste management facilities or disposal sites, and operators of systems or sites. (Responsibilities of the latter two are discussed in Section 4.6.)

Generators include operators of commercial and manufacturing facilities that produce waste as well as operators of waste transfer, bulking, treatment or processing facilities that forward materials off-site for subsequent management.

Generators are responsible for the wastes they generate until the wastes are accepted at an approved disposal site. At that point, the owners or operators of the disposal site become responsible for the wastes. This provides generators with some protection from long-term liabilities associated with disposal. That is an important difference between Ontario legislation and many jurisdictions in the United States where generators are held liable for these wastes even after proper disposal.

Every facility where subject waste is generated must be registered with the MOE by submitting a completed Generator Registration Report. There are two approved ways to submit a report. One is to complete a paper version of the report prepared by the MOE for this purpose (see Figure 4.2). Full size originals can be obtained from the MOE. A photocopy is acceptable, provided it contains an original signature. The second way to register is via the Hazardous Waste Information Network (HWIN), which can be accessed at the MOE Web site at **www.ene.gov.on.ca**. Note that registration is not required in anticipation of wastes that may be generated through spills. Emergency registration procedures are available for spill situations.

Once registered, the MOE will provide the facility a generator registration number. It is an offence to store, dispose and transport subject waste unless a generator registration number has been obtained. Out-of-province waste generators that transport or dispose of subject waste in Ontario must register. There are three parts to the Generator Registration Report.

Part 1 requires the company's name and address, the appropriate North American Industry Classification System (NAICS) code(s) for the site and the name of the company contact and the contact's signature. NAICS codes are provided in the *Registration Guidance Manual for Generators of Liquid Industrial and Hazardous Waste* (MOE, 2001) or can be obtained at **www.statscan.ca/english/subjects/standards/index.htm**. Consultants who act on behalf of the company cannot sign Part 1.

Part 2 requires a description of the waste, the generating process and the quantity of waste. In addition, information on the primary and secondary characteristics of the waste, and the laboratory that analyzed the waste may also be required. Space also is provided for a hazardous waste number that can be found in the federal TDGA, 1992 regulations (TDGR).

Part 3 of the report covers some waste management aspects, notably the intended receiver and carriers. Generator Registration Reports must be submitted by February 15 each year, along with an application fee. Merely submitting the report and the fee does not constitute a completed registration. The true indication that the information has been reviewed and acknowledged by the MOE is when the MOE posts a generator registration document on the HWIN. The MOE no longer mails acknowledgement letters to generators.

If a registration expires, subject waste cannot be taken to a waste management system until the generator renews the registration and it is posted on the HWIN.

If, during a year, a generator produces wastes not declared in the original report, the generator must update its registration. Other reasons for updating a registration include a change in company name, address, contact or telephone numbers, as well as a change in a waste characteristic or waste class. Generators are responsible for keeping their registrations up-to-date and correct.

The annual fee is the sum of three components: an annual flat fee of $50, $5 per manifest used during the year and $10 per tonne of hazardous waste generated during the year. There are options for making fee payments during the year as opposed to making full payments with the filing of the registration report. Details can be found at the HWIN. The only generators exempt from the fees are municipal household hazardous waste facilities operated by or for a municipality, contaminated sites where subject waste is being generated by remediation activities and sites requiring emergency generator registration (typically in response to a spill).

Some of the information provided in a Generator Registration Report is accessible to the public. The MOE maintains a public information database that provides information on the type of waste, the volumes generated and the generator's name and address.

Regulation 347 also requires that waste generators must notify the MOE if subject wastes in quantities greater than the appropriate SQEs are stored on property for longer than three months. This necessitates some type of waste inventory system to track the duration of waste storage.

The generator of a waste is responsible for ensuring that the disposal site receiving the waste has the appropriate Cs of A. Generators often ask for a copy of the C of A or the certificate number prior to entering into an agreement with a disposal site operator.

While a waste is being transported, both the generator of the waste and the carrier of the material are liable should there be an incident involving the material. Generators can make it a contractual condition that ownership of the waste is transferred when the carrier picks up the material.

If the disposal site refuses to receive a waste, the carrier can take the waste back to the generator. The generator is legally required to accept the material back. If the waste is returned to the generator, the generator is then classified as a receiver and must follow the appropriate procedures (see Section 4.7.4).

4.6   Managing Wastes

4.6.1   Definitions

Part V of the EPA refers to waste management facilities as either being waste disposal sites or waste management systems. Disposal sites include any land where waste is deposited, disposed, handled, stored, transferred, treated or processed plus any operation or equipment used in connection with those activities. Disposal sites include landfilling sites, incineration sites, transfer stations, packing and baling sites, dumps, used tire sites and organic soil conditioning sites. Waste management systems include any facilities or equipment used to collect, handle, transport, store, process or dispose of waste and may include one or more disposal sites.

Section 42 of the EPA outlines several conditions that apply to the transfer of responsibility for disposal of a waste from a waste generator to the owner or operator of a disposal facility or transfer station:

> (1) The ownership of waste that is accepted at a waste disposal site by the operator of the site is transferred to the operator upon acceptance.
> (2) Where waste is deposited but not accepted at a waste disposal site, the ownership of the waste shall be deemed to be transferred to the operator of the site immediately before the waste is deposited.
> (3) Subsections (1) and (2) apply only in respect of a waste disposal site for which a certificate of approval or a provisional certificate of approval is in force.
> (4) Subsection (1) applies only in the absence of a contract to the contrary.
> (5) Subsections (1) and (4) do not relieve any person from liability except liability as owner of a waste that is delivered to and accepted by the operator of a waste disposal site in accordance with law including an applicable certificate of approval or provisional certificate of approval.
> (6) Where the operator of a waste disposal site is not the owner of the land on which the site is located, subsections (1) and (2) do not prevent the ownership of waste that is accepted or deposited at the site from being transferred to the owner of the land.

This transferring of responsibility away from waste generators is a major difference between the approach to waste management in Ontario and that used in some other jurisdictions, notably the United States, where generators essentially are responsible for their wastes forever (Saxe, 2001).

4.6.2   Approvals

Section 27 of the EPA specifies that no person shall use, operate, establish, alter, enlarge or expand a waste management system or waste disposal site unless a C of A or a provisional C of A has been issued by the MOE and except in accordance with any conditions set out in such certificate. Operating a waste system without a certificate is an offence.

Section 39 allows an MOE director to impose conditions on a disposal site or management system that are "in the public interest" (but does not expand upon that phrase).

Section 40 prohibits any person from depositing waste without a C of A "upon, in, into or through any land or land covered by water or in any building that is not a waste disposal site" for which a C of A has been issued, except if in accordance with the terms and conditions of such a certificate. Since Section 27 applies to owners and operators of waste management systems, it appears that this section applies to everyone else.

Section 41 limits the equipment and facilities that may be used in conjunction with a waste management system in that "[n]o person shall use . . . any facilities or equipment for the storage, handling, treatment, collection, transportation, processing or disposal of waste that is not part of a waste management system" for which a C of A has been issued and except in accordance with the terms and conditions of such a certificate.

The MOE has prepared various documents to assist in the preparation of a C of A for waste disposal sites and waste management systems, including the *Protocol for Updating Certificates of Approval for Waste Management* in 2002. Copies can be obtained at **www.ene.gov.on.ca/envision**.

4.6.3   Hearings

An MOE director must order public hearings prior to deciding whether a C of A should be issued for any sites that will receive liquid industrial waste, hazardous waste or any other waste equivalent to the domestic waste of more than 1,500 people. In other cases, the director can decide whether or not a public hearing should be held.

Public hearings convened under the EPA are limited in scope to the technical applicability of the proposal. Hearings also can be convened under the Environmental Assessment Act, in which case the scope must be expanded to consider the need for the proposed undertaking, alternatives to the proposed undertaking and alternative methods of carrying out the undertaking. The evaluation of options is a cornerstone of the Environmental Assessment Act and also applies to the selection of technologies and sites.

There also are cases in which waste disposal proposals are subject to hearings under both the EPA and Environmental Assessment Act. The Consolidated Hearings Act permits the two hearings to be held together.

Whether convened under the EPA, the Environmental Assessment Act or the Consolidated Hearings Act, the types of information that are likely to be reviewed include:

- the purpose for the undertaking
- the reasons behind the proposed project
- a description of how the project will affect the environment
- a description of the actions proposed to prevent, change, mitigate or remedy environmental effects
- an evaluation of advantages/disadvantages to the environment of the project

Sections 30 through 36 of the EPA describe various aspects of the hearing process. For example, s. 31 authorizes an MOE director to dispense with public hearings where an emergency situation exists. Section 36 authorizes a director to require a hearing to determine whether an existing or proposed disposal site or management system can be exempt from municipal by-laws.

4.6.4   Restoring or Re-using Disposal Sites

Section 43 provides MOE directors with the statutory authority to order an occupant or the person having charge and control of land or a building to remove wastes and restore the site to a condition satisfactory to the director if the location has not been approved as a waste disposal site. Section 44 authorizes the director to require a waste management system or operator to comply with the requirements of the EPA.

Once a site has been classified as a waste management site, certain restrictions are imposed upon it. Section 46 of the EPA states that if a land site or land covered by water was used for disposal of waste, a period of 25 years from the year in which the waste disposal ceased is required before the land can be used for another purpose unless approval has been obtained from the Minister.

4.7   Manifest System For Subject Wastes

4.7.1   Overview

Regulation 347 describes a manifest system that must be used during the transportation, transfer and disposal of all subject wastes. The system is compatible with the one required under the TDGR, so that only one piece of governmental paperwork needs to be filled out (see Figure 4.3) for transportation within Canada. Additional paperwork is required for international shipments. The TDGR are described in Section 4.9.

Manifests are used to track each shipment of the waste as it passes from generator, to carrier and eventually to the receiver. The MOE oversees the whole system to make sure every aspect is being followed properly.

Carriers that have received approvals from the MOE to transport waste will provide paper copies of manifests. Numbered manifests are available electronically on HWIN. When using the electronic option, the generator needs to provide the carrier with access to the manifest so that the carrier can provide its information. The generator also is responsible for transmitting the manifest to the MOE.

The six-copy manifest is divided into three sections:

- Section A: completed by the consignor (generator)
- Section B: completed by the carrier
- Section C: completed by the consignee (receiver)

Figure 4.4 illustrates how the six copies become distributed as waste is transferred from generator, to carrier, to receiver. That distribution also is described in Sections 4.7.2 through 4.7.4.

4.7.2   Responsibilities of the Generator

A generator must fill out section A of the manifest, including the name of the intended receiver. In addition, the generator must ensure that the carrier completes section B of the manifest. The generator must then remove the first copy and forward it to the director of the Waste Management Branch within three working days of the transfer. The same schedule applies if an electronic manifest is used.

The second copy is removed and retained on file by the generator for a period of two years. The four remaining copies of the manifest are given back to the carrier.

Within two weeks of the transfer of waste, the generator should receive the sixth copy of the manifest from the receiver of the wastes. If this does not happen, the generator should contact the receiver to ensure that the waste did in fact reach its intended destination. If unable to trace the waste within four weeks of the waste transfer, the director of the Waste Management Branch must be notified. These time periods may differ if sending waste outside Ontario (MOE, 2001).

4.7.3   Responsibilities of the Licensed Carriers

A licensed carrier must obtain a C of A and comply with all of the transportation requirements imposed by Regulation 347 and any applicable requirements of TDGA, 1992 (see Section 4.9). The carrier must carry the C of A in the vehicle at all times and only transfer those wastes specified in the certificate.

The carrier must complete section B of the manifest and retain the third through sixth copies of the manifest during the transfer of the waste. The carrier must only transport the wastes to a transfer station or disposal site that is operating under a C of A and transport only those wastes specified in section C of the manifest. Prior to leaving the site of a waste transfer, the carrier must obtain the fourth copy of the completed manifest from the receiver of the waste and retain the document for a period of two years.

Carriers may enter into contracts that offer various forms of assistance to waste generators. For example, a carrier can arrange for samples of the waste stream to be collected and analyzed, fill out necessary forms at the disposal site, arrange for the disposal codes to be put in place, and arrange the scheduling at the receiving site, be it a transfer site or a final disposal site.

If the disposal site refuses to receive a load, the carrier is responsible for either finding an alternative approved receiver or taking the waste back to the generator. The generator is legally required to accept the material if returned.

4.7.4   Responsibilities of the Receiver

A receiver can only accept those waste classes for which the receiver has a C of A. The operation of the site must be in accordance with the conditions specified in the certificate.

The receiver must complete section C of the four remaining copies of the manifest. The third copy of the manifest must be sent to the MOE, the fourth copy is returned to the carrier, the fifth copy is retained by the receiver for two years, and the sixth copy is sent to the generator within three working days after the waste transfer.

If the receiver refuses the waste, a load refusal report must be prepared outlining the manifest number, generator registration number and the reason for the refusal. The report must be sent to the director of the Waste Management Branch within three working days after the refusal.

4.8   Transporting Wastes inside Ontario

4.8.1   Carriers

To transport wastes regulated under Regulation 347, a carrier must have a provisional C of A for a waste management system. The carrier must make an application to the MOE and satisfy certain requirements with respect to driver training and insurance prior to obtaining a C of A.

A C of A will identify the vehicles that the carrier can use, the waste classes that the carrier can transport and the disposal sites or transfer stations that the carrier can use. In addition to the conditions of the certificate, Regulation 347 imposes certain standards upon the carrier:

- waste collection vehicles and waste carriers must be of a design that ensures safe transfer of waste without leakage, emission of offensive odours or falling or blowing of waste material from the vehicle

- a vehicle used for transporting liquid industrial waste or hazardous waste must be clearly marked or placarded in accordance with TDGA, 1992 and display the name and number from the C of A (refer to Section 4.9)
- driver training must be provided for the operation of the vehicle and waste management equipment, relevant legislation, environmental and safety concerns and emergency management procedures
- a copy of the C of A that authorizes the transport must be kept in the vehicle when liquid industrial waste or hazardous waste is being transported
- the driver, generator or receiver must be present whenever liquid industrial waste or hazardous waste is being transferred

4.8.2   Transfer Stations

The transfer station is an intermediate point between the original generator and the ultimate disposal site of the waste. For a transfer station to receive waste, it must have one of two types of certificates: a C of A for a waste disposal site (bulking), or a C of A for a waste disposal site (processing). A public hearing may be required to obtain the necessary certificates (see Section 4.6.3).

4.8.3   Transferring Responsibility or Ownership of Wastes

As noted in Section 4.3, s. 42 of the EPA allows responsibility for a waste to be transferred from the generator to a third party (a disposal site operator, the owner of a disposal site or the owner of a transfer station). While a waste is being transported, both the generator of the waste and the carrier are liable should there be an incident involving the material. Contractual conditions can be used to transfer all liabilities to the carrier. Such a transfer of ownership may not be recognized in court if it is seen as trying to circumvent liability specified under Part X of the EPA.

4.9   Transporting Wastes under the Transport of Dangerous Goods Act, 1992

4.9.1   Overview of the New TDGR

The TDGA, 1992 and associated regulations pertain to marine, air, rail and inter-provincial and international road transportation. TDGA, 1992 is administered by Transport Canada. Environment Canada provides technical advice and recommendations for regulatory initiatives or matters related to dangerous goods.

A revised, new "clear language" version of TDGR came into force in August, 2002. The following is a summary of the key differences between the old TDGR and the clear language regulations. Further details can be obtained at **www.tc.gc.ca/tdg**.

1. UN recommendations – The lists of dangerous goods in both versions are based on the UN Recommendations on the transportation of dangerous goods: the original regulations refer to the 1991, 7th revised edition with the exception of safety marks where the 6th edition is referenced. The clear language TDGR refer to the 11th revised edition, 1999 and incorporate numerous changes adopted by the UN from 1991 to 1999.

2. Layout –The previous TDGR were divided into 16 parts and three schedules. The clear language TDGR are divided into 13 parts and 14 schedules. The subjects of some parts have changed. For example, shipping document requirements were addressed in Part IV, but now are in Part 3 of the clear language version.

3. Variations by mode of transport – The TDGR sometimes permit or require compliance with other legislation, such as the International Maritime Dangerous Goods Code, the International Civil Aviation Organization technical instructions for the transportation of dangerous goods, Title 49 of the American Code of Federal Regulations and the United Nations Recommendations. In the clear language TDGR, these requirements are identified under specific parts. For example, the additional requirements for marine transport are included in Part 11. In the previous regulations, requirements for a specific mode of transport were found in several different parts. For example, the reciprocity provisions for road transport between Canada and the United States were found in Parts IV, V, VII and IX. In the clear language version, the reciprocity provisions for road transport between Canada and the United States are found in Part 9.

4. Waste dangerous goods – The previous regulations had specific requirements related to dangerous goods that are wastes, such as the use of waste manifests. These requirements are no longer addressed in the clear language version.

5. Permits – The clear language version has incorporated some of the more common permits that were issued under s. 31 of TDGA, 1992.

4.9.1   Classifications (Part 2 of TDGR)

A substance is a dangerous good when it is listed in Schedule 1 of the TDGR or when it is not listed but meets the criteria in TDGR's s. 2.1 for inclusion in one of the following nine classes of dangerous goods:

- explosives
- gases

- flammable and combustible liquids
- flammable solids
- oxidizing substances and organic peroxides
- poisonous (toxic) and infectious substances
- radioactive material
- corrosives
- miscellaneous dangerous goods

Each type of dangerous good is assigned to one of the nine classes as its primary classification. In addition, one or more subsidiary classifications may be assigned. The primary classification describes the main hazardous properties of a particular dangerous good. A subsidiary classification describes other hazardous properties. These properties are considered to be of a secondary concern in transportation when compared with the main hazardous properties of the good. For example, Class 2 – Gases, includes several divisions such as Division 2.1 – Flammable Gas; Division 2.2 – Non-flammable Gas; and Division 2.3 – Poisonous Gas.

In the clear language version, there no longer are divisions for Group 9. Marine pollutants, genetically modified micro-organisms, elevated temperature materials, leachable toxic substances for disposal and environmentally hazardous substances for disposal are included in Class 9 (TDGR, s. 2.44). Marine pollutants, a recent addition to this group, are described in s. 2.7 of the TDGR.

The consignor must determine the appropriate classification. The consignor is usually the shipper. If the material is an infectious substance, the director, Office of Biosafety, Health Canada or director, Biohazard Containment and Safety, Canadian Food Inspection Agency may classify the material (see TDGR, s. 2.2).

Key revisions in the clear language regulations related to classification include:

1. The list of classes that take precedence over the table has been amended to reflect changes based upon the UN Recommendations 11th edition (1999).
2. The term "compatibility group" is now described in Appendix 2 of Part 2.
3. A new formula has been introduced to determine the LC_{50} of mixtures.
4. Liquids that fall in the flash point range are not considered flammable liquids if they do not sustain combustion, have a fire point >100 °C and are a water-miscible solution with a water content >90% by mass.
5. There are criteria for Class 4.2 packing Group 1, new testing criteria for Class 4.2 and new criteria for liquids of Class 5.1

6. The formula for LC_{50} or LD_{50} value for a poisonous substance, mixture or solution has changed.
7. The list of infectious substances is included in Appendix 3 of Part 2 as opposed to a separate schedule.
8. Radioactive materials are described as substances with an activity greater than 780 kBq/kg.
9. As noted above, there are no longer any divisions for Class 9 and other pollutants, organisms and substances have been added.

The consignor is responsible for rectifying any errors in classification. The carrier who notices an error must advise the consignor and stop transporting the dangerous goods. The consignor must then verify and correct the classification (see TDGR, s. 2.2).

4.9.2 Shipping Documents (Part 3 of TDGR)

The basic responsibilities of the consignor (generator), carrier and consignee (receiver) are described in Part 3 of the clear language version. The consignor is responsible for ensuring shipping documents are obtained before allowing the carrier to take possession. The consignor can provide an electronic copy of the shipping document if the carrier agrees; the carrier must print out these documents before transporting. The last carrier must provide the consignee a document that identifies the dangerous goods.

The following are key requirements for shipping documents under the clear language version of TDGR:

1. For dangerous goods with special provision 16, the technical name of the most dangerous substances related to the primary class must be provided.
2. Using the heading "class" is optional.
3. The words "subsidiary class" may precede the subsidiary class and the order of the subsidiary class in the description may differ if transported by marine or air.
4. The quantity of dangerous goods must be provided in SI units.
5. The 24-hour number for obtaining technical information must be provided.
6. Written permission must be obtained from CANUTEC to use its 24-hour number.
7. The words "Residue-Last Contained" can be used instead of quantity, if the quantity is less than 10% of the maximum fill limit. This does not apply for Classes 2 and 7.
8. The document must indicate the flash point for Class 3 material transported by ship.
9. The words "marine pollutant" and the name and concentration of the most active substances must be provided if the material is a pesticide.

4.9.3   Dangerous Goods Safety Marks (Part 4 of TDGR)

Dangerous goods safety marks are presented in Part 4. The safety marks illustrated in the UN Recommendations may be displayed in accordance with this part.

Key requirements of labels and placards under the clear language version include:

1. The size of a label can be reduced due to the size or irregular shape of a small means of containment.
2. Each side of a placard must be at least 250 mm in length, and except for the "Danger" placard, have a line running 12.5 mm inside the edge.
3. Dangerous goods safety marks must remain on a means of containment until the contents have been neutralized.
4. Specific requirements are presented for "small means" and "large means" of containment.

4.9.4   Means of Containment (Part 5 of TDGR)

All the standard requirements for means of containment are found in Part 5 of the TDGR. A person must not handle, offer for transport, transport or import dangerous goods unless the means of containment is permitted as per s. 5.1 of TDGR.

In 2003, the requirement to use UN performance packaging for small means of containment will become mandatory.

The following topics that were addressed formerly in Parts 7 and 8 are no longer addressed in the revised, clear language version of the regulations: notification requirements, incompatible substances, discharge emission or escape of dangerous goods from packages and small containers, PCB requirements, and specific requirements for rail.

4.9.5   Training Requirements (Part 6 of TDGR)

Any person who handles, offers for transport, or transports dangerous goods must be trained or be directly supervised by a trained person. The training must be relevant for the types of duties performed and must address shipping names, use of Schedules 1 and 2, shipping documents and emergency response assistance plan requirements.

Following the completion of training, employees must be given a certificate of TDGA training that provides the name and address of the place of business of the employer, the employee's name and the expiry date.

The certificate is valid for three years and must be issued to each employee; except for transportation by air and then it is valid for 24 months as per the International Civil Aviation Organization requirements. Once a certificate is invalid, the employee must be retrained. A trained person may be required to produce the certificate upon request by an inspector.

4.9.6   Emergency Response Assistance Plans (Part 7 of TDGR)

Part 7 of the TDGR requires that a person must have an approved emergency response assistance plan (ERAP) before they can offer to transport or import certain dangerous goods. Sections 7.15 to 7.19 and Schedule XII of TDGR prescribe the dangerous goods and the concentration or quantity for which an ERAP is required. In broad terms, these dangerous goods are more harmful than others, and may present wide spread hazards in the event of an accident.

Emergency response to accidents involving these types of goods may require special equipment (such as fully encapsulated chemical response suits or transfer equipment) or specially trained and qualified personnel. Explosives, toxic gases, flammable gases, multiple hazards and poisons are examples of such dangerous goods. Details about ERAPs are provided in Section 12.2.

4.9.7   Immediate Reporting (Part 8 of TDGR)

An immediate report is mandatory when dangerous goods are released in quantities that exceed prescribed quantities or that pose a danger to the public safety. The quantities are listed in Table 12.1.

The information to be provided during the immediate report is specified in s. 8.2 of the TDGR while the contents of a 30-day follow-up report are outlined in s. 8.3 of the TDGR.

4.10   Import/Export of Hazardous Wastes

As a signatory to the Basel Convention on Control of Transboundary Movements of Hazardous Wastes and Their Disposal, which came into force in 1992, Canada has developed SOR/92-637 (Export and Import of Hazardous Wastes Regulations) under CEPA, 1999. These regulatory requirements help ensure that all shipments of hazardous wastes entering into, leaving or passing through Canada are tracked and controlled. The regulations are administered by the Hazardous Waste Management Division, Office of Waste Management of Environment Canada.

4.11   Reducing, Re-Using and Recycling Wastes

4.11.1   General

The ever-increasing costs of waste management, transportation and disposal are incentives for all waste generators to seek ways to minimize the amounts of waste they create. Increasing numbers of organizations and agencies are emphasizing the three Rs: reduction, re-use, recycling. Recovery is sometimes included as the fourth R.

Reduction can be achieved several ways. The amount of materials needed at a facility often can be reduced by better process control or by implementing changes to processes.

Ongoing inventory tracking of wastes should be used to identify ways to achieve reductions and monitor their effectiveness. Examples of waste reduction approaches and potential waste exchanges are presented in Table 4.3 and Table 4.4.

The Ontario Waste Materials Exchange (OWME) provides a listing and brokerage service for generators and users of industrial by-product materials. By providing this service, waste products of one industry can become the raw materials of another. The OWME is operated by the Ontario Centre for Environmental Technology Advancement.

The OWME can assist companies in various ways:

- link users of industrial wastes to generators
- assist in finding sources of wastes as alternative raw materials
- provide recycling industry contacts
- conduct research on selected hazardous waste streams
- provide technical assistance in waste reduction

The types of materials handled at the OWME include plastics, chemicals, rubber, off-spec and surplus materials, wood, cardboard, paper, construction and demolition debris, and batteries.

Details about the OWME and links to other organizations involved with recycling and waste reduction can be found at **www.owe.org**.

Some examples of other Ontario organizations involved in recycling and waste reduction are the Recycling Council of Ontario, Corporations Supporting Recycling, the Canadian Waste Materials Exchange, and the Waste Diversion Organization. These often have sector-specific mandates. For example, the Waste Diversion Organization is a joint initiative of the MOE, municipalities and industries to assist municipalities in Ontario with blue box programs and other waste diversion initiatives. The Canadian Waste Materials Exchange provides links to waste exchanges in other countries.

The Ontario Waste Diversion Act, 2002 establishes the Waste Diversion Organization as a permanent non-government organization to develop, implement and fund waste diversion programs. The Act also allows the MOE to designate waste materials in regulations and require a diversion program to be implemented for that waste. Waste diversion programs established under the Act must be approved by the MOE. Public consultation is required.

4.11.2   Waste Reduction Regulations

In 1994, a series of waste reduction regulations under the EPA were promulgated. Also referred to as the 3Rs regulations, they set out various requirements for waste reduction, re-use and recycling schemes.

1. Recycling and composting of municipal waste – O. Reg. 101/94 requires municipalities to implement composting and blue box recycling programs, and exempts certain types of recycling and composting sites from approval requirements.
2. Waste audits and waste reductions work plans – O. Reg. 102/94 requires that designated waste generators from the industrial, commercial and institutional (ICI) sectors conduct waste audits and formulate waste reduction plans.
3. Industrial, commercial and institutional source separation programs – O. Reg. 103/94 requires large establishments in the ICI sectors to set up facilities for the separate collection and removal of commonly recycled materials such as fine paper, newspaper, cardboard, wood, glass, plastic, aluminum and steel.
4. Packaging audits and packaging reduction work plans – O. Reg. 104/94 requires manufacturers and importers in the food and beverage sector, the paper and allied products sector, and the chemicals and chemical products sector to conduct audits of packaging practices and to set up reduction work plans.

"Waste audits" (really the preparation of waste inventories) are used to determine the amount, nature and composition of wastes being generated as well as to study the process or activities that generate the wastes. The audits also should critically examine the influence on waste generation of decisions or policies such as the processes used to procure supplies and equipment. O. Reg. 102/94 recommends that the audits be performed annually.

Waste reduction plans respond to the findings of the waste audits. A plan should identify ways to reduce, re-use and recycle waste and should be modified and updated annually. The plan must be communicated to employees. Work plans and waste audits should be kept on file for at least five years.

O. Reg. 103/94 identifies several types of waste generators that need to separate various waste materials at the source. These include:

1. Owners of retail shopping establishments or complexes with total floor area of 10,000 m^2 or greater will separate corrugated cardboard, fine paper, newsprint, and food and beverage containers made of glass, aluminum or steel.
2. Manufacturing establishments with more than 100 full-time employees (or equivalent part-time employees) will separate corrugated cardboard, fine paper, newsprint, wood, steel, aluminum, glass and plastic.
3. Construction or demolition projects involving 2,000 m^2 or more of floor area will separate corrugated cardboard (construction projects only), drywall (construction projects only), wood, steel, concrete and brick.
4. Owners of office buildings with total floor area of 10,000 m^2 or greater will separate corrugated cardboard, fine paper, newsprint, and food and beverage containers made of glass, aluminum or steel.
5. Restaurants that employ more than 100 full-time employees (or equivalent part-time employees) will separate corrugated cardboard, fine paper, newsprint, and food and beverage containers made of glass, aluminum, polyethylene terephthalate (PET) or steel.
6. Hotels, motels, inns, resorts or hostels with more than 75 units will separate corrugated cardboard, fine paper, newsprint, and food and beverage containers made of glass, aluminum, PET or steel.
7. Hospitals will separate corrugated cardboard, fine paper, newsprint, and food and beverage containers made of glass, aluminum or steel.
8. Educational institutions enrolling more than 350 students annually will separate corrugated cardboard, fine paper, newsprint, and food and beverage containers made of glass, aluminum or steel.

In 1996, the MOE proposed several amendments. For O. Reg. 101/94, the list of suitable blue box materials would be slightly changed; for O. Reg. 102/94, consideration was given to reducing the amount of documentation associated with waste audits and work plans; and for O. Reg. 104/94, minor amendments were under consideration. As of late 2002, none of these changes have been made and the regulations stand as promulgated in 1994.

4.12　Summary

In Ontario, waste management is addressed primarily in Part V of the EPA and the associated Regulation 347. The identification and proper disposal of waste is clearly the responsibility of waste generators. Identification should be done in accordance with the classification system set out in Regulation 347. Every generator of subject and registerable wastes must file a Generator Registration Report and be issued a generator registration number.

The transportation of hazardous waste on Ontario roads is governed by Ontario's Dangerous Goods Transportation Act. Transportation of wastes by other modes and between provinces must comply with the requirements of the federal TDGA, 1992 and associated regulations. Transportation of hazardous wastes into, out of, or through Canada must comply with TDGA, 1992 and CEPA, 1999.

Regulation 347 requires a six-copy manifest system to track the movement of waste materials. Carriers must be licensed in accordance with provincial and federal requirements. Similarly, owners/operators of waste management systems, disposal sites and transfer stations must comply with provincial requirements.

The costs of waste management, transportation and disposal are constant incentives for waste generators to seek ways to minimize the amount of waste they create. Various waste exchanges and waste reduction organizations have been established to assist waste generators and municipalities in Ontario.

The 3Rs regulations define mandatory rules and requirements for waste reduction, re-use and recycling activities in Ontario. They require municipalities to implement recycling and composting programs and that major waste generators undertake waste audits and implement waste reduction programs.

References

Ontario Ministry of the Environment (MOE). August 1998. *Criterion for the Management of Inert Fill. Proposed Amendments to Regulation 347.* Waste Management Policy Branch.

Ontario Ministry of the Environment (MOE). December 2001. *Registration Guidance Manual for Generators of Liquid Industrial and Hazardous Waste.*

Saxe, D. 2001. *Ontario Environmental Protection Act Annotated.* Canada Law Book Inc. (looseleaf).

Tricil. Not dated. *The Tricil Guide: Your Guide on What to Know, What to Do, and Where to Find Out about Hazardous Waste Legislation in Ontario.*

Table 4.1

MOE Waste Classes

INORGANIC WASTES
Acid Solutions   **Examples**

111 Spent pickle liquor

Acid solutions of sulphuric and hydrochloric acids containing ferrous salts from steel pickling.

112 Acid solutions, sludges and residues containing heavy metals

Solutions of sulphuric, hydrochloric and nitric acids containing copper, nickel, chromium, zinc, cadmium, tin, lead or other heavy metals; chromic acid waste; acidic emission control sludges from secondary lead smelting.

113 Acid solutions, sludges and residues containing other metals and non-metals

Solutions of sulphuric, hydrochloric, hydrofluoric and nitric acids containing sodium, potassium, calcium, magnesium or aluminum; equipment cleaning acids; cation regenerant; reactor acid washes; catalyst acid and acid washes.

114 Other inorganic acid wastes

Off-specification acids; by-product hydrochloric acid; dilute acid solutions; acid test residues.

Alkaline Solutions   **Examples**

121 Alkaline solutions, sludges and residues containing heavy metals

Metal finishing wastes; plating baths; spent solutions containing metals such as copper, zinc, tin, cadmium; case hardening sludges; spent cyanide destruction residues; dewatered solids from metal and cyanide finishing wastes and cyanide destruction.

122 Alkaline solutions, sludges and residues containing other metals and non-metals, not containing cyanides

Alkaline solutions from aluminum surface coating and etching; alkali cleaner waste; waste lime sludges and slurries; anion regenerants.

123 Alkaline phosphates

Bonderizing waste; zinc phosphates; ferrous phosphates; phosphate cleaners.

Aqueous Salts   **Examples**

131 Neutralized solutions, sludges and residues containing heavy metals

Metal finishing waste treatment sludges containing copper, chromium, zinc, nickel or cadmium; neutral salt bath sludges and washes; lime sludge from metal finishing waste treatment; dewatered solids from these processes

132 Neutralized solutions, sludges and residues containing other metals

Aluminum surface coating treatment sludges; alum and gypsum sludges.

133 Brines, chlor-alkai sludges and residues

Waste brines from chlor-alkali plants; neutralized hydrochloric acid; brine treatment sludges; dewatered solids from brine treatment.

134 Wastes containing sulphides

Petroleum aqueous refinery condensates.

135 Waste containing other reactive anions

Waste containing chlorates; hypochlorite; bromate or thiosulphate.

Miscellaneous Inorganic   **Examples**
Wastes and Mixed Wastes

141 Inorganic waste from pigment manufacturing

Wastewater and sludges from the production of chrome yellow, molybdate orange, zinc yellow, chrome green and iron pigments; dewatered solids from these sources.

142 Primary lead, zinc and copper smelting wastes

Slurries, sludges and surface impoundment solids; treatment plant sludges; anode slimes and leachate residues; dewatered solids from these sources.

143 Residues from steel making

Emission control sludges and dusts; precipitator residues from steel plants; dewatered solids from these sources.

144 Liquid tannery waste sludges	Lime waste mixtures; chrome tan liquors; dehairing solutions and sludges.
145 Wastes from the use of paints, pigments and coatings	Paint spray booth sludges and wastes; paper coating wastes; ink sludges; paint sludges.
146 Other specified inorganic sludges, slurries or solids	Flue gas scrubber wastes; wet fly ash; dust collector wastes; metal dust and abrasives wastes; foundry sands; mud sediment and water; tank bottoms from waste storage tanks that contained mixed inorganic wastes; heavy sludges from waste screening/filtration at transfer/processing sites not otherwise specified in this table.
147 Chemical fertilizer wastes	Solutions, sludges and residues containing ammonia, urea, nitrates and phosphates from nitrogen fertilizer plants.
148 Miscellaneous waste inorganic chemicals	Waste inorganic chemicals including laboratory, surplus or off-specification chemicals, that are not otherwise specified in this table.
149 Landfill leachate	Surface run-off and leachate collected from landfill sites.
150 Inert inorganic wastes	Sand and water from catch basins at car washes; slurries from the polishing and cutting of marble.

ORGANIC WASTES

Non-halogenated Spent Solvents	**Examples**
211 Aromatic solvents and residues	Benzene, toluene, xylene solvents and residues.
212 Aliphatic solvents and residues	Acetone, methylethylketone and residues, alcohols, cyclohexane and residues.
213 Petroleum distillates	Varsol, white spirits and petroleum distillates.

Fuels	**Examples**
221 Light fuels	Gasoline, kerosene, diesel, tank drainings/washings/bottoms, spill clean-up residues.
222 Heavy fuels	Bunker, asphalts, tank drainings/washings/bottoms, spill clean-up residues.

Resins and Plastics	**Examples**
231 Latex wastes	Waste latexes, latex crumb and residues.
232 Polymeric resins	Polyester, epoxy, urethane, phenolic resins, intermediates and solvent mixtures.
233 Other polymeric wastes	Off-specification materials, discarded materials from reactors

Halogenated Organic Wastes	**Examples**
241 Halogenated solvents and residues	Spent halogenated solvents and residues such as perchloroethylene, halogenated still bottoms; residues and catalysts from trichloroethylene and carbon tetrachloride (dry cleaning solvents); halogenated hydrocarbon manufacturing or recycling processes.
242 Halogenated pesticides and herbicides	2,4-D, 2,4,5-T wastes, chlordane, mirex, silvex, pesticide solutions and residues.
243 Polychlorinated biphenyls (PCB)	Askarel liquids such as Aroclor, Pydraul, Pyranol, Thenninol FR, Inerteen, and other PCB contaminated materials.

| **Oily Wastes** | **Examples** |
| 251 Waste oils/sludges (petroleum based) | Oil/water separator sludge; dissolved air flotation skimming; heavy oil tank drainage; slop oil and emulsions. |

252 Waste crankcase oils and lubricants	Collected service station waste oils; industrial lubricants; bulk waste oils.
253 Emulsified oils	Soluble oils; waste cutting oils; machine oils.
254 Oily water/waste oil from waste transfer/ processing sites	Waste oil and oily water limited to classes 251, 252 and 253 that have been bulked/blended/processed at a waste transfer/processing site.

Miscellaneous Organic Wastes and Mixed Wastes	**Examples**
261 Pharmaceuticals	Pharmaceutical and veterinary pharmaceutical wastes other than biologicals and vaccines; solid residues and liquids from veterinary arsenical compounds.
262 Detergents and soaps	Laundry wastes.
263 Miscellaneous waste organic chemicals	Waste organic chemicals including laboratory surplus or off-specification chemicals that are not otherwise specified in this table.
264 Photo processing wastes	Photochemical solutions, washes and sludges.
265 Graphic arts wastes	Adhesives; glues; miscellaneous washes; etch solutions.
266 Phenolic waste streams	Cresylic acid; caustic phenolates; phenolic oils; creosote.
267 Organic acids	Carboxylic or fatty acids; formic, acetic, propionic acid wastes; sulphamic and other organic acids that may be amenable to incineration.
268 Amines	Waste ethanolamines; urea; tolidene; Flexzone waste; Monex waste.
269 Organic non-halogenated pesticide and herbicide wastes	Organophosphorus chemical wastes; arsenicals; wastes from MSMA and cacodylic acid.
270 Other specified organic sludges, slurries and solids	Tank bottoms from mixed organic waste bulking tanks at waste transfer sites; mixed sludges from waste screening/filtration at waste transfer/processing sties not otherwise specified in this table.

Processed Organic Wastes from Transfer Stations	**Examples**
281 Non-halogenated rich organics	Blended/bulked non-halogenated solvents, oils and other rich organics prepared at transfer/processing sites for incineration.
282 Non-halogenated lean organics	Blended/bulked aqueous wastes prepared at transfer/processing sites for incineration and contaminated with non-halogenated solvents, non-halogenated oils and other non-halogenated organics.

Plant and Animal Wastes	**Examples**
311 Organic tannery wastes	Fleshings; trimmings; vegetable tan liquors; Bate solutions.
312 Pathological wastes	Human anatomical waste; infected animal carcasses; other non-anatomical waste infected with communicable diseases; biologicals and vaccines.

Explosive Manufacturing Wastes	**Examples**
321 Wastes from the manufacture of explosives and detonation products	Wastewater treatment sludges; spent carbon; red/pink waters from TNT manufacturing; residues from lead base initiating compounds.

Compressed Gases	**Examples**
331 Waste compressed gases, including cylinders	Methane (natural gas); nitrous or nitric oxide; propane; butane.

Table 4.2

Schedule 4 — Leachate Quality Criteria

LEACHATE QUALITY CRITERIA

Contaminant	Concentration (mg/l)
Aldicarb	0.9
Aldrin + Dieldrin	0.07
Arsenic	2.5
Atrazine + N-dealkylated metabolites (Weedex)	0.5
Azinphos-methyl	2.0
Barium	100.0
Bendiocarb	4.0
Benzene	0.5
Benzo(a)pyrene	0.001
Boron	500.0
Bromoxynil	0.5
Cadmium	0.5
Carbaryl/Sevin/1-Naphthyl-N methyl carbamate	9.0
Carbofuran	9.0
Carbon tetrachloride (Tetrachloromethane)	0.5
Chlordane	0.7
Chlorobenzene (Monochlorobenzene)	8.0
Chloroform	10.0
Chlorpyrifos	9.0
Chromium	5.0
Cresol (Mixture - total of all isomers, when isomers cannot be differentiated)	200.0
m-Cresol	200.0
o-Cresol	200.0
p-Cresol	200.0
Cyanazine	1.0
Cyanide	20.0
2,4-D / (2,4-dichlorophenoxy)acetic acid	10.0
2,4-DCP (2,4-Dichlorophenol)	90.0
DDT (total isomers)	3.0
Diazinon/Phosphordithioic acid, o,o-diethyl o-(2-isopropyl 6-methyl-4-pyrimidinyl) ester	2.0
Dicamba	12.0
1,2-Dichlorobenzene (o-Dichlorobenzene)	20.0
1,4-Dichlorobenzene (p-Dichlorobenzene)	0.5
1,2-Dichloroethane (Ethylene dichloride)	0.5
1,1-Dichloroethylene (Vinylidene chloride)	1.4
Dichloromethane (also see - methylene chloride)	5.0
Diclofop-methyl	0.9

Contaminant	Concentration (mg/l)
Dimethoate	2.0
2,4-Dinitrotoluene	0.13
Dinoseb	1.0
Dioxin & Furan	0.0000015*
Diquat	7.0
Diuron	15.0
Endrin	0.02
Fluoride	150.0
Glyphosate	28.0
Heptachlor + Heptachlor epoxide	0.3
Hexachlorobenzene	0.13
Hexachlorobutadiene	0.5
Hexachloroethane	3.0
Lead	5.0
Lindane	0.4
Malathion	19.0
Mercury	0.1
Methoxychlor/1,1,1-Trichloro-2,2-bis (p-methoxyphenyl) ethane	90.0
Methyl ethyl ketone / Ethyl methyl ketone	200.0
Methyl Parathion	0.7
Methylene chloride / Dichloromethane	5.0
Metolachlor	5.0
Metribuzin	8.0
NDMA	0.0009
Nitrate + Nitrite (as Nitrogen)	1,000.0
Nitrilotriacetic acid (NTA)	40.0
Nitrobenzene	2.0
Paraquat	1.0
Parathion	5.0
PCBs	0.3
Pentachlorophenol	6.0
Phorate	0.2
Picloram	19.0
Pyridine	5.0
Selenium	1.0
Silver	5.0
Simazine	1.0
2,4,5-T (2,4,5-Trichlorophenoxyacetic acid)	28.0
2,4,5-TP/ Silvex/ 2-(2,4,5-Trichlorophenoxy) propionic acid	1.0
Temephos	28.0
Terbufos	0.1
Tetrachloroethylene	3.0
2,3,4,6-Tetrachlorophenol /(2,3,4,6-TeCP)	10.0

Contaminant	Concentration (mg/l)
Toxaphene	0.5
Triallate	23.0
Trichloroethylene	5.0
2,4,5-Trichlorophenol (2,4,5-TCP)	400.0
2,4,6-Trichlorophenol (2,4,6-TCP)	0.5
Trifluralin	4.5
Uranium	10.0
Vinyl chloride	0.2

* Toxic equivalent (TEQ)

Reference: Regulation 327, Sch. 4.

Table 4.3

Approaches to Waste Reduction

Waste Abatement
Substitute new low-waste process

– Replace sulphuric acid in for an old process to eliminate or steel pickling with reduce the quantity of waste. hydrochloric acid

– Replace liquid paints with powder coatings

– Replace solvent-based adhesives   with water-based adhesives

Waste Minimization
Reduce the quantity of waste through good house keeping practices or use technologies to produce more concentrated wastes.

– Separate waste streams to permit recovery
– Use countercurrent rinsing to minimize volume of discharge

Reduce the amount of hazardous waste by in-plant treatment.

– Neutralize then use precipitation to reduce waste sludge volume
– Fix leaky taps and nozzles

Waste Re-use
Directly re-use a waste as is, or with a minor modification.

– Re-use surplus chemicals
– Use oil sludges in asphalt manufacturing
– Use blast furnace slag as aggregate
– Use solvents from electronics industry in paints manufacture
– Use refinery spent caustic in wood pulping
– Use electronic circuit manufacturing plating baths in regular plating shops

Waste Recycle

Reclaim the value of waste material by reprocessing, distillation, etc.

– Oil re-refining
– Solvent distillation
– Recover iron salts such as from pickle liquor
– Recover heavy metals from sludges
– Recover and re-use spent foundry sands
– Recover scrap metal
– Regenerate and re-use activated carbon

Table 4.4

Waste Exchanges

Materials	Possible Receivers
Acids	
nitric acid	metal reclaimer
sulphuric acid	metal reclaimer
Alkalis	
calcium hydroxide	broker to new business
potassium hydroxide	chemical company
Other Inorganic Chemicals	
alumina	abrasives manufacturer
foundry sands	asphalt manufacturer
Organic Chemicals	
ink wastes	recycling and recovery
paints	organizations that build or maintain assisted housing
latex materials	manufacturer for re-blend
oils (PCB free)	oil recycler
Plastics and Rubber	
polystyrene	plastic recycler
plastic drums	municipal composting programs

Figure 4.1

Waste Identification Flowchart

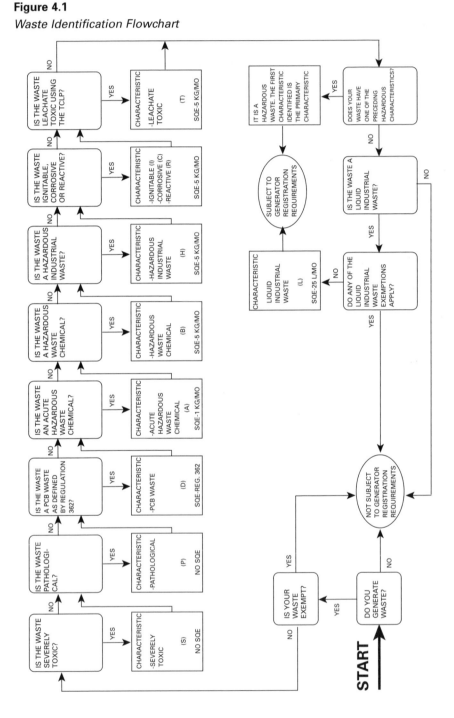

Figure 4.2

Blank O. Regulation 347 Generator Registration Report

Ontario | Ministry of the Environment | Ministère de l'Environnement | **Generator Registration Report** / *Rapport d'inscription du producteur*

Please Mail completed form to:
Environmental Monitoring & Reporting Branch, Ministry of the Environment, 135 St. Clair Avenue West, AREA "M"
Toronto, ON M4V 1P5

Veuillez faire parvenir le formulaire dûment rempli à l'adresse suivante:
Direction de la surveillance environnementale, Ministère de l'Environnement, 135 avenue St. Clair ouest, SECTEUR M
Toronto, ON M4V 1P5

Part 1 - Generator Identification / *Partie 1 - Identification du producteur*

This report is / *Le présent rapport constitue :*

Generator Registration Number
N° d'inscription du producteur

1. ☐ an initial generator registration report /
un premier rapport d'inscription du producteur

or / *ou*

2. ☐ a revision - enter Ontario Generator Registration Number /
une révision - veuillez donner le numéro d'inscription du producteur (Ontario)

3. For generators located outside of Ontario, enter Registration/Notification number assigned by your local environmental authority. / *Si vous êtes un producteur de l'extérieur de l'Ontario, veuillez inscrire le numéro d'inscription / d'identification attribué par les autorités locales en matière d'environnement.*

Name of Generator (Enter the corporate name or, if a proprietorship, the name of the principal(s). If the generator intends to carry on business under a separate name or style, this should also be entered.) / *Nom du producteur (Veuillez inscrire la dénomination sociale ou, s'il s'agit d'une société en nom collectif ou d'une société à propriété unique, le nom du (des) principal (principaux) propriétaire(s). Si le producteur envisage d'exploiter une entreprise sous une dénomination ou un nom distinct, veuillez également le noter.)*

4. Name / *Nom*

5. Address / *Adresse*

6. Municipality / *Municipalité* | Province / State *Province / Etat* | Postal Code / *Code Postal*

7. Site location / *Emplacement des installations*

8. Municipality / *Municipalité* | Province / State *Province / Etat* | Postal Code / *Code Postal*

9. Name of Contact / *Nom de la personne à contacter* | Telephone No. / *N° de tél.*

10. North American Industry Classification System (NAICS) Codes for Site noted in Section 7. / *Le Système de classification des industries de l'Amérique du Nord (SCIAN) pour les installations dont l'adresse figure au N° 7*

11. Total number of wastes to be registered with this report /
Nombre total de déchets à inscrire dans ce rapport

12. Name of Company Official / *Nom du représentant de la compagnie* | 13. Position / *Poste*

14. Signature / *Signature* | 15. Date / *Date*

16. **Ministry Use Only /**
Réservé au Ministère

County Code / *Code de comté*

Regional/ District Code / *Code de région/district*

Municipal Code / *Code de municipalité*

Inter City Tie Line / *Ligne privée interurbane*

Figure 4.2

Blank O. Regulation 347 Generator Registration Report (cont'd)

Part 2 - Waste Identification / *Partie 2 - Nature des déchets*
(Please complete one page 2 for each waste stream to be registered)
(*Veuillez photocopier cette page et en remplir une pour chaque flux de déchets à inscrire.*)

1. Description of Waste / *Description des déchets*

2. Description of generating process / *Description du procédé de production*

3. Waste quantity generated or accumulated / *Quantité de déchets produite ou accumulée*

   Kg/mo. / *kg/mois*

4. Primary characteristic / *Caractéristique principale*

   Analytical data (if applicable). If the data has been estimated, attach separate sheet outlining basis for the estimate. / *Données analytiques (le cas échéant). Si les données sont estimatives, veuillez annexer une feuille à part pour décrire sur quoi reposent les estimations.*

   Name of laboratory (if applicable). / *Laboratoire (le cas échéant)*

   | Waste Class *Catégorie de déchets* | Hazardous Waste Number *Numéro de déchets dangereux* | Specific Gravity *Masse volunique* | Physical State (Solid - S, Liquid - L, Gas - G) *État (Solide -S, liquide - L, gaz - G)* |

   For Ministry Use only / *Réservé au Ministère*

5. Secondary Characteristic / *Caractéristique secondaire*

   Analytical data (if applicable) / *Données analytiques (le cas échéant).*

Part 3 - Waste Management / *Partie 3 - Gestion des déchets*

1. Principal Intended Receiver / *Réceptionnaire principal prévu*

   Company name and address / *Nom et adresse de la compagnie*   Receiver No. / *N° du réceptionnaire*

   Municipality: *Municipalité:*   Province/State: *Province / Etat:*   Postal Code: *Code Postal:*

2. Principal Intended Carrier / *Transporteur principal prévu*

   Company name and address / *Nom et adresse de la compagnie*   MOE Carrier No. *N° du MEO du transporteur*

   Municipality: *Municipalité:*   Province/State: *Province / Etat:*   Postal Code: *Code Postal:*

Figure 4.3

Blank O. Regulation 347 Manifest

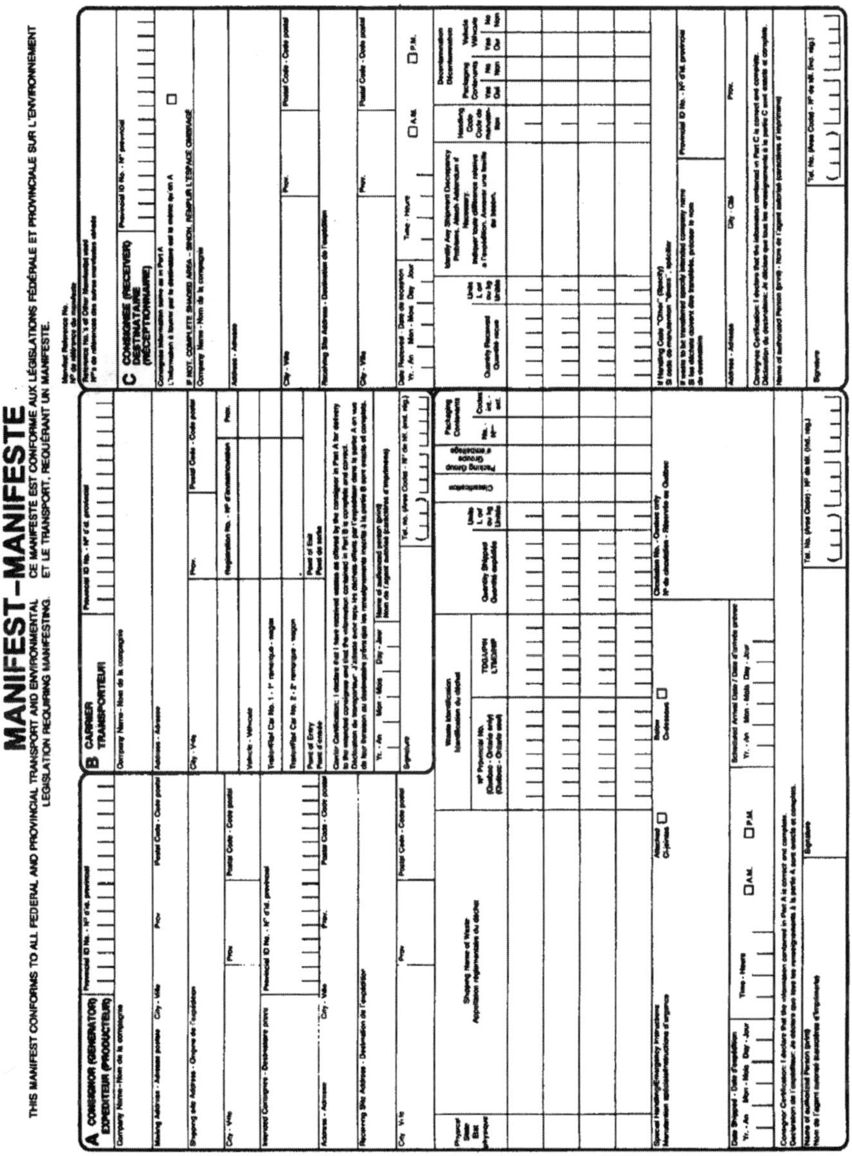
MANIFEST-MANIFESTE

THIS MANIFEST CONFORMS TO ALL FEDERAL AND PROVINCIAL TRANSPORT AND ENVIRONMENTAL LEGISLATION REQUIRING MANIFESTING.

CE MANIFESTE EST CONFORME AUX LÉGISLATIONS FÉDÉRALE ET PROVINCIALE SUR L'ENVIRONNEMENT ET LE TRANSPORT, REQUÉRANT UN MANIFESTE.

Figure 4.4
Distribution of Manifests

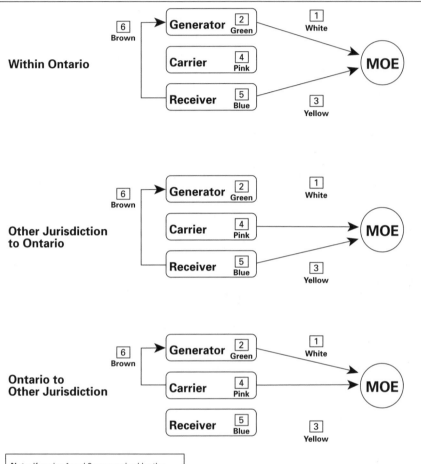

Note: If copies 1 and 3 are required by the other jurisdiction, photocopies are acceptable for submission to the MOE.

5

Assessing and Remediating Contaminated Property

5.1 Overview

There are thousands of properties in Ontario where soil or ground water quality can be considered to be contaminated. When and how these properties are remediated will be influenced strongly by how the goals of remediation are defined and how the remediating responsibilities are distributed.

Numerous factors have contributed to the growing importance of contaminated property issues over the past 20 years. Some originate in regulatory requirements. Legislation like the Environmental Protection Act (EPA) make various persons responsible for removing, treating, remediating or otherwise managing contaminated soil and ground water. These persons include current owners, previous owners, site managers, current tenants and past tenants. If a corporation owns a contaminated site, or if a site becomes contaminated as a result of the actions of a corporation, the individual officers of the corporation can be liable.

Other factors stem from changes in social expectations. These reflect the general heightened awareness of the public to contaminated property issues. For industrial and commercial properties, environmental assessments have become standard parts of the process of changing land ownership. Increasingly, other types of properties are being assessed prior to being sold.

Contamination issues pose challenges to buying, selling, leasing, remediating and redeveloping property. They also clearly illustrate the need for current property owners to protect the environmental quality of their sites. While some aspects of this issue have improved substantially in recent years in Ontario, it is anticipated that issues related to contaminated property will continue to be hampered by uncertainty for some time to come.

5.2 Regulatory Framework

5.2.1 Federal Acts, Regulations and Initiatives

Federal agencies participate in contaminated property issues in several ways. Federal agencies are responsible for facilities such as international airports, harbours, federal research facilities and defence facilities including any property assessment or remediation that may be required.

Federal agencies are responsible for establishing clean-up requirements for radioactive materials associated with nuclear fuel cycles and specifically regulated substances such as polychlorinated biphenyls (PCBs).

Another role is developing numerical criteria for chemicals in soil or ground water. This typically is done in conjunction with the provinces through bodies such as the Canadian Council of Ministers of the Environment (CCME). Environmental quality criteria developed by the CCME are described in Section 5.3.5.

The National Round Table on the Environment and the Economy (NTREE) is an advisory council to the prime minister. It has established a Brownfield Redevelopment Strategy Task Force that is devising a national strategy for promoting redevelopment of contaminated lands that are unused or unproductive (brownfields). It has published several background reports that pertain to brownfield redevelopment. More information can be obtained at **www.ntree-trnee.ca.**

Both NTREE and the Prime Minster's Caucus Task Force on Urban Issues have recommended that the governments should investigate the use of tax incentives for brownfield redevelopment, as well as co-ordinate regulations and provide support to the private sector.

5.2.2 Provincial Acts and Regulations

While all three levels of government are involved in contaminated property issues, this is largely a matter of provincial jurisdiction. Since contamination can cause several of the adverse effects noted in the EPA, the EPA is the key provincial statute for contaminated property issues.

When Part X (Spills) of the EPA came into effect in 1985, it created two categories of contaminated sites. One category includes all sites where the contamination has been caused by spills that began after November 29, 1985. For those cases, s. 93 of the EPA clearly imposes a duty to restore the environment upon the owner or the person having charge of the pollutant that is spilled. Other sections in Part X of the EPA authorize the Minister to order clean-ups, take whatever remedial actions are necessary, and to sue the person(s) responsible for the costs.

The second category includes all sites contaminated by events other than spills and all sites contaminated before November 29, 1985. There is no automatic obligation to remediate contaminated sites that fall into this second category unless the contamination is causing an adverse effect. It is possible for a site to be contaminated for many years before an owner or the MOE is aware of the contamination. (The likelihood of this occurring increases where the contamination is relatively immobile, present at only mild levels of contamination, and has not migrated on to neighbouring lands.)

The MOE does not have the resources to seek out or investigate potentially contaminated properties except when prompted by circumstances that suggest an adverse effect has occurred such as the contamination of a water supply or when the MOE becomes aware of a documented adverse effect. Despite these limitations, large numbers of properties in Ontario have been investigated over the past two decades, and substantial sums have been paid to clean up some properties. These actions largely have been in response to the growing desire of owners, purchasers, lenders and insurers to avoid the potential liabilities that contaminated property can pose.

In addition to these market forces, the MOE can become aware of contaminated property via some regulatory processes:

1. When a change in land use or redevelopment is proposed, a Record of Site Condition (RSC) may be submitted to the MOE for its acknowledgement.
2. The MOE may decide to assess contamination related to a spill, or be asked to assess such a situation by a municipality.
3. Some conservation authorities and federal agencies will ask the MOE for opinions of environmental suitability of proposed projects and applications.
4. If a project is the subject of the environmental assessment process, the MOE may recommend that site assessment and/or remediation be required as conditions to the approval or exemption of the project.

When the MOE becomes aware of soil or ground water contamination at a property, it can take action if it determines that the contamination has caused or may cause an adverse effect. For example, s. 17 of the EPA authorizes MOE directors to issue remedial orders for contaminated sites regardless of when the contamination occurred. Such an order can be issued to the "person who causes or permits the discharge of a contaminant into the natural environment". Responsible persons can include past and present owners, tenants, site managers and operators of properties.

While the intent of the EPA is to have those responsible for contamination to restore the environment (often referred to as the "polluter pays" principle), the EPA does not provide clear rules or definitions for determining when clean up is necessary and when clean-up activities should be terminated. These limitations are addressed to some degree by the maximum acceptable concentrations of chemicals (the "criteria") that appear in MOE guidelines (see Section 5.3).

In 2001, the Brownfields Statute Law Amendment Act was enacted in Ontario. The intention of the Act is to bring more certainty to site assessment and remediation. The Act amends sections of the EPA, the Municipal Act, and the Planning Act among others. Amendments to the EPA create an environmental site registry, where RSCs will be filed. While the RSC was first introduced in 1996, proposed regulations under this Act will give it a more formalized role. Future regulations will prescribe land use changes where an RSC will be mandatory, set MOE standards for Phase I and II environmental site assessments (see Section 5.4), define a process for certifying "qualified persons" to undertake site assessments, set a policy for the disposal of excess soils and codify (give force of law to) the MOE guideline that contains soil and groundwater criteria (see Section 5.3). As of late 2002, none of these regulations had been enacted. Other regulations under the Act will protect municipalities, secured creditors, receivers and trustees in bankruptcy from some types of EPA orders.

The Technical Standards and Safety Act, 2000 replaced several pieces of provincial legislation that addressed the remediation of contaminated sites where petroleum products have been released from storage tanks. The Act itself does not address contamination issues, but O. Reg. 217/01, Liquid Fuels, promulgated under the Act, describes spills, leaks or other losses of petroleum products as events that are regulated. The regulation provides no further details but refers to the Liquid Fuels Handling Code. Section 7 of the Code requires environmental restoration when an underground storage tank is being removed or replaced, when there has been a spill or leak of a petroleum product to the environment, and when storage tank systems are decommissioned. The Code does not provide further details but refers to the *Environmental Management Protocol for Operating Fuel Handling Facilities in Ontario* (GA 1/99). The Protocol makes frequent reference to the soil and ground water criteria recommended by the MOE (see Section 5.3).

At fuel handling facilities that are to remain operational, the Protocol offers the option of leaving petroleum-related contamination in place provided that various conditions are met (the foremost being that immediate corrective action is not necessary). In this case, the facility operator can develop and submit a contamination management plan, which provides for on-going monitoring of the contaminants until remediation is undertaken (that is, when the facility is to be closed permanently). The Protocol outlines the minimum topics to be addressed in a contamination management plant.

The Protocol indicates that immediate corrective action is needed to:

- eliminate all liquid phase-separated product evident on the surface or in the subsurface
- eliminate any potential explosion hazards in enclosed spaces
- eliminate the potential for off-site migration of petroleum product and related contaminants

The Act, regulation, code and protocol are administered by the Technical Standards and Safety Authority. Where a release of petroleum product may cause an adverse effect or impact a drinking water supply, the regulatory lead over environmental remediation will be transferred to the MOE. More information about the Technical Standards and Safety Authority can be obtained at **www.tssa.org**.

5.2.3   Municipal Involvement

Municipalities enforce the Ontario Building Code Act, 1992. The Building Code requires that a building permit must be issued by the chief building official before a building is constructed or demolished. A permit will not be issued if the proposed construction or demolition contravenes the Building Code or "any other applicable law". This wording encompasses many statutes and regulations, but it is not clear how it relates to contaminated property issues, except in those infrequent cases where there clearly were adverse effects as defined in the EPA.

Until 1996, most municipal government agencies turned to the MOE for assistance with site assessment and remediation issues that arose in development applications submitted to the municipality under the Planning Act. There were some exceptions. For example, the Environmental Protection Office of the Toronto Health Department reviewed site assessment studies and remediation plans.

In 1996, the MOE reduced its advisory role and shifted responsibility for environmental reviews to municipalities. This put more of the onus on municipalities to assess the environmental suitability of sites during the planning review process. Some municipalities had the internal resources to fill this role, many decided to use outside experts as peer reviewers (and had development applicants pay for these reviews), and others decided to rely on MOE acknowledgement of RSCs. Overall, this has proven to be a challenge for most municipalities.

Some muncipalities have created or adopted screening procedures and protocols to determine if a development proposal should be subject to an environmental review and/or RSC submission and acknowledgement. These protocols generally seek to obtain, review and evaluate information on historic land uses, potential on-site contaminant sources and proximity to potential off-site contaminant sources.

The Brownfields Statute Law Amendment Act likely will make it mandatory for the chief building official to consider environmental matters prior to issuing a building permit if a RSC or a certificate of property use is needed (see Section 5.5.4) or other circumstances as may be prescribed in regulations under the Act. For example, it has been proposed that regulations will make an RSC mandatory whenever an industrial or commercial property is to be converted to residential or park use.

A chief building official may issue a conditional permit at any stage of construction. This can help applicants combine excavating for environmental purposes (that is, remediation) with excavating for construction purposes, rather than excavating, backfilling and then excavating again.

For many years, municipalities in Ontario have used community improvement plans as a mechanism for providing incentives for redevelopment such as reduced or waived planning and development fees, and the easing of various planning requirements. Recently, some communities have adopted the community improvement plan model for redeveloping derelict or contaminated lands. One example is the Environmental Remediation and Site Enhancement (ERASE) program at the City of Hamilton. It provides grants to help defray the costs of environmental studies. Hamilton also is the first municipality in Canada to introduce tax incremental financing to promote brownfield redevelopment by granting back a portion of the property tax increase generated by a brownfield redevelopment project to the developer to help pay for environmental remediation, demolition and infrastructure upgrading costs.

5.3 Environmental Quality Criteria and Guidelines

5.3.1 General Approaches to Setting Criteria

Most discussions of contaminated property eventually turn to "the numbers", the maximum acceptable concentrations of chemicals in soil or ground water that regulatory agencies have recommended or established.

In Ontario, there have been "numbers" since the early 1970s. At first, these numbers addressed mostly metals, general indicators such as pH and only a few chemicals. A few more chemicals were added in the 1980s and again in 1991. A major expansion occurred in 1996 when the MOE issued the *Guideline for Use at Contaminated Sites in Ontario* (MOEE, 1997) along with several supporting documents.

The Guideline contains criteria for more than 110 chemicals in both soil and ground water for three categories of land use and other combinations of factors including depth below the surface, soil texture and whether or not ground water is a potable water supply. In addition to the numbers, the Guideline and supporting documentation explain the process used to set the numbers.

The Guideline outlines three basic approaches to setting criteria;

- set them equal to background or ambient concentrations
- determine the types of adverse effects a chemical can cause, then set generic criteria low enough to avoid those effects in most environmental settings

- use site-specific information to determine criteria appropriate for site-specific conditions

Criteria recommended by the MOE are not appropriate for all properties in Ontario. For lands under federal jurisdiction, it may be more appropriate to use criteria recommended by federal agencies. These are discussed in Section 5.3.5.

5.3.2   The Role of Ambient (Background) Concentrations

The Guideline defines the background approach as one where the goal is to restore conditions to those found in the natural environment, or to conditions that existed prior to contamination. This approach can be used at any site in Ontario.

In 1991, the MOE initiated a program to establish Ontario Typical Range (OTR) values to describe background concentrations. OTR values are intended to represent the expected distribution of chemicals due to natural processes and normal human activity at locations remote from known point sources of emissions.

The locations sampled by the MOE subsequently were organized into two land use categories: rural parkland and urban parkland. The data showed variability (as expected) and the decision was made to define the OTR values as equal to the 98th percentile value of each distribution. (Modifications to this procedure were made for arsenic and DDT.) From a statistical perspective, the 98th percentile is equivalent to the mean plus two standard deviations of a normally distributed population. While it means that OTR values should not be exceeded at 98% of unimpacted sites, it also means that there is a 2% likelihood that samples from unimpacted locations will exceed the OTR values.

The OTR values became the basis for the criteria presented in Table F of the Guideline. Separate criteria are presented for agricultural land and the broad category of all other land uses. Table F criteria have been developed for soil only.

If the background approach is to be used, but the chemicals of interest are not included in Table F, or if the criteria in Table F are not appropriate because of a local or regional geology, alternative background criteria may be developed. The data gathering process used must be consistent with the methods used by the MOE to establish the OTR values. To establish a background criterion for a chemical not in Table F, sampling must involve not less than 30 separate sampling sites from at least 10 different locations. The sample sites must be in areas that have not been affected by local point sources of air or land pollution, by local roads or highways, or by other known sources of contamination. The Guideline indicates that suitable sites include parks, schoolyards, cemeteries, forests, wood lots or large undeveloped areas. While the provincial OTR values are based on the 98th percentile of the distributions, local background values are to be set at the 90th percentiles and cannot (despite what the distribution may be) be set above generic criteria provided in Table A (discussed below) if the Table A criteria are based on health effects. In the years since the Guideline was issued, seldom have new or local background concentrations been determined to the satisfaction of the MOE.

The Guideline does not provide background criteria for chemicals in ground water. In the past, the MOE has suggested that background levels in water should be defined according to the objectives outlined in the document entitled *Water Management: Policies, Guidelines, Provincial Water Quality Objectives for the Ministry of the Environment and Energy* (MOEE, 1994). That document does not provide much guidance for evaluating ground water quality, except perhaps where ground water enters surface water.

5.3.3   Setting Generic or Effects-Based Criteria

As noted in Section 5.3.1, the second approach used in Ontario to identify acceptable concentrations of chemicals in soil or ground water is to set criteria low enough to avoid the types of adverse effects a chemical can cause. These determinations are made by the MOE using a methodology (the generic approach) to identify generic or "effects-based" criteria.

The generic approach considers many factors. These include: potential impacts to human health; potential impacts to ecological receptors; background concentrations; detection limits; the environmental behaviour of chemicals, their solubilities in water and the potential to cause odours. Detailed descriptions of the generic approach are provided in the Guideline and supporting documents (MOEE, 1997; 1996a; 1996b; 1996c).

The generic process considers five broad categories of land use: agricultural, residential, parkland, commercial and industrial. Generally, more stringent clean-up requirements are needed for agricultural, residential and parkland because it is assumed that there are more opportunities for people and ecological receptors to come into contact with chemicals in soil or ground water than at commercial and industrial sites.

The generic process generally identifies more stringent criteria for sites where ground water is or may be used as a source of potable water than for sites where ground water is not used as a source of potable water. Similarly, the process can identify more stringent criteria for coarse-textured soil than for medium- to fine-textured soil, since chemicals tend to adhere more strongly to medium- to fine-textured soils. Coarse-textured soil is defined as material having more than 70% (by dry weight) of particles equal to or greater than 50 μm in diameter (that is, more than 70% sand).

The generic process also differentiates between surface soil (the upper 1.5 m) and subsurface soil (deeper than 1.5 m). Criteria for the former tend to be more stringent because it is assumed that there are more opportunities for people and ecological receptors to come into contact with chemicals in the shallower soil.

The Guideline provides generic criteria for more than 110 parameters in soil and ground water a series of four tables:

- Table A provides criteria for surface soils and ground water at sites with potable ground water conditions
- Table B provides criteria for surface soils and ground water at sites with non-potable ground water conditions
- Table C provides criteria for subsurface soils at sites with potable ground water
- Table D provides criteria for subsurface soils at sites with non-potable ground water conditions

The Guideline also provides sediment quality criteria (in Table E) and the background concentrations (in Table F) noted in Section 5.3.2.

Table 5.1 and Table 5.2 illustrate how generic criteria can vary among these tables using a small subset of the chemicals addressed in the Guideline. To develop generic criteria that would have broad applicability, the MOE made several assumptions about typical site conditions. For example, the generic criteria are based upon the assumption that there is at least 2 m of soil above rock, and that the soil pH lies in the range of 5 to 9. To ensure that the generic criteria are sufficiently protective at the majority of sites in Ontario, the generic approach incorporates numerous conservative elements.

Despite incorporating conservative elements into the generic approach, the Guideline cautions that the generic criteria may not be sufficiently protective at some sites. The MOE refers to these as "potentially sensitive sites" and the Guideline describes several conditions at a site that can render the generic criteria inappropriate. Examples include a site that is a nature reserve, an area identified as environmentally sensitive by a municipality or conservation authority, a fish habitat, a significant wetland, a provincial park, a site with less than 2 m of soil or overburden, and soil outside the pH range of 5 to 9.

5.3.4   Setting Site-Specific Criteria

As noted in Section 5.3.1, the third approach used in Ontario to identify acceptable concentrations of chemicals in soil or ground water is to use site-specific information to determine criteria appropriate for site-specific conditions. The MOE describes this as a site-specific risk assessment (SSRA) approach. This can be done when it is felt that generic criteria are inappropriate (because they are based on conditions not representative of those at a specific site), if the chemical of interest is one for which generic criteria have not been established, or if the generic criteria are not adequately protective (for example, at a potentially sensitive site).

While the Guideline and a supporting document (MOEE, 1996b) describe the SSRA in general terms, details are left to those who wish to pursue this option and the results are then reviewed by the MOE.

Obviously, the onus is on the proponent to undertake whatever effort is required and to demonstrate that the proposed site-specific criteria values are sufficiently protective of the environment and human health. Efforts in that regard should consider the following issues:

- environmental mobility of the contaminants of concern (for more details see Chapter 9)
- the pathways by which people may come into contact with the contaminants of concern
- the potential risks posed to human health
- the potential risks posed to non-human site users such as terrestrial or aquatic species of plants and animals
- anticipated future uses of the site
- surrounding land uses
- possible synergistic or antagonistic effects of the contaminants of concern
- possible exposures to off-site receptors by pathways such as diet or water supplies
- physical features and environmental conditions of the site such as soil type, ground water regime, local meteorology
- background concentrations of contaminants
- possible aesthetic considerations
- compatibility with other relevant environmental guidelines or criteria

Risk assessment and management are major components of efforts to develop site-specific guidelines. Such techniques appear to offer a more defensible alternative than some of the methods that historically have been used such as basing criteria on analytical detection limits, background conditions or technological capabilities. Failure to consider risks can result in limited resources being used to reduce already low levels of risk at one situation while higher risks go unattended at another.

While it would be misleading to present SSRA and risk management as panaceas for resolving complex issues, it is equally true that risk management is critical to identifying cost-effective actions and ensuring that resources are used efficiently. More details about risk assessment and risk management are presented in Chapter 11.

5.3.5   Using Soil Criteria Developed in Other Jurisdictions

The Guideline indicates that when the MOE has not established criteria for a specific chemical of interest, a proponent may choose to adopt criteria recommended by other regulatory agencies. In such cases, human health, ecological health and protection of the natural environment must be considered. The process used — including reference information — must be fully documented and submitted to the MOE for review.

Examples of other sources of criteria include: the CCME, environment departments in other provinces, state and federal environmental agencies in the United States (most notably the U.S. EPA), and environment departments in other countries. There also are a few notable examples of multi-stakeholder organizations that have developed criteria such as the Total Petroleum Hydrocarbons Criteria Working Group (TPHCWG).

Criteria from other jurisdictions require careful review to determine their applicability to conditions found in Ontario. Criteria from jurisdictions where predominant geology, weather, building construction methods, types of agriculture or the types of organisms used to assess ecological toxicity are clearly different from those found in Ontario. Criteria also tend to evolve or be updated frequently, so it is important to know that the criteria being considered reflect the current deliberations of that agency. Proponents need to review any statements from the developers of the criteria concerning their application or interpretation before values are proposed for a specific site.

Among the many possible sources of criteria from other jurisdictions, those looked to for guidance most often in Ontario include those from the CCME, a few of the states (Massachusetts and Michigan are two examples) and British Columbia. In addition, the U.S. EPA and the TPHCWG are frequently cited as sources of human health and environmental fate information that are key components to setting or revising criteria.

There have been instances where the MOE has expressed reservations about the criteria recommended by other jurisdictions. Contacting the MOE with the specifics is recommended before pursuing criteria from other jurisdictions.

Table 5.3 lists various sources of other jurisdictions' soil and ground water criteria.

5.3.6   Some Limitations to Current MOE Criteria

The Guideline indicates that criteria may need to be adjusted in response to advances in environmental sciences, research and analytical techniques, however, no updates have been issued since 1996 (except for some errors in the original Guideline documents that were amended in 1997).

In 2002, the MOE awarded several contracts to identify possible improvements to the Guideline and the criteria it contains. These suggestions are to be considered in preparation for incorporating the Guideline into proposed regulations to the Brownfields Statute Law Amendment Act. The regulations are not expected before 2003. The following examples illustrate some of the areas where the Guideline might be improved.

Total Petroleum Hydrocarbons

The underlying rationale for the total petroleum hydrocarbons (TPH) criteria is not science-based. The generic criteria for TPH in soil are described by the MOE as "management numbers" used primarily to protect against adverse aesthetic effects. The generic criterion of 1 mg/L for TPH in potable ground water is based on a Massachusetts document that describes it simply as a default value. Massachusetts subsequently replaced that value and does not require TPH to be monitored in drinking water. Criteria have not been established for non-potable ground water in Ontario.

These are serious limitations, particularly for a contaminant that is encountered as frequently as TPH. Many agencies are replacing TPH with better-defined fractions and setting criteria based on the potential adverse effects those fractions can cause. Two important examples are the efforts of the TPHCWG, which subsequently have been adapted by the Maritime provinces, and the recently developed Canada-wide standards for petroleum hydrocarbons in soil by the CCME. Similar changes seem inevitable in Ontario.

Chlorinated Volatile Organic Compounds

Chlorinated volatile organic compounds (VOCs) make up another group of chemicals that is frequently encountered in the environment. The generic criteria in the Guideline pivot around vinyl chloride and on the assumption that other chlorinated VOCs always degrade to vinyl chloride. Aside from that assumption being overly simplistic, two of the chemicals that can degrade to vinyl chloride (1,2-dichloroethane and 1,1-dichloroethylene) have been left out of this scheme. Furthermore, the toxicological assessment of vinyl chloride used by the MOE has subsequently been revised (to be somewhat less stringent) by the U.S. EPA. Currently, there is some scientific evidence and actions in other jurisdictions that suggest the criterion for trichloroethylene might need to be lowered.

Beryllium

Beryllium is the only element for which the generic criteria are based on background concentrations. In preparing generic criteria, the MOE assessed the human health aspects of beryllium using information from occupational exposures to elemental beryllium. This led to the conclusion that acceptable risks correspond to concentrations that are less than background concentrations in soil. To avoid setting criteria that cannot be achieved, criteria for beryllium were set equal to the background levels. But the health risks likely were overestimated. In the natural environment, most beryllium occurs in silicate and phosphate minerals, not as elemental beryllium. As a result, there are higher concentrations of beryllium in products such as rock wool insulation than the generic criteria allow in soil.

Sodium Absorption Ratio

The sodium absorption rate (SAR) is an indicator of the presence of calcium, sodium and magnesium in soil. In the 1950s, the U.S. Department of Agriculture was concerned that frequent irrigation of farmland in the arid southwest was causing salinity to increase and injure commercial crops — specifically fruits and nuts. Research led to creating the SAR as a way of expressing these chemical characteristics of soil. The generic criteria used in Ontario are intended to protect these sensitive crops and to prevent soil permeability from being affected. There seems to be little to tie the SAR criteria to the types of plants typically grown in Ontario. One wonders if the current criteria for SAR are relevant in the Ontario context and if criteria are needed for both it and similar parameters such as electrical conductivity. Conversely, if both parameters are needed, are the criteria compatible with one another?

Polycyclic Aromatic Hydrocarbons

The basis of the criteria for polycyclic aromatic hydrocarbons (PAHs) at industrial and commercial lands is not clear. The Guideline contains the vague comment that the criteria are "derived from the Shell/Texaco criteria". In 1988/89, criteria were set for two petroleum refinery properties in Oakville and Port Credit. The criteria included a value of 1.2 μg/g for benzo[a]pyrene for a residential land use scenario complete with backyard gardens. Another report by the team that originally calculated the 1.2 μg/g value indicated that the value for various types of industrial and commercial land should be eight to 30 times higher. But the MOE chose to use only the 1.2 μg/g value in the Guideline and then multiplied it by 1.57 to create the industrial/commercial criterion of 1.9 μg/g. The multiplier was set to mimic ratios between some benzo[a]pyrene criteria recommended by Massachusetts. The generic criteria for benzo[a]pyrene are particularly important in the Guideline because criteria for several other PAHs are multiples of benzo[a]pyrene. Examples include benzo[b]fluoranthene, benzo[k]fluoranthene, chrysene, dibenzo[a,h]anthracene, and indeno[1,2,3-cd]pyrene. The net result is that the scientific basis for the generic criteria for all of these PAHs is poorly explained in the Guideline.

Statistical Analysis of Soil and Ground Water Data

The development of statistical methods to interpret soil and ground water data is proving to be elusive in Ontario. Instead, there is a tendency to assess data by merely comparing measured concentrations to criteria. While the MOE has endorsed a simple statistical treatment of soil data from excavations at petroleum storage tanks, no similar approach exists for any other soil or ground water assessment activities. It has been reported that one of the frequent reasons RSCs fail MOE audits is because the MOE is rigorous in its interpretation that it is unacceptable to exceed any criterion anywhere on a site, regardless of the amount of the excess or its statistical significance (Engineering Dimensions, 2001).

Depth to Water Table

Around 1999, representatives of the Standards Development Branch (SDB) at the MOE indicated that there was an aspect of the Guideline that needed to be clarified. The Guideline indicates that sites with less than 2 m of overburden are potentially sensitive sites (and therefore generic criteria may not be appropriate at such sites), however, the SDB indicated that the intention was that sites where there is less than 2 m to the water table may be potentially sensitive. This has important implications because it would mean that there are many more potentially sensitive sites than previously thought. Since that time, requests for clarification have been asked of several district and regional offices of the MOE. While some disagree with the SDB interpretation, most offer no comment.

Human receptors for VOCs

The Guideline relies heavily on some concepts and equations in a 1991 article by Johnson and Ettinger to estimate exposures to vapours that migrate into buildings. While it is obvious that not all buildings are houses with full basements, and not all houses have basement walls and floors with an overabundance of cracks and openings, the MOE sees deviations from those default assumptions by Johnson and Ettinger as reasons for a risk assessment to be labelled as using a Level 2 approach to risk management and thereby trigger additional (and often undesirable) administrative consequences. There clearly is a need for this aspect of risk assessment to evolve beyond the 1991 article. As a minimum, a standard industrial/commercial building should be defined. Furthermore, organizations such as Canada Mortgage and Housing Corporation (CMHC) have produced better "Canadian" values for building characteristics that should be used.

Ecological Receptors in Urban Settings

How much attention do ecological receptors merit when assessing urban settings, particularly those covered by buildings, parking lots, roads and sidewalks? This question often arises when an SSRA is to be prepared. MOE guidance indicates that species that exist on or near the site must be assessed. But the same MOE document indicates that a screening level assessment often is sufficient and that such an assessment should focus on receptors of major importance or so-called valued ecosystem components (VECs). Some SSRAs conclude that VECS are not present and therefore further ecological assessment is not required. Other SSRAs go further and try to assess organisms that are present but hardly seem to be VECs. Examples include soil micro-organisms and small organisms such as the earthworm or wood louse. The situation is further complicated when one considers that the MOE chose to ignore terrestrial receptors when setting generic criteria for soils deeper than 1.5 m below the surface. The net result is a highly inconsistent approach to assessing ecological receptors.

5.4   Assessing Site Conditions

5.4.1   Standardizing Terminology

As contaminated property issues grew in importance in the late 1980s and early 1990s, people and organizations looked for ways to minimize the chance of unknowingly taking on the responsibilities of owning or financing contaminated land. Advisors to potential owners (including lawyers, lenders and property managers) started to expect or require that certain types of environmental information be reviewed as a prerequisite to closing property transfers.

By the early 1990s, many terms were being used to describe this gathering of environmental information including site inspection, site assessment, site evaluation, site characterization, site audit, property audit, real estate audit, historical audit and transaction audit. Out of the resulting confusion, efforts were made in Canada and elsewhere to standardize the terminology and tasks for various types of site assessment studies. From those efforts, the terms "transaction screen" and "environmental site assessment" (ESA) gained wide acceptance. While standard definitions are now well-established, buyers and sellers of site assessment services still should ensure that there is a clear understanding of the work to be done prior to undertaking such studies.

5.4.2   Transaction Screen

A transaction screen usually takes the form of a questionnaire about current and past activities at a site. The questions are phrased so that simple "yes", "no" or "don't know" answers will suffice. The screens often are developed by financial institutions for use by in-house staff. The results indicate whether a transaction should proceed or stop or whether it requires further assessment by environmental professionals.

Examples of questions that might be addressed in a transaction screen are presented in Table 5.4. A positive response to any of the questions should raise a red flag and hence require a more detailed assessment (that is, a Phase I or Phase II ESA). The more red flags, the greater the need to consult with environmental specialists.

5.4.3   Phase I Environmental Site Assessment

A Phase I ESA typically consists of four tasks:

- gathering information about past and present activities and uses of the site
- inspection of the site by an environmental professional, usually accompanied by someone familiar with the property (for example, a site manager, plant manager or long-time employee)
- reviewing environmental files maintained by the site owner and regulatory agencies
- preparing a report that identifies existing and potential sources of contamination on the property

A Phase I ESA report may also identify the types of studies needed to confirm or deny suspected conditions. It is important to note that Phase I ESAs typically do not include collecting any samples for analysis or taking any measurements.

Review of Historical Information

The review of historical information can offer clues as to the types of contaminants that might be at a site and locations where contamination is most likely be found. Some types of industrial activities are recognized as probable sources of contamination. These include coal gasification works, salvage yards, rail spurs and maintenance yards, petroleum refineries, fuel storage and distribution facilities, metal foundries, metal-plating operations, wood-preserving facilities, leather tanneries, chemical producers, pesticide manufacturers, and paint and ink manufacturers.

Transfers of ownership complicate the review of historical information. With each new owner, the clues by which contamination can be identified may become more difficult to uncover. The title search that often is a legal prerequisite for a transfer of ownership can identify previous owners of a property and the types of activities that may have occurred on the site, but legal title searches often extend back 40 years. This may not be sufficient for a Phase I ESA where historical information sources (such as aerial photographs, fire insurance plans and municipal directories) should be traced back to the point where there is a long period of agricultural use or non-use.

Fire insurance plans and reports can be reviewed at some libraries, but not copied. These plans and maps can be obtained from the Insurers' Advisory Organization Inc. for a fee. Further information and documents can be found at **www.iao.ca**.

Ground water is a special concern wherever it is used as a drinking water supply or where aquatic habitat is at risk. When reviewing historical information be aware of the possibility of ground water contamination. The legal and financial consequences of contamination migrating beyond property boundaries are much greater and more complicated than on-site contamination.

Site Inspection

The overall objectives of a site inspection are to develop a general appreciation for site layout and activities. At properties where chemicals are used or stored, the site assessor should come away with a general understanding of all chemicals used, the ways in which chemicals may be released to the environment (that is, emission release points), and all chemical handling and waste management procedures.

Even at properties where activities do not involve the use, storage or production of chemicals, there are numerous conditions that need to be noted during a Phase I ESA. The topography or physical layout of the property should be checked for obvious signs of past and present waste disposal practices such as lagoons, pits, ponds or standing water. Note should be made of areas with distressed, sparse or absent vegetation as a possible indicator of contaminants.

Facilities should be checked for the presence of any hazardous materials (as defined by Regulation 347) or designated substances (as defined by the Ontario Ministry of Labour). All PCB-containing equipment such as transformers and capacitors should be identified. Any equipment no longer in service must be stored in properly designed facilities (see Section 7.1). Asbestos (a generic term that applies to naturally occurring hydrated mineral silicates) was used extensively for ceiling and floor tiles, pipe insulation, cement and insulating materials (see Section 7.2). The presence of asbestos-containing materials will complicate and increase the costs of renovation or demolition efforts.

Storage tanks and piping can be the source of leaks or spills and the subsequent contamination of soil and/or ground water. Vent or filler pipes can often indicate underground storage tanks. Leaking tanks can go undetected for years. Various sources have estimated that 5 to 20% of all underground storage tanks have leaked. There are several hundred thousand underground storage tanks in Canada.

Finally, any activities on nearby properties that have the potential to adversely affect the surrounding environment should be noted. Examples include facilities that handle large volumes of chemicals in liquid form, facilities that generate or store large amounts of waste materials, locations where materials have been stored in tanks for many years, locations where large amounts of material have been used to raise grades, and landfills.

Review of Company Files and Records

This next step can provide information on site conditions and the environmental performance of a facility. Files to review include previously prepared ESA reports or environmental audits, reports of incidents such as spills or unscheduled releases, and waste manifests and records of chemical use. Company files can provide an account of how well a company has complied with environmental requirements, but may ignore significant environmental risks at a facility (for example, inadequate spill containment or prevention at a discharge adjacent to a sensitive fish habitat). Reviews of past documents must consider the terms of references of those who prepared the documents.

A list of the chemicals used, stored or produced at the facility can be prepared using records of waste manifests and chemical inventories/purchases.

Under the Freedom of Information Act, any citizen has the ability to request and review MOE files. Information regarding processes/chemicals, the names of individuals registered on complaints and information used in a current court proceeding may be confidential.

An MOE "approval" file may contain Cs of A for air emissions, water discharges, and waste treatment and/or disposal systems. The certificates for subject waste stored on-site should be critically reviewed for the types and volumes of chemicals stored, and the technology used to prevent ground water contamination. An approval file should also contain violation notices, control orders or tickets issued under the EPA (see Section 8.5.2). The file may contain information concerning unscheduled releases of contaminants onto the property.

The MOE Spills Action Centre should be contacted for information about spills on the subject property and adjacent properties (see Chapter 12).

Another potential source of information is the MOE Public Information Dataset, which contains summarized information about the generators, carriers and receivers of subject waste. The Dataset is updated annually.

Various environmental databases of information collected by federal and provincial agencies can be obtained for a fee by contacting EcoLog Environmental Risk Information Services Limited. Additional information can be obtained at **www.ecologeris.com**. Municipalities can be contacted for information about sewer use by-law infractions and spills.

Interviews with knowledgeable personnel can provide information regarding on-site contaminants that may not have been recorded. Informal discussions with long-time plant personnel can identify past practices or unusual events, such as spills that should be investigated further. Interviews with neighbours may be useful, as they may have noticed odours, dust, noise, or other conditions that could be linked to on-site activities. Some judgments, however, may be required to separate facts from fiction.

Reports

Phase I ESA reports should identify existing and potential sources of contamination on the property, document all findings, indicate the relative degree of uncertainty associated with evidence of potential contamination, describe steps that could be taken to confirm, refute or delineate contamination (Phase II ESA activities), and provide preliminary estimates of the costs associated with those steps. Facts should be clearly distinguished from opinion.

Several organizations have published guidelines or outlines for Phase I ESA reports. These include the Canadian Standards Association (CSA, 2001), the Canada Mortgage and Housing Corporation (CMHC, 1994), and the American Society for Testing and Materials (ASTM, 2000). There also is a guide to the environmental assessment of sites and organizations adopted by the CSA (2002) from the International Organization for Standardization. The MOE also is developing a Phase I ESA guideline, but has not published it as of late 2002.

5.4.4   Phase II Environmental Site Assessment

While Phase I ESAs rely on existing information, the focus of Phase II ESAs is to gather new information about a property. Phase II ESA tasks can include:

- collecting and analyzing samples
- directly measuring conditions such as noise levels or radiation
- comparing the analytical data and measurements to environmental criteria to determine the types of contaminants present, the physical extent of contaminants and the potential volumes of material that may need to be treated, removed or otherwise managed
- using environmental fate or transportation models to evaluate the potential migration of the contamination

Initial sampling efforts should be based on information (typically gathered during a Phase I ESA) about historical and current operations at the site, observations made during a site inspection, the results of previous sampling and the future use of the site. Numerous sample-gathering techniques are available. Factors such as the type(s) of material to be sampled (such as soil, ground water, etc.), the accessibility of the materials to be sampled (for example, whether they are located at or near the surface, at substantial depths, beneath buildings, etc.), and the type(s) of analyses to be performed will influence the selection of sample-gathering techniques.

Boreholes are one of the most commonly used ways to investigate the physical and chemical quality of soil. As soil samples are brought to the surface, they are inspected for signs of impacts such as unusual colour, staining, odours, vapours or materials such as debris, cinders, wood or glass.

Test pits also can be used for soil sampling and can provide a better picture of local stratigraphy, but are best suited when depths of interest are shallow (less than 3 m), and where the surface disruption associated with test pits is not a concern.

Monitoring wells can be installed in boreholes and then used to study water levels, ground water flow directions and ground water quality.

Geotechnical investigations conducted primarily for construction purposes may provide qualitative information regarding the extent of contamination at a site. For example, geotechnical borehole logs may note stains, odours or debris. Similarly, observations made during excavation for building additions, installing utilities or removing pipelines or underground structures can indicate the locations and types of contamination present.

Non-disruptive investigations using geophysical techniques such as near-surface conductivity surveys and magnetometer surveys can be used to delineate buried metallic objects such as tanks, drums, pipelines and debris. All of the anomalies identified by such a survey should be thoroughly reviewed prior to taking further action since survey anomalies can be induced by nearby objects such as buildings, fences and surface debris.

While most sampling programs generate samples that are taken off-site for analysis, there are a growing number of options for on-site monitoring. It is common practice to use an organic vapour meter when investigating volatile organic compounds. Vapour levels can be measured quickly in the headspace of soil or ground water sample containers, or by placing the meter close to soil samples as they are obtained from boreholes or test pits.

Other portable measuring devices include X-ray fluorescence devices (measuring lead in paint), similar devices for arsenic, and gas chromatographs (measuring organic compounds in soil or water). These devices require specialized training and care to operate and tend to be used when large numbers of samples need to be analyzed at a site.

Other on-site monitoring techniques include test kits developed for certain parameters. These typically involve adding a small amount of a reagent to the sample that will react with the chemical of interest to cause a colour change. The resulting colour can be translated into a concentration.

For the initial phase of sampling programs, it may be cost-effective to analyze samples for indicator parameters such as pH, electrical conductivity, total chromium, total petroleum hydrocarbons, or total dioxins and furans. The results of these analyses should indicate if further sampling is required or if specific individual chemicals need to be investigated. There may need to be more than one campaign of sample gathering and analysis with subsequent campaigns used to improve the delineation of contamination.

Off-site sampling of adjacent surface water may be appropriate where a facility has a direct discharge to surface water or its storm water is directed to surface water. This can include locations that are upstream, at, or downstream of a facility. In addition, surface water may be sampled at locations where ground water emerges.

Phase II ESA reports should document all findings and criteria used to define contamination. Assessors and clients may agree to expand the scope of a Phase II ESA to include estimates of volumes of contaminated materials, preliminary options for managing the contamination or the next steps to developing a remediation plan. In all cases, the methods used should be described.

Some organizations have published guidelines or outlines for Phase II ESA reports (CSA, 2000; ASTM, 1998). The MOE also is developing a Phase II ESA guideline, but it had not been published as of late 2002.

Although remediation planning sometimes is referred to as a Phase III ESA (with implementation of the plan being called Phase IV ESA), that terminology is somewhat misleading and not as helpful as the preferred terms "remediation planning and implementation".

5.5   Remediation

5.5.1   When Is Remediation Necessary?

Remediation can be broadly defined as the actions taken to manage contamination. These actions can include physical actions (such as the destruction, removal, isolation or immobilization of contaminants) or administrative actions (such as restricted site access, signs, registration on title and the "holding" designation in municipal zoning by-laws).

The need for remediation usually is prefaced upon the findings of an ESA, environmental audit or in response to an accidental release of a substance. The evidence most often cited as demonstrating the need for remediation is one or more substances (usually in site soil or ground water) at concentrations that exceed whatever guidelines or criteria are appropriate to use.

The reasons for remediation can include the need to improve conditions as a prerequisite for changing land use; as a means of preventing contamination from migrating off-property (and avoiding the various types of liability that such a condition can pose); and the desire to improve property value and/or the ability to show that a property is "clean".

The *Guideline for Use at Contaminated Sites in Ontario* (MOEE, 1997) includes remediation in the four basic steps of site assessment:

Step 1 – the initial site assessment consists of the activities described in Section 5.4.3.

Step 2 – detailed site assessment consists of the activities described in Section 5.4.4.

Step 3 – remediation work planning consists of the activities described below in Sections 5.5.2 and 5.5.3.

Step 4 – completion of the site assessment and remediation process consists of the activities described in Section 5.5.4.

5.5.2   Remediation Planning

The kinds of information needed to plan remediation activities typically concern the types of contaminants present and their locations; site conditions such as soil type(s) and depth to ground water table; the clean-up objectives to be achieved; and the capabilities of the option(s) being considered.

At relatively simple sites, one remediation option may be adequate to address contamination. At more complex sites, multiple options may be needed. It may also be necessary to clean certain portions of a site or specific contaminants before other problems can be addressed. The considerations of timing, sequencing, logistics and combining (either in series or in parallel) various technologies all need to be taken into account in planning remediation.

As noted in Section 5.5.1, remediation plans can include options other than those that remove contaminants. Other types of options can include isolating contaminants (so that they are immobile in the environment), on-site or off-site storage, or pairing future use to clean-up efforts (for example, in areas where a clean-up to meet residential requirements is difficult, it may be designated for industrial or commercial use).

Remediation planning often will begin by reviewing numerous options and identifying those that appear to be best suited for application to a specific site. Several dozen options are available for soil and a similar number of options are available for ground water (see Section 5.6.2). Factors that may need to be considered when selecting options include the ability of technologies to meet clean-up guidelines; the time required to implement and carry out the clean-up; the risks to workers and neighbours during clean-up; and costs.

Procedures such as environmental fate modelling, pathways analysis, risk assessment, and financial analysis may be used to evaluate aspects of a site remediation plan. In some cases (for example, in response to an order or when there are concerns about potential adverse effects), the results of remediation planning may need to be reviewed by regulatory agencies such as the MOE. For some remediation technologies, bench-scale and/or pilot-scale investigations will be needed to demonstrate their effectiveness, to develop detailed plans for full-scale implementation, and/or gather data necessary to obtain MOE approval.

5.5.3   Implementation

While implementation is the phase of remediation in which much of the physical work gets done, it also is a relatively straightforward phase if site assessment and remediation planning have been done carefully. There are numerous companies (also referred to as contractors or vendors) that provide treatment, removal and disposal services. Independent inspectors may be used while remediation efforts are underway.

Before a plan can be implemented, environmental approvals may be required. The approvals will not be for the plan *per se* but for any activities of the plan that release contaminants to the environment. Approvals may be needed that address the atmospheric emissions of activities, the treatment or disposal of waste waters or residues, and noise levels along the site boundary. Before approvals are issued, sufficient information characterizing the releases must be submitted to the MOE and reviewed.

Obtaining approvals for various types of remediation technologies is an important area of uncertainty surrounding the remediation of contaminated sites in Ontario.

Toward the completion of the site remediation plan, soil and/or ground water monitoring can be used to verify that clean-up objectives have been met. This information should adequately demonstrate that the remediation has been successful. The types of samples, number of samples and the parameters tested often are left to the assessor's judgment. In one of the supporting documents to the Guideline, the MOE provides a guide to the minimum number of samples for excavations around underground tanks used to store petroleum products (MOEE, 1996c).

5.5.4   Documentation/Registration

At the conclusion of implementing the site remediation plan, a final comprehensive report should be prepared that describes the goals of the remediation, all remediation activities and the information that verifies the goals have been met. Reports prepared during site assessment or remediation planning can be used to facilitate the final report's production.

If remediation occurred in advance of redevelopment, the verification information will need to be shared with the municipality. If remediation is a response to an MOE order or concern, verification information will need to be shared with the MOE. In some instances, it may be necessary to share the verification information with both. In relatively rare circumstances, the MOE may want to conduct its own verification testing.

When the Guideline was issued in 1996, it introduced a form called the Record of Site Condition (RSC). An RSC may be submitted to the MOE whenever an applicant wants to provide a summary of site conditions (after remediation, if any remediation has been done). The RSC includes a list of the environmental reports prepared for the site, summaries of the maximum concentrations of chemicals that remain on site, and statements signed by the property owner and the consultant.

Once received and checked for accuracy and completion, the MOE will return the RSC with Part 7 (acknowledgement of receipt) signed and dated. As the name implies, this is strictly a receipt and does not signify that the MOE has reviewed or concurred with the contents of the RSC.

In certain circumstances, an RSC must be filed with the MOE. These conditions include using the stratified approach to assess or remediate a site, and the use of a Level 2 approach to risk management at the site. When the Brownfields Statute Law Amendment Act comes into force, it also will become necessary to submit an RSC whenever an industrial or commercial property is converted to a higher use (for example, agriculture, residential or park land).

If the RSC indicates that a stratified approach has been used or that a Level 2 approach to risk management has been used, before the RSC is acknowledged the MOE may issue an order to the property owner that a Certificate of Prohibition must be registered on title. The purpose of the certificate is to provide a means of notifying any person who may acquire an interest in the site. A stamped duplicate copy of the registered certificate must be returned to the MOE. Once it has received confirmation of the registration, the MOE will acknowledge receipt of the RSC.

If site conditions improve to the extent that the reason for the Certificate of Prohibition no longer exist, the property owner can submit an environmental update report to the MOE for review. If the MOE concurs, it will issue an order to the property owner that a Certificate of Withdrawal of Prohibition must be attached to the title of the property.

Under the Brownfields Statute Law Amendment Act, it is proposed to rename the Certificate of Prohibition and the Certificate of Withdrawal of Prohibition with a Certificate of Property Use.

Some of the RSCs submitted to the MOE are subjected to a more thorough review. The MOE has established a central audit team to perform this function. RSCs are selected primarily on a random basis, although other factors are taken into consideration. To prepare an audit, the MOE typically requests copies of all the reports listed in the RSC. These audits have resulted in fairly high percentages of failures. Failures can be referred to the Investigation and Enforcement Branch (see Chapter 8) although this is relatively rare. Failures also can prompt the MOE to undertake its own site assessment investigations, but this too happens rarely.

5.5.5   Remediation beyond Government Criteria

In Ontario, particularly since the *Guideline for Use at Contaminated Sites in Ontario* (MOEE, 1997) was issued, most remediation efforts have been evaluated by comparing post-remediation information to the generic criteria in the Guideline. While this approach is seen as addressing regulatory interests, it does not insulate property owners from common or civil law. This fact was highlighted by court decisions in 2001 and by a subsequent decision from the Ontario Court of Appeal in 2002 in the case of *Tridan Developments Limited v. Shell Canada Products Ltd.*

A Shell gasoline station impacted soil on the adjacent Tridan property. Shell proposed to remediate the Tridan property so that generic criteria recommended by the MOE were no longer exceeded. Tridan argued that it was entitled to have its property remediated to "pristine" conditions (those that existed before the impacts occurred). Tridan also argued that to do otherwise would stigmatize its property and lower its value. The court ruled in favour of Tridan. It is unclear from the decision how "pristine" is to be defined. In the case of total petroleum hydrocarbons, pristine might be equated to less than detection limits. (The MOE has not established background concentrations for total petroleum hydrocarbons. See Table 5.2.)

5.6   Remediation Options

5.6.1   Options for Remediating Soil

Many ways to manage contaminated soil have been developed in response to the many combinations of contaminants and site conditions that are possible. Continuing research and development efforts are expected to expand the number of available options. Discussions of these options can be simplified by grouping the options according to common, basic characteristics. Six categories are used in the following discussion. The discussion only briefly touches upon clean-up technologies that are available or are in the process of being developed. There is extensive documentation for virtually each of the technologies and the amount of literature grows rapidly as demonstrations and actual applications are published.

Off-site Disposal

This category involves removing material from a property without prior treatment. Excavation and off-site disposal of contaminated soils has been the most commonly used option in Ontario. The type of facility to which the excavated soil should be taken is determined by the types of contaminants present and their concentrations and leachability (ability to solubilize and be transported by ground water). For example, material classified as "hazardous waste" using the Regulation 347 leachate test must be disposed at a hazardous waste management facility. The only such facility currently in Ontario is located near Sarnia. Less contaminated materials can be suitable for disposal at municipal landfills.

Destruction

The options in this category destroy organic contaminants. The options usually involve converting the contaminants to other, simpler forms such as water, carbon dioxide and acid gases. Most destruction methods generate intermediate products (vapours, ash or wastewater) that need to be managed.

Many of the methods can be implemented in place (*in situ*), in reactor vessels (typically large tanks), or off-site. Some are limited to taking place off-site at treatment facilities.

Bioremediation refers to the destruction of organic contaminants in soil and ground water through microbial degradation. Indigenous micro-organisms can be used to degrade or detoxify most organic materials, as can non-indigenous bacteria specifically designed to metabolize particular compounds or classes of compounds. Bioremediation options can be applied to soil *in situ*, in a prepared soil bed (for example, land farming or a biopile), or in a reactor tank.

All bioremediation options require the careful monitoring and adjusting of nutrients (phosphorus, nitrogen, sulphates, trace minerals, etc.) and oxygen, as well the control of water content, pH and soil temperature. These factors are more easily controlled in reactor piles or tanks than *in situ*. Bioremediation options have the potential to create noxious or hazardous gases and/or leachable biodegradation products.

Chemical destruction options employ the chemical reactivity of the contaminant to alter the form of the material or destroy it completely and render it inert or less reactive.

Thermal destruction options use extreme temperature to destroy organic soil contaminants through combustion. The emissions are chiefly gases — such as carbon dioxide — water and acid gases such as sulphur dioxide and nitrous oxides.

Separation

This category of options involves separating the contaminants from the soil by transferring the contaminants to another media such as air, liquid or solid. Most separation options take advantage of the physical properties of contaminants such as their volatility, solubility or density. The media that receive the contaminants usually need to be managed.

Volatilization transfers the contaminant from the soil or the ground water to air spaces present in the soil, which subsequently is collected for release or treatment.

Solubilization transfers the contaminant from the soil to a liquid (often ground water), which subsequently is collected for treatment. Adsorption separates contaminants from soil by physically binding the contaminants to an inert material that can be subsequently removed. The contaminants can be stripped from the inert material so as to regenerate and recycle the inert material.

Electrokinetics applies an electrical field to soil, causing a complex set of phenomena to occur. The two principal effects are electro-osmosis (fluids and dissolved species are forced through stationary media) and electrophoresis (solid particles are induced to move through a stationary fluid).

Immobilization

The options in this category use physical or chemical processes that permanently solidify or immobilize the contaminants present in soil. These processes prevent the contaminants from migrating into the ground water or air.

Solidification usually involves the production of solid blocks of material in which the contaminants are permanently, physically locked. Stabilization involves adding materials that ensure the contaminants are converted and maintained in their least mobile form. Encapsulation methods coat or enclose contaminants with a non-permeable substance.

Isolation

In this category, contaminated soil is neither treated nor removed from the site. The soil can be excavated and placed in surface storage facilities or left in place and isolated from the environment by the use of physical barriers such as impermeable liners, caps and walls.

Re-use

This category involves either the conversion of contaminated soil into a useful product (such as asphalt) or to use the soil as fill material where it is acceptable to do so.

5.6.2   Options for Remediating Groundwater

Many of the technologies used to remediate soils have analogous versions for treating ground water. Several of these options involve bringing the water to the surface and having it flow through a container or tank where a treatment process occurs. Types of treatment processes can include bioremediation, adding chemicals that destroy contaminants, air stripping (a form of enhanced volatilization), adsorption and the use of ultraviolet light (to break down organic contaminants by photolysis). Collectively, these are referred to as "pump and treat" approaches.

Treated water may be discharged to the sewer system or surface water if its quality is acceptable. Alternatively, treated water may be re-injected into the ground to assist with the recovery of affected ground water. Atmospheric emissions from the treatment component may need to be monitored, and/or require treatment of the off-gases before release.

Pump and treat approaches work best in permeable soils. Non-homogeneous conditions in the subsurface can prevent clean-up of pockets of contamination. Even in homogeneous soils, residual contamination held in place by interstitial forces can act as a long-term source of contamination. As a result, there are many cases of pump and treat efforts continuing for several years without reaching clean-up targets.

There also are techniques for treating ground water in place. Such techniques typically involve injecting chemicals to destroy contaminants or to encourage indigenous bacteria to degrade contaminants. *In situ* techniques are subject to many of the same limitations as pump and treat options.

Finally, there are techniques that can be used to contain impacted groundwater, deflect impacted ground water toward collection or treatment devices, or prevent impacted ground water from reaching selected locations. These techniques include the use of interceptor wells and subsurface barriers.

5.6.3   Current Status of Soil Remediation

Table 5.5 lists more than 20 remediation options. Their use in Ontario is ranked as either very common, rare, or three intermediate categories.

Only two of the options are described as very common in Ontario. Both involve excavation and off-site disposal at two types of facilities.

Four of the options are described as "regularly used". Both bioremediation and low temperature thermal desorption are used for organic contaminants (mainly petroleum hydrocarbons). On the other hand, surface storage and containment are used for various types of chemicals.

Two options are described as "occasionally" used. These include chemical oxidation and dual phase extraction (where high vacuum pumps collect ground water and soil vapours from wells). Both are used to manage organic chemicals and both can be used to treat both soil and ground water simultaneously.

Three options are described as being used on a "limited" basis. These include biological seeding (used for resistant organic compounds such as PAHs), dechlorination (to treat PCBs in transformer oil) and metal chelation (to remove various heavy metals from soil or ground water).

The remaining options are described as being "rarely" used in Ontario. Several of these options would need to overcome opposition from regulatory agencies and/or the public before they can be widely used. Examples include options that use thermal destruction (due to concerns about atmospheric emissions), immobilization (due to concerns about the long-term integrity of immobilized materials) and subsurface containment (due to concerns about the long-term integrity of the containment features).

The costs of the options listed in Table 5.5 range from $25/m^3 to more than $1,000/m^3.

Since excavation and off-site disposal is so widely used, its cost of $100/m^3 to $150/m^3 acts as a benchmark for assessing the costs of other options. Although disruptive and not feasible where impacts extend beneath buildings and around buried utilities, it permanently eliminates the impacted material and can be implemented relatively quickly.

Most of the other options that destroy, remove or immobilize contaminants in soil have costs in the range of $150/m^3$ to $300/m^3$.

All of the options shown in Table 5.5, except those classified as "very common" or "regularly used" face the cyclical challenge that the lack of local experience and/or proven effectiveness makes it difficult to provide the information the MOE needs to issue the required permits to use the option. While there were federal and provincial programs in the 1990s to develop and demonstrate site remediation technologies, those programs have ended.

One of those initiatives at the federal level was the Development and Demonstration of Site Remediation Technology Program. Although the program stopped in 1995, the results of demonstrations projects are available at the Environment Canada Web site (**www.ec.gc.ca**).

There was also the national Groundwater and Soil Remediation Program (GASReP), which was jointly sponsored by government and the petroleum industry. GASReP focused on the development and assessment of technology for cleaning ground water and soil contaminated by petroleum hydrocarbons. It appears that the program was terminated in the later 1990s, but GASReP projects still appear in listings of remediation technology summaries.

From time to time, the MOE has indicated a willingness to assist remediation technology vendors to obtain approvals for pilot projects that demonstrate new or innovative technologies.

Many organizations have published reviews of remediation technologies or provide reviews and information on Web sites. Table 5.6 identifies a few such sources.

5.7   Legal Aspects of Buying or Selling Contaminated Property

5.7.1   The Role of Common Law

Notwithstanding the use of the transaction screens and ESAs described in Section 5.4, there remains the basic tenet of common law that the onus is on buyers to protect themselves by an express warranty that premises are fit for their purpose and free from specified defects. The *caveat emptor* principle applies when a buyer purchases a property that has a defect that could have been discovered during a careful inspection. If the defect could not be discovered, the onus may be on the vendor, however, the purchaser must prove that the vendor knew of the latent defect. At many sites, it is difficult to prove that the owner knew of contamination. Examples can include ground water contamination resulting from a leaking tank or historical disposal activities that occurred before the owner took possession.

The *caveat emptor* principle may not apply if misrepresentation has occurred during the purchase of the contaminated property, however, the misrepresentation must be actual fraud to rescind the contract and recover out-of-pocket losses. If an innocent misrepresentation occurs (that is, the owners believed they were acting in good faith), the contract may be rescinded but recovery of out-of-pocket losses may not be allowed. Several other torts (civil wrongs excluding breach of contract) may be applicable during a property transfer. These include negligence, nuisance and trespass. (Section 8.8 discusses common law as it pertains to torts.)

Where responsibility for contamination is complicated by changes in ownership, the MOE can elect to issue an order to all parties that can be legally associated to the cause of the contamination. Because of the broad definitions in the EPA of "owner" and "operator", liability for past contamination can extend through unsuspecting parties.

5.7.2   The Role of Contract Law

To maximize protection from assuming a property that poses environmental liabilities, a purchaser's offer should contain the following provisions (Ruderman, 1988; McAree 2001):

1. The purchaser and their advisors have the right to access files, documents, etc. pertinent to the property, inspect the property, interview staff and conduct environmental tests during a conditional period. The conditional period should be long enough to complete these activities.
2. If not satisfied with the inspection, the purchaser has the right to terminate the agreement or proceed with the agreement such that it is not deemed to be a waiver of non-compliance with any of the vendor's warranties.
3. The vendor must provide an absolute warrant that there are no noxious, dangerous, toxic substances or conditions on the property, nor has the vendor received notice or knowledge of any judicial or administrative action related to the presence or discharge of noxious, dangerous or toxic substances on the property or adjacent lands.
4. The vendor must provide an absolute warrant that all necessary licenses, certificates and registrations to operate the business have been obtained and the business is in compliance with all government laws and regulations.
5. The purchaser must be informed promptly in writing if the vendor has knowledge of any event or likely event that may result in any of the warranties no longer being true.
6. A statement that the warranties will survive closing.

7. The purchaser has the right to terminate if warranties are found to be untrue and potential recovery of inspection costs.
8. The vendor is required to rectify the breach of warranty prior to closing and a portion of the purchase price is held back as security on the completion of the work.
9. The agreement of purchase and sale should be contingent upon the purchaser meeting the requirements of its lender, insurer and muncipality.

In recent years, the rights of a purchaser to claim damages arising from non-disclosure of facts relating to the condition of the property have expanded. Therefore, if a vendor has any knowledge of potentially hazardous conditions, the vendor may not be protected by an offer that contains no warranties. The knowledge portion of this liability can extend to a corporation's officers, directors and agents.

The vendor's counter-offer may contain the following (Ruderman, 1988; McAree, 2001):

- Agreement that the vendor will disclose all latent defects of which the vendor has knowledge.
- A requirement that the purchaser acknowledge the previous uses of the property that may have resulted in the existence of hazardous, noxious or toxic conditions or substances in soil, ground water and structures.
- Agreement that the purchaser can conduct an independent assessment at its own expense and that the purchaser is buying on an "as is" basis.
- If the purchaser does not terminate the agreement during the assessment or inspection period, the purchaser is deemed to have accepted the conditions of the property.
- The purchaser is to indemnify the vendor during and after closing from all claims, liabilities and obligations respecting such substances or conditions.
- If the purchaser terminates the agreement following the assessment or inspection, the purchaser will be responsible for all costs and will not be entitled to make any claim for damages arising out of breach of warranty.

Such clauses will not protect the vendor from liability if there is an actual concealment of conditions relating to the property, a direct false statement about the condition of the property or an intentional withholding of facts.

The above counter-offer terms may conflict with those described in the offer to purchase and may scare off a purchaser. It is best to have a lawyer knowledgeable in the field of environmental law involved in the transaction to ensure that the appropriate steps are being taken.

Note that the various types of professional advisors (such as engineers, lawyers, real estate agents, etc.) who may participate during the transfer of ownership have a duty to act with a reasonable degree of care. A person under contractual duty to make an inspection may be liable for breach of contract and/or negligence by his or her failure to perform the tasks properly.

5.8   Summary

Owners, prospective buyers and sellers of property need to be aware of environmental legislation in Ontario that influences site assessment, remediation, redevelopment and the assignment of liabilities. At greatest risk are individuals who deal with properties that have a history of industrial or commercial use. Past practices have resulted in conditions at many sites that do not meet the environmental criteria now in place.

The Ontario Environmental Protection Act clearly places responsibility for making a site acceptable upon the current owner, but previous owners or site operators can also be held responsible. Where responsibility for contamination is complicated by changes in ownership, the MOE can elect to issue an order to all parties that can be legally associated to the cause of the contamination. Where the transfer of property is concerned, the adage "let the buyer beware" is more applicable today than ever before.

The MOE publication *Guideline for Use at Contaminated Sites in Ontario* is often cited as the basic reference when proponents consider site remediation in Ontario. As of late 2002, the Guideline does not have force of law but the MOE has indicated it likely will be incorporated into regulation.

Many methods for remediating sites or addressing various types of site contamination are in the developmental and demonstration stages, but very few of the methods are used with any regularity in Ontario.

Collectively, theses conditions pose numerous obstacles to buying, selling, remediating and redeveloping contaminated property. They also clearly illustrate the need for current property owners to protect the environmental quality of their sites.

References

American Society for Testing and Materials (ASTM). 1998. *ASTM Standards Related to the Phase II Environmental Site Assessment Process.*

American Society for Testing and Materials (ASTM). 2000. *Standard Practice for Environmental Site Assessments: Phase I Environmental Site Assessment Process.* ASTM E 1527-00.

Canada Mortgage and Housing Corporation (CMHC). 1994. *Phase I Environmental Site Assessment Interpretation Guidelines.*

Canadian Standards Association (CSA). March 2000. *Phase II Environmental Site Assessment.* CSA Publication Z760-00.

Canadian Standards Association (CSA). November 2001. *Phase I Environmental Site Assessment.* CSA Publication Z768-01.

Canadian Standards Association (CSA). March 2002. *Environmental Management - Environmental Assessment of Sites and Organizations (EASO).* CAN/CSA-ISO 14015:02.

Engineering Dimensions. September/October 2001. *Practice Bulletin for Engineers: Completing RSCs for Site Assessment and Remediation.* p. 6.

McAree, M. November 2001. "The Legal Perspective on Managing Risk in Contaminated Land Transactions". Presented at Managing Contaminated Land and Brownfield Redevelopment. Waterloo, Ontario.

Ontario Ministry of Environment and Energy (MOEE). July 1994. *Water Management: Policies, Guidelines, Provincial Water Quality Objectives of the Ministry of Environment and Energy.*

Ontario Ministry of Environment and Energy (MOEE). July 1996a. *Rationale for the Development and Application of Generic Soil, Groundwater and Sediment Criteria for Use at Contaminated Sites in Ontario.* ISBN 0-7778-2818-9.

Ontario Ministry of Environment and Energy (MOEE). July 1996b. *Guidance on Site Specific Risk Assessment for Use at Contaminated Sites in Ontario.* ISBN 0-7778-4058-8.

Ontario Ministry of Environment and Energy (MOEE). July 1996c. *Guidance on Sampling and Analytical Methods for Use at Contaminated Sites in Ontario.* ISBN 0-7778-4056-1.

Ontario Ministry of Environment and Energy (MOEE). February 1997. *Guideline for Use at Contaminated Sites in Ontario.* ISBN 0-7778-4052-9.

Ruderman, J.C. 1988. "Negotiating the Agreement of Purchase and Sale to Reduce Risk of Environmental Hazards." Presented at the Environmental Real Estate Transaction Conference, Toronto. 19 September.

Table 5.1

Selected Examples of MOE Criteria

	Arsenic	B[a]P	Copper	Lead	TCE	VCM
Table A						
AG, soil	25/20	1.2	200/150	200	3.9/1.1	0.0075/0.003
R/P, soil	25/20	1.2	300/225	200	3.9/1.1	0.0075/0.003
I/C, soil	50/40	1.9	300/225	1,000	3.9/1.1	0.0075/0.03
Groundwater	25	0.01	23	10	50	1.3/0.5
Table B						
R/P, soil	25/20	1.2	300/225	200	3.9/1.1	0.0075/0.003
I/C, soil	50/40	1.9	300/225	1,000	3.9/1.1	0.0075/0.003
Groundwater	480	1.9	23	32	50	1.3/0.5
Table C						
R/P, soil	50/40	1.9	2,500	1,000	3.9	0.25/0.094
I/C, soil	NV	7.2	2,500	NV	3.9	0.25/0.094
Table D						
R/P, soil	50/40	1.9	2,500	1,000	3.9	0.25/0.094
I/C, soil	NV	7.2	2,500	NV	3.9	0.25/0.094
Table E						
Sediment	6.0	0.37	16	31	NV	NV
Table F						
AG, soil	14	0.10	56	55	0.004	0.003
Other, soil	17	0.49	85	120	0.004	0.003

Notes

- Soil and sediment values in µg/g on dry weight basis.
- Soil criteria for inorganics apply only when soil pH is 5.0 to 9.0.
- Soil criteria in Table C and D apply to subsurface soils (> 1.5 m below surface).
- Where two values are listed the first applies to locations where soil is medium to-fine textured. The second applies to soils that are coarse-textured.
- Ground water values are in µg/L.
- Sediment values are described as lowest effect levels.

NV = no value set   AG = agriculture
R/P = residential/parkland   I/C = industrial/commercial
B[a]P = benzo[a]pyrene   TCE = trichloroethylene
VCM = vinyl chloride

Table 5.2

More Examples of MOE Criteria (Petroleum Hydrocarbons)

	Benzene	Toluene	Ethyl Benzene	Xylenes	TPH(g/d)	TPH(ho)
Table A						
AG, soil	0.24	2.1	0.28	25	100	1,000
R/P, soil	0.24	2.1	0.28	25	100	1,000
I/C, soil	0.24	2.1	0.28	25	100	1,000
Ground water	0.005	0.024	0.0024	0.3	1	1
Table B						
R/P, soil	25/5.3	150/34	500/290	210/34	1,000	1,000
I/C, soil	25/5.3	150/34	1,000/290	210/34	2,000/1,000	5,000
Ground water	12/1.9	37/5.9	50/28	35/5.6	NV	NV
Table C						
R/P, soil	0.24	2.1	0.28	25	100	1,000
I/C, soil	0.24	2.1	0.28	25	100	1,000
Table D						
R/P, soil	63	1,000/510	1,000	1,000/460	5,000	5,000
I/C, soil	230/89	2,500 /510	2,500 /460	2,500	10,000 /5,000	10,000 /5,000
Table E						
Sediment	0.32	NV	NV	NV	NV	NV
Table F						
AG, soil	0.002	0.002	0.002	0.002	NV	NV
Other, soil	0.002	0.002	0.002	0.002	NV	NV

Notes

- Soil and sediment values in µg/g on dry weight basis.
- Soil criteria for inorganics apply only when soil pH is 5.0 to 9.0.
- Soil criteria in Table C and D apply to subsurface soils (> 1.5 m below surface).
- Where two values are listed the first applies to locations where soil is medium- to fine-textured.
- The second applies to soils that are coarse-textured.
- Ground water values are in mg/L.
- Sediment values are described as lowest effect levels.
- NV = no value set.

Table 5.3

Other Sources of Soil and Ground water Criteria

In British Columbia, the *Criteria for Managing Contaminated Sites in British Columbia* (1995) can be obtained at **www.gov.bc.ca/wlap/**.

In Alberta, the *Tier I Criteria for Contaminated Soil Assessment and Remediation* (1994) can be obtained at **www3.gov.ab.ca/env**.

In Quebec, the *Generic Criteria for Soils and Ground Water* can be obtained at **www.menv.gouv.qc.ca**.

For Atlantic Canada, criteria for petroleum hydrocarbons can be found at **www.atlanticrbca.com**.

The Canadian Council of Ministers of the Environment (CCME) has a Web site — **www.ccme.ca** — that provides access to CCME documents including the *Canada-Wide Standards for Petroleum Hydrocarbons (PHCs) in Soil* (2000) as well as earlier documents including *A Protocol for the Derivation of Environmental and Human Health Soil Quality Guidelines* (1996).

The journal *Contaminated Soil Sediment & Water* produces an annual issue that provides a state-by-state review of criteria for hydrocarbons in soil and water. The listing for 2002 was published in the January/February issue. The journal is published by the Association for the Environmental Health of Soils (AEHS).

The U.S. EPA Web site **www.epa.gov/iriswebp** provides access to the toxico-logical information in the Integrated Risk Information System.

The Part 201 criteria established by the Michigan Department of Environmental quality can be obtained at **www.michigan.gov/deq**.

The criteria contained in the Massachusetts Contingency Plan can be found at **www.state.ma.us/dep/bwsc**.

The American Petroleum Institute (API) Web site at **http://api-ep.api.org** pro-vides access to documents that describe how risk-based criteria can be derived for petroleum hydrocarbons.

Table 5.4

Examples of Transaction Screen Questions

Q1. Were significant volumes of hazardous chemicals used, produced or stored at the site?
Q2. Have waste materials been disposed on-site or nearby?
Q3. Are there any tanks or pipelines (above or below ground) on the site? Are there records or indications that any of the tanks or pipelines have leaked?
Q4. Are there unexplained earthworks or vegetation damage on the site?
Q5. Are there signs of stained or discoloured soils or building surfaces? Are oily sheens present on surface waters (including puddles)?
Q6. Are there facilities in the vicinity that may pose a health or environmental risk or nuisance problem because of current or past operation?
Q7. Are PCBs, asbestos or other designated substances present in on-site equipment or structures?
Q8. Is ground water used by nearby residents and/or does ground water discharge to a nearby water course?

Table 5.5

Examples of Remediation Technologies

Category/Option	Extent of Use*	Contaminants Addressed
OFF-SITE DISPOSAL		
Hazardous Waste Facility	very common	all**
Municipal Landfill	very common	all
DESTRUCTION, BIOREMEDIATION		
Biostimulation	regularly used	best on BTEX, some PAHs
Biological Seeding	limited	heavy oils, some PAHs
DESTRUCTION, CHEMICAL		
Oxidation	occasionally used	BTEX, chlorinated organics
Dechlorination	limited	PCBs
DESTRUCTION, THERMAL		
Fluidized Bed Incineration	rarely	BTEX, PAHs, cyanides
Rotary Kiln Incineration	rarely	BTEX, PAHs, phenols
Pyrolysis	rarely	organics and volatile metals
SEPARATION, VOLATILIZATION		
Soil Vapour Extraction	rarely	BTEX, solvents, some PAHs
Low Temperature Thermal	occasionally used	BTEX, some PAHs
Dual or Multi-Phase Extraction	occasionally used	BTEX, solvents

SEPARATION, SOLUBILIZATION

Soil Washing	rarely	most organics, most metals
High Pressure Soil Washing	rarely	all

SEPARATION, ADSORPTION

Metal Chelation	limited	most metals, cyanides

SEPARATION, ELECTROKINETICS

Electrokinetics	rarely	most metals

IMMOBILIZATION, SOLIDIFICATION

Cement Solidification	rarely	metals, PCBs
Lime-Silicate Solidification	rarely	metals, oils, solvents
Glassification	rarely	all

IMMOBILIZATION, ENCAPSULATION

Thermoplastic	rarely	metals, some organics

ISOLATION

Surface Storage	regularly used	PCBs
Subsurface Containment	rarely	all

REUSE

Asphalt Batching	rarely	BTEX, PAHs

Notes

* extent of use in Ontario; some options rarely used in Ontario are commonly used elsewhere.

** except PCBs and radionuclides (this note applies to each entry shown as "all").

BTEX - benzene, toluene, ethylbenzene, and xylenes
PAHS - polycyclic aromatic hydrocarbons

Table 5.6

Sources of Information about Remediation Technologies

Contaminated Sites Management Working Group, 1997. *Site Remediation Technologies: A Reference Manual*. Prepared by Water Technology International Corp. Available at **www.ec.gc.ca/etad/csmwg**.

Profiles of selected technologies and case studies can be found at the Web site **www.aboutREMEDIATION.com**.

REMTEC is computer software that contains more than 500 entries describing technologies for remediating soil, sediment, water and off-gases. The technologies range from those at bench scale to commercial use. For more information can be obtained at **www.scisoftware.com**.

The MOE produced two reviews of remediation technologies in 1996:
 Site Remediation Technologies Used in Ontario. Prepared by U. Sibul. ISBN 0-7778-4838-4.
 Treatment Technologies For Contaminated Groundwater (A Literature Review). Prepared by J.E. Mulira. ISBN 0-7778-4837-6.

Various U.S. EPA offices have web sites that focus on remediation technologies. The Technology Innovation Office has **www.clu-in.org**, which includes detailed descriptions of technologies, citizen's guides to technologies, annual reports of tests of technologies and tools for screening or selecting technologies. The Superfund Innovative Technology Evaluation (SITE) Program has **www.epa.gov/ORD/SITE**, which describes the technologies used at hazardous waste sites. The Remediation Technologies Development Forum (RTDF) is designed to demonstrate how government and industry can work together to improve remediation technologies. Its Web site, **www.rtdf.org**, describes the activities of several teams, each investigating a specific approach or remediation challenge. The Groundwater Remediation Technologies Analysis Centre (GWRTAC) compiles and analyses information on innovative ground water remediation. Its Web site, **www.gwrtac.org**, has descriptions of technologies and vendor information.

The Contaminated Sites Management Working Group (CSMWG) is a federal, interdepartmental committee established in 1996 to develop a common federal approach to managing contaminated sites under federal custody. One of its first projects was to create a reference manual of site remediation technologies that can be accessed at **www.ec.gc.ca/ etad/csmwg**.

CHAPTER 6

Noise and Vibration

6.1　Overview

Noise is defined as unwanted or disturbing sound. Vibrations also can be disturbing and cause damage. In the context of environmental compliance, the types of situations where sound becomes noise or generates complaints about vibrations often involve commercial operations, industrial facilities or transportation corridors in proximity to residential areas.

Both noise and vibration involve fluctuating or oscillating motions. Noise consists of energy waves, usually travelling through the air. Vibrations are similar waves of energy, usually travelling through solids such as the ground or buildings. Noise can be heard but only felt at very high levels, while vibrations cause sound and can be felt.

Both sound and vibration are defined as environmental contaminants in the Environmental Protection Act (EPA) and should be treated no less seriously than chemicals released into the environment. In Ontario, residents can file complaints about noise and vibration with various agencies including the MOE and municipal authorities.

In Ontario, the courts have stated that when a person with average sensibilities is bothered by noise impinging on their property, the person is entitled to complain and the courts will assist in having the bothersome noise checked. The courts also have indicated that one cannot acquire a right to inflict noise on another's property unless the other property owner is actually there, suffering the noise and taking no action to prevent it (Working Group on Noise Control, 1989).

The potential for people to complain about sound and vibration will increase when the following conditions occur (Thumann and Miller, 1986):

- sources such as outdoor cooling towers, vents, stacks, etc. are clearly visible
- sound or vibrations go through obvious changes or fluctuations
- sound occurs as pure tones or in discrete frequencies
- sound or vibration interfere with sleep or communications
- background levels are unusually low
- low frequency sound induces vibrations (for example, rattling windows) in residences

- sound occurs as impulses or is startling
- sound conveys displeasing information (for example, glass breaking)

The units of measurement for sound and vibration span several orders of magnitude when used to describe conditions typically found in the environment. For example, the frequencies of sounds in the environment can range from near zero to more than 1,000,000 hertz (cycles per second), although the normal range of human hearing is approximately 20 to 20,000 hertz. Even larger ranges can be encountered for sound intensity and sound pressures. As a result, logarithmic scales are used for several sound and vibration parameters (Beranek, 1971).

The logarithmic parameter most commonly associated with acoustics is the decibel (dB). The logarithmic nature of sound is illustrated in Table 6.1, which provides examples for several common sources. Table 6.2 shows similar information for vibration.

6.2   Regulatory Framework

6.2.1   Federal

The federal government establishes guidelines for noise related to interprovincial and international transportation systems including aircraft, trains and ships. Health Canada provides advice on health-related aspects of noise to assessments undertaken by or involving other federal departments (Health Canada, 1998). Overall, federal agencies play a secondary role in managing most common sources of noise or vibration.

6.2.2   Provincial

The roles of the province include approving activities or processes that emit sound or vibration into the environment, setting numerical limits for acceptable levels of sound and vibration, assisting with the development of municipal noise by-laws, and helping municipalities consider sound and vibration issues in land use planning.

The main piece of provincial legislation that addresses sound and vibration is the Environmental Protection Act (EPA). Sound and vibration are defined as contaminants in s. 1(1) of the EPA. As noted in previous chapters, the EPA contains a general prohibition against the discharge of a contaminant into the natural environment that causes or is likely to cause an adverse effect. In the context of sound and vibration, key parts of the EPA definition of "adverse effect" include:

- impairment of the natural environment for any use that can be made of it

- harm or material discomfort to any person
- loss of enjoyment of normal use of property
- interference with the normal conduct of business

In keeping with the concept that sound and vibration are two types of contaminants, the requirements to obtain a Certificate of Approval (C of A) from the MOE for sources of sound and vibration are essentially the same as those described in Section 2.2 for emissions of other types of contaminants to air. Obtaining approvals is addressed further in Section 6.3.

The EPA contains no numerical criteria for sound and vibration. Neither are there regulations that specify numerical limits. These details along with descriptions of monitoring equipment specifications, mathematical models to estimate noise levels and other aspects of sound and vibration are contained in various MOE documents including Noise Pollution Control (NPC) publications, guidelines for minimizing sound and vibration incompatibilities during land planning exercises, and bulletins described later in this chapter.

6.2.3   Municipal By-Laws and Guidelines

Section 178(1) of the EPA provides local municipalities with the power to pass sound and vibration by-laws, subject to the approval of the Minister. Municipal by-laws can address various aspects of sound and vibration:

- regulate or prohibit the emission of sounds or vibrations
- provide for the licensing of persons, equipment and premises with respect to the emission of sounds or vibrations
- prescribe maximum permissible levels of sounds or vibrations that may be emitted
- prescribe procedures for determining the levels of sounds or vibrations that are emitted
- exempt specific people, equipment, activities or premises from other provisions of the by-laws.

To assist municipalities develop noise by-laws and provide a consistent basis for noise control across Ontario, the MOE issued the *Model Municipal Noise Control By-Law* in 1978. Since that time, many municipalities in Ontario have passed noise by-laws. Some have qualitative prohibitions while others prescribe specific limits. Most describe the penalties that can be imposed for contravening by-law provisions.

6.3   Obtaining Approvals

As outlined in Section 9 of the EPA, a C of A must be issued by the MOE to any person who intends to construct, alter, extend or replace any plant, structure, equipment, apparatus, mechanism, process, rate of production or thing that may emit or discharge sound or vibration into any part of the natural environment other than water. Failure to obtain a C of A, especially for an industrial operation, may result in prosecution.

Exemptions to the C of A requirements with respect to sound and vibration include:

- routine maintenance carried out on any plant, structure, equipment, apparatus, mechanism or thing
- any equipment, apparatus, mechanism or thing in or used in connection with a building or structure designed for the housing of not more than three families where the only contaminant produced by such equipment, apparatus, mechanism or thing is sound or vibration
- any plant, structure, equipment, apparatus, mechanism or thing used in normal agricultural practice
- any motor or motor vehicle that is subject to the provisions of Part III of the EPA

General guidance on the information required by the MOE to assess C of A applications for noise and vibration sources is provided in a supplementary document entitled *Guide to Applying for Approval (Air): Noise and Vibration* (MOE, 1995a). There also is a supplementary publication, NPC-233, which describes the information to be submitted to the MOE for approval and/or audit of stationary sources of noise or vibration.

The requirements apply to all new sources of sound and vibration as well as the expansion, alteration or conversion of existing sources. The types of mandatory information to be submitted include a description of the equipment and the facility where it is to operate, the hours of operation, land use zoning designations in the surrounding area, the locations and distances to points of reception, relevant architectural and mechanical drawings, and details of any noise and vibration control measures. For complex or multiple sources, the MOE may require an acoustical report, an acoustic audit report, a vibration report, or a vibration audit report. The types of information to be provided for these reports are described in NPC-232.

6.4 Numerical Limits and Criteria

6.4.1 Noise

The MOE has issued numerous NPC documents in support of the Model Municipal Noise By-law. These are listed in Table 6.3. Several provide sound level limits from specific types of sources or in specific situations.

NPC-205 addresses sound level limits for stationary sources in urban areas. In general, the limits are set equal to the minimum background one-hour equivalent sound level (L_{eq}) at the time when the source in question (often an industrial activity) will be producing sound. For urban areas, road traffic is an important contributor to background sound levels.

NPC-205 provides specific limits for some types of sound sources including industrial metal working operations, gun clubs, infrequent impulsive sound sources and pest control devices. It also excludes sources that do not exceed 45 to 50 dBA (A-weighted decibels). The upper limit applies to time from 7:00 A.M. to 11:00 P.M. The lower limit applies to times from 11:00 P.M. to 7:00 A.M.

NPC-232 addresses sound level limits for stationary sources in rural areas, including communities of less than 1,000 people, agricultural areas, recreation areas and wilderness areas. These limits are expressed in terms of the L_{eq} and/or the one-hour 90th percentile sound level (L_{90}). In general, the limits are defined by the background sound level measured within 30 m of a dwelling or camping area.

NPC-232 provides specific limits for some types of sound sources including industrial metals working operations, gun clubs, infrequent impulsive sound sources and pest control devices. It also excludes sources that do not exceed 40 to 45 dBA, the lower limit applying to times from 7:00 P.M. to 7:00 A.M.

NPC-115 sets out sound limits for various items of construction equipment according to the date of manufacture. It includes limitations for excavation equipment, dozers, loaders, backhoes, pneumatic pavement breakers, portable air compressors and tract drills.

As noted in Section 6.2.2, one of the roles of the MOE is to help municipalities prevent or minimize incompatibilities that may arise during land use planning or from the operation of specific types of facilities. MOE assistance with such matters is outlined in documents such as Guideline D-1 (land use compatibility) and Guideline D-6 (compatibility between industrial facilities and sensitive land uses). In support of those two guidelines, the MOE issued LU-131, which specifies sound level limits for the assessment of noise-sensitive land uses (for example, residential) adjacent to relatively noisy facilities such as airports, transportation corridors and industrial facilities. The maximum acceptable limits, described as "criteria" in LU-131, are recommended for various scenarios.

For road and rail transportation facilities:

L_{eq} = 55 dBA   at outdoor living areas during the day (7:00 A.M. to 11:00 P.M.)

For road facilities:

L_{eq} = 45 dBA   living and dining areas at homes, hospitals, schools, etc. during the day

L_{eq} = 40 dBA   sleeping quarters during the night (11:00 P.M. to 7:00 A.M.)

For rail facilities:

L_{eq} = 40 dBA	living and dining areas at homes, hospitals, schools, etc. during the day
L_{eq} = 35 dBA	sleeping quarters during the night

For air traffic:

NEF/NEP = 30	all outdoor areas
NEF/NEP = 5	living and dining areas at homes, hospitals, schools, etc.
NEF/NEP = 0	sleeping quarters

For stationary sources (industrial and commercial activities):

L_{eq} = 50 dBA	during the day (7:00 A.M. to 11:00 P.M.) in Class 1 urban areas
L_{eq} = 50 dBA	during the day (7:00 A.M. to 7:00 P.M.) in Class 2 urban areas
L_{eq} = 45 dBA	during the evening (7:00 P.M. to 11:00 P.M.) in Class 2 urban areas
L_{eq} = 45 dBA	during the night (11:00 P.M. to 7:00 A.M.) in all urban areas

An "outdoor living area" is an outdoor area that is easily accessible from the building and is designed for quiet enjoyment. Examples include front and back yards, gardens, terraces, patios, balconies and passive recreation parks.

Noise Exposure Forecast/Noise Exposure Projection (NEF/NEP) values for major Ontario airports are specified on maps available from the Ministry of Municipal Affairs and Housing.

A Class 1 urban area is one where the background sound level is dominated by urban hum (largely traffic noise) most or all of the time. A Class 2 urban area is one that experiences urban hum during the day, but not in the evening (7:00 P.M. to 11:00 P.M.) or night (11:00 P.M. to 7:00 A.M.). A rural area is defined by the absence of significant road traffic.

6.4.2   Vibration

Relative to noise, there are fewer numerical values for vibration. In part, this is due to vibration being more complicated to measure and predict than noise.

Acceptable levels of vibration are based upon perceptibility. One method of evaluating the perceptibility of vibration has been developed by the International Organization for Standardization (ISO) and adopted by Canadian National Railroad (CNR) and Canadian Pacific Railway (CPR). The approach is based on the premise that vibrations with frequencies between 4 and 200 hertz have the potential to cause annoyance at a residential receptor location if the vertical velocity of the vibration exceeds 0.14 mm/s during an averaging time of one second. Below this speed, vibration is not a concern; above it, consideration should be given to mitigation measures. The MOE has adopted the same values in draft form but has not yet included them in official guidelines.

Stronger vibration levels are usually needed to produce annoyance or inconvenience in commercial or industrial areas than at residential locations, however, there are no guidelines for acceptable vibration levels in commercial or industrial spaces. If complaints arise, each situation is assessed individually according to the activities occurring in the space and the perceptibility of the vibration intrusion.

LU-131 recommends a peak particle velocity of 1 cm/s as the criterion for vibration from blasting operations at a mine or quarry.

The Federal Railroad Administration in the United States has recommended vibration criteria for high speed trains (FRA, 1998). For frequent events, the criteria range from 65 to 75 vibration decibels (VdB). For infrequent events, the criteria range from 65 to 83 VdB. For both ranges, the lower values apply to buildings where vibrations would interfere with interior operations such as manufacturing or using sensitive equipment or at concert halls and recording studios. Mid-range values would be used for residences and other buildings where people normally sleep. Upper values would be used for institutional uses with primarily daytime occupancy.

6.5   Land Use Compatibility

When a change in land use will place a sensitive land use within the possible influence area of other facilities or activities, municipal authorities and developers are expected to prevent conflicts and incompatibilities. Nevertheless, the MOE has an interest in seeing that such incompatibilities are eliminated or minimized.

MOE Guideline D-1 provides a general description of the types of studies the MOE may want to review and comments on methods of preventing or minimizing potential incompatibilities between land uses. The types of incompatibilities include noise and vibration (MOE, 1995b).

As noted in Section 6.4.1, LU-131 describes the noise criteria recommended by the MOE for land use planning. A companion document to LU-131 provides technical details for assessing noise impacts, identifies when noise studies may be required and appropriate assessment methods and procedures.

During the planning stages of sensitive developments, a noise study may be required to demonstrate the feasibility of meeting the recommended sound level criteria, and to specify control measures if they are needed. LU-131 describes how these feasibility studies should be conducted. A detailed study then should be prepared to determine the appropriate layout, design and implementation of control measures.

Control measures may include site planning (such as orientation of buildings and other outdoor recreational areas), set back distances, acoustical barriers (such as berms, walls), architectural design (such as placement of windows, balconies), construction methods (such as acoustical treatment of windows and doors) and installing noise control devices on industrial sources.

One frequently used control measure is to separate sources and receptors. Set back distances are provided in LU-131 for roads and rail lines. The MOE cautions that these are guidelines only, and the effectiveness of distance as a control measure will vary according to site specific conditions, particularly topography.

A noise feasibility study may not be needed if the lands of interest are:

- more than 100 m from a freeway right-of-way or more than 50 m from a provincial highway right-of-way;
- more than 100 m from a principal railway line right-of-way or more than 50 m from a secondary railway line right-of-way.

A detailed noise study may not be needed if the lands of interest are:

- more than 500 m from a freeway right-of-way, more than 250 m from a provincial highway right-of-way, or more than 100 m from the right-of-ways of other roads;
- more than 500 m from a principal railway line right-of-way, more than 250 m from a secondary railway line right-of-way, or more than 100 m from other railway lines.

MOE Guideline D-6 recommends minimum separation distances between sensitive land uses (residential areas, hospitals, churches, schools, daycare facilities, camp grounds or recreational areas deemed sensitive) and three classes of industrial facilities (MOE, 1995c). Class I facilities are generally small, daytime operations that are small sources of noise, odour, dust or vibration. The other two classes are progressively larger and more frequent sources of emissions. Based on MOE experience, industrial facilities can cause incompatibilities at distances of: 70 m for Class I; 300 m for Class II; and 1,000 m for Class III. While mitigating measures may be taken to reduce incompatibilities, the recommended minimum separation distances are: 20 m for Class I; 70 m for Class II; and 300 m for Class III.

Both CNR and CPR require the measurement of vibration at all proposed residential sites within 75 m of rights-of-way for main and branch lines.

The provincial Aggregate Resources Act recommends set back distances to avoid incompatibilities of noise and dust. In 1997, the distances were revised to be 500 m for quarries and 150 m for pits.

A report for the U.S. Federal Railroad Association recommends screening distances for vibration assessments of trains. Where distances exceed those that are recommended, vibrations should not become a concern. For trains with steel wheels that travel less than 160 km/h (100 mph) the recommended distances are 20 to 40 m for residential areas (the lower distances used if there are less than 70 passbys per day) and 7 to 30 m for institutional uses (FRA, 1998).

Another noise control measure is to prevent windows from opening at the receptor's location. LU-131 describes requirements for central air conditioning systems in such circumstances. It also addresses the use of "warning clauses" or restrictions that notify potential purchasers or tenants of conditions such as the potential for noise levels that may occasionally exceed criteria, or the presence of central air conditioning as a noise control measure. If the source of the noise is industrial activity, additional control at the source is recommended over the combination of sealed windows and air conditioning.

LU-131 concludes with a set of tables that summarize the recommended criteria, conditions where control measures are required and conditions that require warning clauses.

Table 6.4 lists various guidelines and policy documents issued by the MOE and other provincial agencies that address noise.

6.6   Monitoring

6.6.1   Noise

Sound pressure levels are measured in A-weighted decibels (dBA). The term "A-weighted" indicates that factors are assigned to specific frequencies to approximate the relative sensitivity of the normal human ear to different frequencies (pitches of sound). Sound pressure level is defined according to the following equation (Thumann and Miller, 1986):

Equation 6.1

$$Lp = 20 \log (P/Po)$$

where Lp = sound pressure level (dBA)
        P = sound pressure (micropascals)
        Po = reference pressure = 20 micropascals

The reference pressure is the pressure equivalent to the threshold of human hearing.

Sound levels fluctuate with time, thus a single measurement of sound level is insufficient for an assessment. The statistical method that is most commonly used to describe environmental noise is the energy equivalent sound level (L_{eq}). The L_{eq} is a time-weighted, mean square, A-weighted sound pressure.

To present the random fluctuation of community noise, especially where traffic is present, values for L90, L50, L10, and L1 may be used. These are the noise levels that are exceeded 90% of the time (also referred to as background or ambient), 50% (median), 10% (intrusive), and 1% (the "peak" noise levels), respectively.

Sound measuring devices typically consist of a microphone, electronic amplifier, filters and a readout meter. When measuring noise, the microphone should be located at least 1.2 m above the ground and 3 m away from significant sound-reflecting surfaces. Average wind speeds of greater than 15 km/h may invalidate the data. Short-term (less than 5 minutes) wind gusts up to 20 km/h are acceptable. Steady precipitation can also invalidate monitoring.

It is essential that the device be calibrated in the laboratory and in the field. Field calibration should be conducted prior to and after taking each set of sound level readings.

MOE publication NPC-102 describes the instrumentation requirements for sound meters, while NPC-103 describes the preferred procedures for measuring or calculating sound levels.

Measured or calculated sound levels may need to be adjusted to take into account characteristics of the sound such as intermittence, tonality, and cyclic variations, or if it is quasi-steady impulsive sound. NPC-104 describes these adjustments.

Table 6.5 lists helpful publications concerning the measurement of sound from other agencies including the Canadian Standards Association, American Society for Testing and Materials, and the ISO.

6.6.2   Vibration

Vibration usually is described in terms of velocity or acceleration (of the object receiving or experiencing the vibration).

The velocity refers to rate of change in the amplitude of the vibration. The peak particle velocity (PPV) is the maximum instantaneous peak (either positive or negative) of a vibration signal. While peak velocity is a useful measure of the potential of vibration to damage a building, human perception takes place over longer periods of time (typically one second). Over such periods, the root mean square (RMS) amplitude is used to describe the average velocity.

PPV and RMS velocities typically are expressed as m/s. These can span wide ranges. Logarithmic notation can be used to compress these ranges. Velocity decibels (VdB) are defined as:

Equation 6.2

$$L_v = 20 \log (v/v_{ref})$$

where L_v = vibration level (VdB)
v = the RMS velocity (m/s)
v_{ref} = reference velocity (m/s)

The reference velocity typically is assigned a value of 1 to 5 x 10^{-8} m/s, although this varies slightly between organizations.

Typical background velocity levels in residential areas are around 50 VdB or lower, well below the threshold for perception by people, which is around 65 VdB (FRA, 1998).

In NPC-233, the MOE recommends that vibration reports should describe sources using peak or RMS vibration velocities in mm/s.

Although vibration is usually a concern inside buildings, measurements of vibration usually are taken outdoors except if the concern relates to vibration-sensitive equipment or activities at indoor locations.

Most vibration studies measure vertical velocities. Transverse vibration often is ignored since it is a minor component of the vibration and not transferred as efficiently into buildings. (Seismic vibration has a strong horizontal component.)

Vibration typically is measured by mounting or gluing transducers or accelerometers onto surfaces at locations of interest. The transducers often are located at several distances between the source and the receptor or along the path of a mobile source such as a rail line. Signals are recorded on an instrumentation quality digital tape recorder. In the lab, a multi-channel spectrum analyzer is used to display the results.

Ambient or background vibration usually is assessed over a period of 10 to 30 minutes and may be expressed as the L_{eq} of the velocity over that period. The maximum RMS velocity over that period may be designated as the L_{max} level.

For very short-lived events such as blasting at quarry or construction site, vibrations often will be described using PPV values.

In situations where the data will be used to understand vibration movement or for other types of detailed analysis, vibration frequencies may be measured, divided into bands of frequencies and analyzed.

The results of vibration surveys can be used to identify locations where remedial actions are warranted or to identify appropriate setback distances to avoid excessive vibration levels.

6.7   Mathematical Models

6.7.1   Noise

To estimate sound pressure levels outdoors, the basic power formula can be used (Thumann and Miller, 1986):

Equation 6.3

$$Lp = Lw + 10 \log [Q/(4 \pi R^2] + 0.02$$

where Lw = sound power level (dBA)

   = $10 \log (W/10^{-12})$

R = distance from the source (m)

Q = directivity factor of the source

W = acoustic power (watts)

10^{-12} = standard reference power (watts)

The parameter Q takes into account the directivity of the source and the reflecting surfaces:

Q = 1 for point sources radiating uniformly in all directions with no reflecting surfaces

Q = 2 for sources radiating from flat surfaces

Q = 3 for noise source radiating from corners

A simple formula can be used to predict noise levels at a plant boundary based on measurements taken adjacent to the source (Thumann and Miller, 1986):

Equation 6.4

$$Lp,X = Lp,Y - 20 \log (X/Y)$$

where X = distance at which noise is being estimated for (m)

Y = distance at which noise monitoring data are available (m)

Equation 6.3 will result in a level decrease of 6 dBA when the distance is doubled (that is, the inverse square law) in free space. The above relationship does not apply directly adjacent to the noise source (also referred to as the near field). Sound measurement should be made at a minimum distance of two machine dimensions and at least one wavelength. If information on the wavelength is unavailable, a good rule of thumb is to monitor at a distance of several times the equipment dimensions (Thumann and Miller, 1986).

When more than one source exists, the overall sound level is not the sum of each sound level (W). Two equal sound sources result in the overall sound level being 3 dBA higher than either source alone.

More sophisticated models have been used to estimate sound levels. Two models developed by the MOE and used frequently in Ontario are:

ORNAMENT - Ontario Road Traffic Noise Prediction Methodology (1989)

STEAM - Sound from Trains Environmental Analysis Method (1990)

ISO has issued standards that specify methods for describing the sound emitted by various types of sources and describing sound levels at outdoor locations. ISO standard 9613 describes a method for predicting noise levels at locations away from sound sources. It covers most of the major mechanisms by which sound levels are attenuated during propagation outdoors. Part 2, the general method of calculation, is cited by the MOE in draft regulations (MOE, 1998) and is likely to be used more frequently in future assessments of noise assessments in Ontario.

6.7.2   Vibration

Predicting vibration levels is not as standardized as predicting sound levels. Vibration propagation through ground is complex and difficult to model. Procedures have been developed for specific types of sources or for specific locations, but often are not well-suited to use elsewhere.

Numerous site-specific factors influence the migration of vibration from a source to a receptor. Important characteristics of the source itself are whether it is located above ground or underground, its speed (if it is a mobile source such as a truck or train), and the effectiveness of any dampening or suspension at the source. Between the source and the receptor, some of these factors include the type of soil, depth to bedrock, depth to the water table and the presence of frost. At the receptor, the type of foundation, building construction techniques, the mass of the building and the effectiveness of any acoustical dampening will be factors (FRA, 1998).

An equation for estimating vibration propagation from construction activities such as pile driving, drilling or operating a jackhammer can be estimated as follows:

Equation 6.5

$$PPV_d = PPV_{ref} (25/D)^{1.5}$$

where PPV_d = the PPV (m/s) at D m from the source
PPV_{ref} = the reference PPV measured 8 m (25 feet) from the source
D   = distance from the source (m)

PPV_{ref} values for various types of equipment have been published (FRA, 1998).

6.8   Future Directions

The MOE is considering exempting various activities from the approval requirements described in s. 9 of the EPA (MOE, 1998). At present, a C of A must be updated when sources of sound and vibration are altered, extended, or replaced, or when processes are altered, or rates of production are altered. The MOE is proposing to exempt such modifications if:

- as a result of the modification, no impulse sounds are to be produced and structures on a noise sensitive site that is beyond the property line of the site of the modification will not vibrate in a manner that can be perceived by a person
- minimum separation distance standards for noise must be met or a professional engineer has used an acoustical assessment to demonstrate that noise standards are met

The minimum separation distances have been set at 500 m for Schedule 1 equipment/processes, 1,000 m for Schedule 2 equipment/processes and other distances are set out in the *Noise Red Flag Tables* (MOEE, 1997).

If these separation distances are not met, a professional engineer must conduct, check or review an acoustical assessment and verify that after the modification the sound level at the nearest sensitive point of reception will not exceed the greater of 40 dBA and the one-hour L_{eq} produced at the nearest noise sensitive receptor before the modification.

Another proposed change is that a start-up notice and a fee would need to be submitted to the MOE when a client chooses to claim such an exemption. The notice and supporting studies or records prepared by a professional engineer would need to be kept on file and available for MOE review. If an acoustical assessment is required, a copy of the professional engineer's written verification that noise impact standards are met would need to be kept on file. Records would need to be maintained for as long as the modification was in operation and for two years after it ceases to operate.

Some types of new sound and vibration sources also may be candidates for the proposed exemptions. The types of sources include standby generators, combustion equipment, ethylene oxide sterilizers and arc welding equipment. Details of the exemption process vary according to the source type, but generally include a written assessment from a professional engineer that noise impact standards are being met, or that minimum separation distances are being met.

6.9   Summary

Sound and vibration should be treated as any other pollutants that may cause the loss of enjoyment of normal use of property. As such, it is important that a company be aware of its contribution to the surrounding noise and vibration levels, and reduce levels if they are having a detrimental impact on local residents.

Noise control for the most part is regulated by municipal by-laws. The MOE has published a Model Noise Control By-Law as a means of providing a uniform basis for noise control across Ontario. The Model By-Law is supported by several Noise Pollution Control documents that set numerical criteria, as well as describe methods to measure, estimate, model, assess and control sound and vibration.

References

Beranek, L.L. (ed.). 1971. *Noise and Vibration*. New York: McGraw-Hill Book Co.

Federal Railroad Administration (FRA), U.S. Department of Transportation. December 1998. *High-Speed Ground Transportation Noise and Vibration Impact Assessment*. Final Draft. Report 293630-1.

Health Canada. 1998. *The Health and Environment Handbook for Health Professionals: Health and Environment*.

Ontario Ministry of the Environment (MOE). August 1978. *Model Municipal Noise Control By-Law*. Final Report.

Ontario Ministry of the Environment (MOE). 1979. *ORNAMENT: Ontario Road Noise Analysis Method for Environment and Transportation*. ISBN 0-7729-6376.

Ontario Ministry of the Environment (MOE). 1990. *STEAM - Sound from Trains Environmental Analysis Method*. ISBN 0-7729-6376-2.

Ontario Ministry of the Environment (MOE). 1995a. *Guide to Applying for Approval (Air): Noise and Vibration*.

Ontario Ministry of the Environment (MOE). July 1995b. *Land Use Compatibility*. Guideline D-1.

Ontario Ministry of the Environment (MOE). July 1995c. *Compatibility Between Industrial Facilities and Sensitive Land Uses*. Guideline D-6.

Ontario Ministry of the Environment (MOE). December 1998. *Guide For Draft Regulation Entitled: "Certificate of Approval Exemptions – Air, Start-Up Notice Required" Under the Environmental Protection Act*. Formerly known as Standardized Approval Regulations.

Ontario Ministry of Environment and Energy (MOEE). 1997. *Noise Red Flag Tables*.

Thumann, A., and R. Miller. 1986. *Fundamentals of Noise Control Engineering*. The Fairmont Press.

Working Group on Environmental Noise. March 1989. *National Guidelines for Environmental Noise Control*. Federal-Provincial Advisory Committee on Environmental and Occupational Health.

Table 6.1

Representative Sound Sources and Levels at Close Range

Source	Sound Pressure Level (in dB)	Power (in Watts)
human breath	10	0.00000000001
rustling leaves	20	0.0000000001
soft whisper	30	0.000000001
small electric clock	40	0.00000001
ventilation fan	60	0.000001
conversation	70	0.00001
shouting	90	0.001
blaring radio	110	0.1
small aircraft engine	120	1
large pipe organ	130	10
turboprop aircraft at takeoff	150	1000

Table 6.2

Representative Vibration Levels at Close Range

Source	Velocity Level (VdB)
typical background	50
15 m from a moving bus or truck	62
threshold of human perception	65
frequent events are annoying in residential areas (i.e. streetcar)	72
infrequent events are annoying in residential areas (i.e train)	80
15 m from heavy construction equipment	92
threshold of minor damage to fragile buildings	100
15 m from blasting at construction project	100

Notes:

Vibration velocities are RMS velocities.

Adopted from: FRA, 1998.

Table 6.3
Publications in Support of the MOE Model Municipal Noise Control By-Law

NPC-101	Technical Definitions
NPC-102	Instrumentation
NPC-103	Procedures
NPC-104	Sound Level Adjustments
NPC-119	Blasting
NPC-205	Sound Level Limits for Stationary Sources in Class 1 & 2 Area (Urban) – replaced NPC-105 in 1995
NPC-206	Sound Levels of Road Traffic - replaced NPC-106 in 1995
NPC-207	Impulse Vibration in Residential Buildings (draft)
NPC-216	Residential Air Conditioning Devices – replaced NPC-116
NPC-232	Sound Level Limits for Stationary Sources in Class 3 Areas (Rural) – replaced NPC-132 in 1995
NPC-233	Information to be Submitted for Approval of Stationary Sources of Sound – replaced NPC-133 in 1995
LU-131	Noise Assessment Criteria in Land Use Planning – replaced NPC-131 in 1997

The following documents are not referenced in recent MOE publications, but may still be helpful sources of guidance:

NPC-115	Construction Equipment
NPC-117	Domestic Outdoor Power Tools
NPC-118	Motorized Conveyances
NPC-134	Guidelines on Information Required for the Assessment of Planned New Uses with Respect to Sound and Vibration Impacts

Table 6.4

Provincial Noise Guidelines and Policies

Ministry of the Environment
- Guideline D-1, Land Use Compatibility (1995)
- Guideline D-6, Compatibility Between Industrial Facilities and Sensitive Land Uses (1995)
- Guideline for Noise Impact Assessment for Off-Site Vehicular Traffic (1988)*
- Guidelines for Quarries (1986)*
- Manual for Noise Assessment in Land Use Planning (1997)
- Noise Assessment Criteria in Land Use Planning: Requirements, Procedures and Implementation (1997); replaced NPC-131 Guidelines for Noise Control in Land Use Planning
- Environmental Noise Guidelines for Installation of Residential Air Conditioning Devices (1994)

Ministry of Municipal Affairs and Housing
- Land Use Policy Near Airports (1987)*
- Guidelines on Noise and New Residential Development Adjacent to Freeways (1979)*

* Some of the documents seldom are referenced and may be no longer used or have been superseded.

Table 6.5

Selected References Concerning Sound Measurement

American Society for Testing and Materials (ASTM). 2000. *Standard Method for Measurement of Outdoor A-Weighted Sound Levels.* ASTM Standard E1014.

Canadian Standards Association (CSA). 1983. *Recommended Practice for the Prediction of Sound Levels Received at a Distance from an Industrial Plant.* CSA Standard Z107.55.

International Organization for Standardization (ISO). 1987. *Description and Measurement of Environmental Noise.* Publication No. 1996. Part 1: Basic Quantities and Procedures. Part 2: Acquisition of Data Pertinent. Part 3: Application to Noise Limits.

International Organization for Standardization (ISO). 1996. *Acoustics – Attenuation of Sound During Propagation Outdoors.* Publication No. 9613.

Ontario Ministry of the Environment (MOE). 1983. *Procedures for the Measurement of Sound.* Publication NPC-103.

7

Special Materials

7.1   Polychlorinated Biphenyls (PCBs)

7.1.1   Overview

There are 209 compounds that are classified as polychlorinated biphenyls (PCBs). Each PCB compound consists of two, fused benzene rings attached to which are one or more chlorine atoms. The physical, chemical and toxicological properties vary widely among PCB compounds, but all are stable at elevated temperatures and sparingly soluble in water. These characteristics made PCBs suitable for various uses including dielectric fluid in electrical equipment and in hydraulic fluids, paints and inks. PCBs came into commercial use in the late 1920s. The major Canadian use was in dielectric fluid for industrial electrical equipment.

In 1973, the Organization for Economic Cooperation and Development (OECD) urged all member countries to limit PCBs to enclosed uses and to develop control mechanisms to eliminate the release of PCBs into the environment. In 1977, PCBs became the first class of substances to be regulated under the Canadian Environmental Contaminants Act, and subsequently most non-electrical uses of PCBs were prohibited in Canada. Since that time, several federal and provincial regulations have been issued that address virtually every aspect of the use, management and disposal of PCBs. These are summarized in Table 7.1.

The same properties that made PCBs well-suited for use as dielectric fluid also have resulted in PCBs becoming widely dispersed in the environment. Airborne transport probably is an important contributor to their environmental distribution.

PCBs can accumulate in living organisms and have been found in animals in remote areas of Canada and at low levels in the fatty tissue and blood of Canadians (CCEM, 1986). Table 7.2 presents physical characteristics for a few PCB compounds. The octanol-water partition coefficient (Kow) is a good indicator of the potential for bioaccumulation.

Birds, aquatic invertebrates and most species of fish are particularly sensitive to PCBs. Effects which have been observed include the reduction in litter sizes of otters and minks, as well as birth defects such as crossed beaks in birds (CCREM, 1986). It is these effects on wildlife, the persistence of PCBs, and their ability to bioaccumulate, which prompted the federal government to regulate PCBs.

PCBs were identified as a suspected human health hazard approximately 25 years ago. More recent research indicates that the risks posed to human health may have been overestimated. The possible effects of long-term exposures to relatively low concentrations are still not well known. The available information comes largely from people exposed to relatively high levels of PCBs in occupational settings. It is clear that PCBs can cause a severe form of acne called chloracne. Other health effects include numbness in limbs, weakness and problems with the nervous system. There have been suggestions that PCBs can cause cancer in people, but according to Health Canada, "there is no proof yet of a definite risk between PCBs and cancer" (Health Canada, 2001). PCBs with relatively high numbers of chlorine atoms are considered capable of causing some cancers in laboratory animals if exposed to high concentrations for prolonged periods. PCBs with high numbers of chlorine atoms are found in small amounts in the PCBs supplied to the electrical industry.

Nevertheless, PCBs must be handled, stored and transported as if they were highly toxic because of the requirements that legislation has imposed. Although PCBs may be less hazardous to humans than previously suspected, PCBs can be converted to dioxins and furans if heated sufficiently. The oily soot produced at fires involving PCBs can be hazardous.

7.1.2   Federal Acts and Regulations

As noted in Section 7.1.1, the first federal statue pertaining to PCBs was promulgated in 1977. The following regulations now fall under the Canadian Environmental Protection Act, 1999 (CEPA, 1999), although most were originally promulgated under other acts:

- Chlorobiphenyls Regulations (SOR/91-152)
- Disposal at Sea Regulations (SOR/2001-275)
- Federal Mobile PCB Treatment and Destruction Regulations (SOR/90-5)
- PCB Waste Export Regulations (SOR/97-109)
- Storage of PCB Material Regulations (SOR/92-507)

The four older regulations have been amended from time to time, and most recently were amended in 2000.

In addition to the regulations, PCBs are designated as toxic substances and therefore subject to various controls and initiatives under CEPA, 1999 (as noted in Section 7.1.3).

The original Chlorobiphenyls Regulations (SOR/91-152) permit the continued use of PCBs for certain electrical applications and prohibit most non-electrical uses.

The Storage of PCB Material Regulations (SOR/92-507) set out requirements for PCBs stored or in use at federal facilities and lands.

The Federal Mobile PCB Treatment and Destruction Regulations (SOR/90-5) set PCB clean-up standards and emission standards for incinerator systems (one of the main PCB treatment technologies when the regulation was passed in 1990) that operate on federal lands or are operated for federal institutions.

The PCB Waste Export Regulations (SOR/97-109) regulate the transboundary movement of PCB wastes. In effect, PCB wastes cannot be exported except to the United States if sent to an approved facility where they will be destroyed in an environmentally sound manner. The regulation clarifies (for PCBs) controls previously described in the Export and Import of Hazardous Waste Regulations.

The Disposal at Sea Regulations (SOR/2001-275) replaced the Ocean Dumping Regulation originally promulgated under the Ocean Dumping Control Act. For wastes that contain less than 0.1 ppm of PCBs, no ecological toxicity tests are required prior to disposal at sea. Wastes that contain more than 0.1 ppm of PCBs must first be tested and shown to be acceptable before they can be disposed at sea.

The Transportation of Dangerous Goods Act, 1992 (TDGA, 1992) and associated regulations (TDGR) apply to the movement of PCB wastes by ship, plane, rail between provinces, and with other countries (see Section 4.9 for additional information on TDGA, 1992 and TDGR). In August, 2002, a new clear language version of TDGR came into effect. There are several revisions in the new regulations that pertain to the transportation of PCBs. These are highlighted in Section 7.1.12.

7.1.3   Other Federal Agreements and Programs

The Canada-U.S. Agreement on Transboundary Movement of Hazardous Waste has been in place since 1986. It sets out various conditions that must be met when transboundary movement of hazardous waste, including PCBs, is to occur.

In 1989, Canada signed the Basel Convention, the international treaty governing the shipping of hazardous waste. Informed consent must be obtained from the country receiving the PCBs if the concentration of PCB in the waste exceeds 50 ppm.

In 1994, Canada and the United States established the Great Lakes Binational Toxics Strategy, the main goal of which is the "virtual elimination" of persistent, toxic substances in the Great Lakes. In turn, this led to the Canada-Ontario Agreement Respecting the Great Lakes Basin Ecosystem. In the Agreement, PCBs are identified as a Tier I substance and therefore require immediate action to eliminate the use, generation or release into the Great Lakes environment.

From that Agreement came the Canadian Challenge for PCBs, which included the goals of decommissioning 90% of high-level PCB wastes (described as materials containing more than 1% PCBs), and destroying 50% of the high-level PCBs in storage in Ontario.

Since PCBs are defined as toxic under CEPA, 1999, they are subject to the federal Toxic Substance Management Plan and therefore scheduled for virtual elimination.

The North American Free Trade Agreement (NAFTA) includes an action plan for PCBs that calls for the phase-out of all PCB equipment by 2007 and destruction of all PCB waste by 2009.

In 2001, Canada signed the Stockholm Convention, aimed at eliminating the release of persistent organic pollutants (including PCBs). This likely will limit destruction and decontamination options for PCBs to those that do not produce unwanted chemicals including dioxins and furans.

7.1.4   Provincial Acts and Regulations

There are two main regulations in Ontario that address PCBs. Both fall under the Environmental Protection Act.

- Mobile PCB Destruction Facilities (R.R.O. 1990, Reg. 352)
- Waste Management – PCBs (R.R.O. 1990, Reg. 362)

Regulation 362 defines various PCB terms and establishes management requirements for the disposal, storage and shipping of PCB waste. Regulation 352 requires mobile destruction facilities to obtain a Certificate of Approval (C of A) from the MOE before the facility can accept PCB waste.

Some of the same definitions in Regulation 362 also are used in Regulation 347 to define PCB waste as a class of subject waste and therefore must be managed in accordance with the requirements set out in Regulation 347 (and described in Chapter 4). For example, generators of PCB waste must be registered as a subject waste generator as described in Section 4.5.

The provincial Dangerous Goods Transportation Act (DGTA) governs the movement of PCB wastes on roads inside Ontario. (See Section 7.1.12 for more information about transporting PCBs and Section 4.9 for more information on DGTA).

7.1.5   Municipal Involvement

Until recently, most PCB waste in Ontario was kept in storage. Some municipalities limited areas where hazardous substances could be stored and others did not allow transfers into their jurisdictions. For example, North York (now part of the City of Toronto) passed a by-law that, with the exception of specified sites, prohibited the storage of PCB waste if the PCBs are generated from lands other than the storage site (Basrur, 2001).

7.1.6   Definitions and Classifications

Provincial and federal regulations define several key terms such as PCB material, PCB liquid, PCB waste and PCB equipment. The two sets of definitions generally are consistent with one another. The following definitions are from Regulation 362 with differences in federal definitions noted as appropriate. In 2002, the MOE proposed numerous changes and additions to definitions of PCB terms. These changes are noted below, but had not been made as of late 2002.

All PCBs have the molecular formula $C_{12}H_{10-n}Cl_n$ where "n" represents the number of chlorine atoms. Regulation 362 defines PCBs as having n>1 while Schedule 1 of CEPA, 1999 (the List of Toxic Substances) defines PCBs as having n>2. The MOE has proposed to amend its definition to be consistent with the federal definition.

PCB material is any material that contains more than 50 ppm by weight of PCB regardless of whether the material is liquid or not.

A PCB liquid is defined as:

(a) liquids, other than liquids used or proposed for use for road oiling, containing PCBs at a concentration of more than fifty parts per million by weight

(b) liquids used or proposed for use for road oiling, containing PCBs at a concentration of more than five parts per million by weight, and

(c) liquids made . . . by diluting liquids referred to in clause (a) or (b).

PCB equipment means equipment designed or manufactured to operate with PCB liquid or to which PCB liquid has been added. It also includes drums or other containers used to store PCB liquid. The federal definition of PCB equipment is similar but does not extend to include drums or containers. The MOE has proposed defining the term "PCB electrical equipment" as any electrical equipment designed or manufactured to operate with PCBs or that is contaminated with PCBs.

Regulation 362 does not define PCB solid or PCB substance (both of which are defined in federal SOR/92-507 as materials that contain more than 50 ppm of PCBs). The MOE has proposed defining "PCB solid" and "PCB mixture" as a solid or mixture containing more than 50 ppm of PCBs. Similarly, the MOE has proposed defining "PCB-contaminated soil" as soil or loose material that contains more than 50 ppm.

Regulation 362 defines PCB waste to include PCB material, PCB liquid and PCB equipment. Exceptions include:

(a) PCB material or PCB equipment that has been decontaminated pursuant to guidelines issued by the MOE or instructions issued by the director.

(b) PCB equipment that is,

(i) an electrical capacitor that has never contained over 1 kg of PCBs;

(ii) electrical, heat transfer, or hydraulic equipment or a vapour diffusion pump that is being put to the use for which it was originally designed or is being stored for such use by a person who uses such equipment for the purpose for which it was originally designed, or

(iii) machinery or equipment referred to in sub-subclause (c)(i).

(c) PCB liquid that,

(i) is at the site of fixed machinery or equipment, the operation of which is intended to destroy the chemical structure of PCBs by using the PCBs as a source of fuel or chlorine for purposes other than the destruction of PCBs or other wastes and with respect to which a C of A has been issued under the EPA

(ii) is in PCB equipment referred to in sub-subclause (b)(ii).

The exemption in sub-subclause (b)(i) also is found in Regulation 347. Although capacitors that have never contained more than 1 kg of PCBs need not be registered as PCB waste, the MOE asks owners of large numbers of such capacitors (as could be found in the ballasts of fluorescent light fixtures manufactured prior to 1980) to collect them voluntarily so that they ultimately can be disposed in an environmentally acceptable manner. In 2001, the MOE proposed that projects involving the replacement of 100 ballasts or more should treat each ballast as if it is PCB waste, while at smaller projects 40 ballasts would need to be accumulated before they are categorized as PCB waste.

Environment Canada has published useful guides that describe ways to determine if capacitors and ballasts likely contain PCBs (Environment Canada, 1988; 1991).

Regulation 362 does not address the length of time that equipment can go unused before it needs to be stored or is considered to be PCB waste. Federal regulation SOR/92-507 indicates that storage is required for equipment shut down for greater than six months. The MOE-proposed definition of PCB electrical equipment would allow equipment to be shut down for up to six months before being deemed PCB waste.

Similarly, Regulation 362 does not define any minimum quantities of material that might be exempt. Federal regulation SOR/92-507 applies to materials not being used daily in quantities of more than 100 L of PCB liquid, more than 100 kg of PCB solids or substance, or anything containing greater than 1 kg of PCBs.

Regulation 362 defines PCB related waste as waste containing low levels of PCBs or waste arising from a spill or clean up of PCB liquid or PCB waste. No further guidance is provided as to what constitutes low levels of PCBs.

Askarel is a generic name for PCB liquids used as electrical insulating materials. Askarels range from crystal clear to pale yellow in colour and are denser than water. The term often is used to describe fluid mixtures that contain PCBs in excess of 30% by weight (CCME, 1995).

7.1.7 Environmental Guidelines and Allowable Release Rates

The federal Chlorobiphenyls Regulations (SOR/91-152) prohibit the release of more than 1 g per day of PCBs from any one piece or package of equipment in the course of operation, servicing, maintenance, decommissioning, transportation or storage of this equipment. It also prohibits the use of oils that contain more than 5 ppm by weight for the application to road surfaces, and the release of PCBs from all sources — except those discussed above — in excess of 50 ppm by weight.

The MOE has issued numerical criteria and guidelines for allowable concentrations of PCBs in water, air and soil, as well as used concentrations as a basis for defining PCB wastes. These values are summarized in Table 7.3.

The MOE provides guidance on the protocols to be used for sampling and testing materials for PCBs (MOE, 2000a). This protocol describes the sampling of contaminated soil, rocks, asphalt, concrete, wood, cables, sludges, PCB materials in drums and stockpiles, the wipe tests for surfaces, and the statistical analysis of the data.

7.1.8 Manufacture, Importation and Use

PCBs were first synthesized in 1881, but not manufactured on a commercial scale until 1929. Most PCBs used in Canada were imported from the United States either in pure form or as askarel. In 1977, production in the United States was terminated voluntarily, while in Canada, most non-electrical uses of PCBs were prohibited under the former Canadian Environmental Contaminants Act and a national inventory of PCB-filled equipment was undertaken by Environment Canada.

While the major Canadian use of PCBs was in dielectric fluid for industrial electrical equipment, they also were used in waxes, adhesives, heat exchange fluids, vacuum pump oil, paints, de-dusting agents, hydraulic fluids, specialized lubricants, painting inks, pesticides, cutting oils, sealants, plasticizer and carbonless copying paper. Some of the trade names under which PCB fluids were sold include Aroclor, Askarel, Chlorinol, Diachlor, Hyvol, Inchlor, Inerteen, Pyranol, and Sovol.

The federal Chlorobiphenyls Regulations (SOR/91-152) prohibit the manufacture, use, sale or import of PCBs in excess of 50 ppm for any of the following commercial, manufacturing or processing uses:

- the operation of any product or equipment other than
  -electrical capacitors and transformers and capacitors
  -heat transfer equipment, hydraulic equipment, electromagnets and vapour diffusion pumps that were designed to use PCBs other than those that were in use in Canada before September 1, 1977
- the operation of electromagnets that are used to handle food, animal feed or any additive to food or animal feed
- as a constituent of any product, machinery or equipment manufactured in or imported into Canada on or after September 1, 1977 other than electrical capacitors and transformers
- as a constituent of electrical capacitors and transformers manufactured in or imported into Canada on or after July 1, 1980
- in the servicing or maintenance of any product, machinery or equipment other than electromagnets and electrical transformers and associated electrical equipment from which PCBs are removed to allow servicing and maintenance
- as new filling or as make-up fluid in the servicing or maintenance of electromagnets or electrical transformers and associated electrical equipment

Exemptions to these prohibitions include the sale of PCB-filled equipment which is a necessary and integral part of a building, plant or structure that is offered for sale; the sale of PCB-filled equipment for destruction or for storage awaiting destruction of the PCBs contained therein; and the importation of PCB-filled equipment for destruction of the PCBs contained therein. The latter exemption was required to develop a reciprocal agreement between Canada and the U.S. for the use of PCB-destruction facilities (CCME, 1989).

Equipment in service that contains PCBs does not need to be registered with the MOE or Environment Canada. Labelling of in-service equipment will assist in inventory control as well as during handling, storage and disposal. Figure 7.1 presents a label that should be used on large pieces of equipment such as transformers. Once the label is affixed, it should only be removed if the equipment has been decontaminated and the MOE and/or Environment Canada are satisfied that the PCB concentration is less than 50 ppm.

For smaller items, the label presented in Figure 7.2 can be used. If several smaller pieces of PCB equipment are found together, one label may be sufficient. The label should have an Environment Canada registration number at the bottom. The registration numbers allow Environment Canada to keep track of the amounts and locations of PCB equipment and liquids.

Figure 7.3 presents a general warning label that should be placed in a clearly visible position at the entrances to locations where PCB equipment is found.

Both the MOE and Environment Canada maintain inventories of all askarel and PCB-contaminated equipment that has been labelled. Both agencies should be informed as to the status of a piece of PCB equipment (for example, if it is taken out of service, relocated, stored, decontaminated or disposed). Environment Canada recommends that owners of PCB equipment retain inventory records for five years after equipment is removed from service.

7.1.9   Decommissioning PCB Equipment

When PCB equipment is to be taken out of service, whether through failure, retrofit or redundancy, it must be carefully decommissioned. The following suggested procedures are taken from guidelines issued by the CCME (1989) and a bulletin issued by Environment Canada when decommissioning PCB equipment at federal facilities in Ontario (Environment Canada, 2001).

Notification and Record Keeping

Prior to decommissioning equipment, the MOE must be notified and both the site and material must be registered. Before the equipment is decommissioned, the following information should be recorded: nameplate data, serial numbers, dates of decommissioning and shipment, destination of equipment, names of decommissioning personnel, as well as the Environment Canada label identification number, if applicable.

Planning

All persons assigned to handle the PCB equipment should be thoroughly instructed in the proposed procedures, particularly with respect to safety precautions, the use of safety equipment and the applicability of federal and provincial regulations. Prior to decommissioning, aspects such as containment, ventilation and working space available should be examined. The type, condition and level of PCBs in the equipment dictate the extent of precautions to be taken.

If the equipment is located in an open area, suitable curbs, barriers and/or metal pans should be provided to prevent the release of PCBs in the case of a spill during handling operations. All floor drains should be plugged and air ducts leading to other parts of the building should be closed. If cracks or leaks are apparent, liquids should be removed from the equipment prior to movement. The area of work should be appropriately identified and unauthorized persons prohibited from entering the area.

Protective Clothing and Apparatus

The required protective clothing will depend upon the individual circumstances, such as concentration, quantity of PCBs and whether the material is in solid or liquid form. If workers are to come into direct contact with askarel, protective clothing impervious to PCBs should be worn.

Procedures

Sealed capacitors should be placed into 205-L drums made of 18-gauge steel (or heavier), fitted with removable steel lids and gaskets made of PCB-resistant material such as nitrile rubber, cork or Teflon. Capacitors should be stored with the terminals up to prevent leakage from the capacitor bushings. As many capacitors as space allows may be placed in each drum. Drums or containers smaller than 205 L may be used when the size or quantity of capacitors does not justify the larger container.

Leaking capacitors should be drained and then placed in heavy-duty polyethylene bags before storing in a drum (one capacitor per bag). The drum should be packed with adsorbent material to adsorb PCBs that may escape from the bags.

Non-leaking capacitors that are too large to fit into a 205-L drum should be wrapped in heavy gauge polyethylene and crated for transfer to a storage area. If the capacitor is leaking, it should be drained and stored in a drip pan containing sufficient adsorbent to adsorb any remaining liquid.

Small transformers may be stored or transported in leak-proof containers, without draining, in a manner similar to that for capacitors. Transformers stored on-site need not be drained as long as they are structurally sound, external parts are protected from the weather and spill containment is provided. Where large askarel transformers are being stored pending transportation or disposal, the askarel should be removed or stored in double-bung drums made of 16-gauge steel (or heavier). PCB liquids may be stored in tanks rather than drums provided the tanks are above ground and are sound, properly labelled, regularly inspected, protected from the weather, and spill containment is provided.

Volume Reduction

Small items such as fluorescent lamp ballasts that contain PCBs may be numerous and can represent large volumes of material to be stored. The PCBs are contained in a capacitor inside the ballasts. The contents of a ballast typically are sealed in asphalt or tar. This makes removal difficult, and it should be done by properly trained individuals. By separating the capacitors from the ballasts, the volume of material to be stored as PCB waste can be reduced by more than 50%. Similar volume reductions can be achieved with power factor condensers.

7.1.10   Decontamination of PCB Equipment

The removal of PCBs from equipment or mineral oil is called "decontamination." There are two general approaches to decontamination: solvent cleaning and retrofilling.

Solvent cleaning is used for metal equipment that contains PCBs. Once the PCB fluid is removed, the metal hulk or carcass is rinsed with an organic solvent (such as hexachlorobenzene) to remove residual PCBs. Double- or even triple-rinsing may be necessary. Cleaned equipment may be suitable for disposal or re-use, or be sold for its scrap metal content. Cs of A have been issued by the MOE to a few sites in Ontario where transformer carcasses with low concentrations of PCBs (<1,000 ppm) are drained, rinsed and then sold for scrap. The MOE recommends that the procedures described in a CCME publication be used for this purpose (CCME, 1995). The scrap yard that receives decontaminated transformers must be advised in writing as to the original PCB content. There is one facility in Ontario where transformers with high concentrations of PCBs can be sent to be drained, rinsed and then sold for scrap.

For transformers that are to remain in service, a non-PCB fluid is used to replace the original fluid. Several weeks after retrofilling, the new fluid should be analyzed to determine whether residual PCBs in the equipment have migrated into the new fluid. Two or more refills may be needed to reach acceptably low concentrations of PCBs in the fluid.

There is also an *in situ* or on-line method of decontamination in which the transformer is connected directly to a processing unit that extracts the PCBs from the fluid. This process continues until the PCB concentration is reduced to an acceptable limit.

Depending upon the level of contamination, size of equipment, transportation regulations and draining, decontamination can take place at the point of removal from service, or at some other location prior to transportation. Decontamination can also take place at a disposal site.

Like decommissioning, decontamination must be undertaken carefully. A C of A specifically applicable to PCB wastes is required to decontaminate liquid PCBs. The following suggested procedures are taken from guidelines issued by the CCME (1989; 1995).

1. **Non-electrical, Askarel-filled equipment** – Non-electrical equipment that contains askarel can be decontaminated by triple-rinsing. Metal components may then be suitable for recovery.

2. **Electrical, Askarel-filled equipment** – For electrical equipment that contains askarel, the procedure is complex and the success of the retrofilling is dependent on the type of transformer.

3. **Contaminated mineral oil equipment** – Equipment that contained mineral oil contaminated with PCBs at concentrations <500 ppm may be drained and refilled with clean oil for re-use. Scrapping for metal recovery is considered acceptable once all free liquid is removed from the hulk by an approved method. The drained oil is a PCB waste.

4. **Containers** – Askarel-contaminated containers, such as drums or tanks, should be decontaminated by triple-rinsing with an appro-

priate solvent. Containers that held PCB-contaminated mineral oil or solvent should be rinsed in a manner appropriate to the degree of contamination and the intended use of the empty container or to the disposal method.

5. **Solvent disposal** – Solvents used for PCB decontamination are classified as PCB waste if they contain more than 50 ppm of PCB by weight. It is acceptable, however, to use rinse solvent contaminated with greater than 50 ppm PCB as the first rinse when more than one PCB article is being decontaminated and the article being rinsed is more highly contaminated than the rinse solvent.

7.1.11   Storage of PCB Waste

As of 2001, the MOE estimated that approximately 99,000 tonnes of PCB waste were stored at 1,000 sites across Ontario. The waste includes approximately 7,000 of high level wastes that contain more than 1% PCBs (MOE, 2001b). These volumes have decreased considerably since the mid-1990s due to efforts by federal and provincial agencies, but still are a telling legacy of a time when the proper management of PCB waste in Ontario usually translated into storage. When Regulation 362 was promulgated in 1990, there were few opportunities to destroy or dispose PCB wastes in Ontario. While Regulation 362 defined all sites used to store PCB wastes as PCB disposal sites, this was misleading since there were few true disposal opportunities. It became commonplace to refer to the storage facilities as being "interim".

Regardless of the terminology used, Regulation 362 describes various administrative requirements for operating a PCB waste disposal or storage site. For example, section 4 of the regulation outlines the responsibilities that a site operator has with respect to keeping records and reporting to the MOE.

Section 5 of the regulation requires that certain activities such as removing and transferring PCB waste can only take place in accordance with written instructions from an MOE director. Section 7 of the regulation requires that every person storing PCB waste shall ensure that the waste is in a safe and secure location so as to prevent waste coming into contact with any person and so that any liquid containing PCBs that may escape can be readily recovered and will not discharge directly or indirectly into a watercourse or ground water.

On the other hand, Regulation 362 provides few details as to how PCB materials should be stored. The MOE directs owners of PCB waste to federal regulation SOR/92-507, which describes proper storage requirements in detail including site security and access, the types of containers to be used, the precautions that should be taken to prevent releases to the environment, the stacking of containers, fire protection and emergency procedures, maintenance and inspection practices, labelling requirements and record-keeping. Table 7.4 presents a list of features that facilities used to store PCB materials should have.

Figure 7.4 shows the label that should be used to identify drums, tanks or packaging where contaminated mineral oils, rinsing fluids or other low-level PCB wastes are stored. The label allows for the entering of the PCB concentration, date of analysis, company name and the signature of an authorized company official.

Drums or other containers that contain PCB liquids in concentrations above 10,000 ppm require special identification to alert people to separate these liquids from low-level wastes in the storage area and that special disposal requirements may be necessary. The label presented in Figure 7.5 can be used.

Environment Canada recommends that owners of PCB wastes retain an inventory record for five years after removal or disposal of the last of their PCBs.

To assist with the proper characterization of PCB materials in storage, the MOE has developed protocols for sampling, analysis and statistical interpretation of the results (MOE, 2000a).

As noted above, Regulation 362 had the effect of creating many PCB storage sites across the province. This could have been problematic because Regulation 347 defines PCB waste as a class of subject waste and Part V of the EPA requires a C of A to operate a facility where such wastes are managed. To avoid issuing approvals to all owners of PCB storage sites in Ontario (estimated to be well over 1,000 by the mid-1990s), PCB storage sites that comply with Regulation 362 are exempt from the C of A requirement of Part V.

It is generally recognized that there are merits to consolidating and centralizing PCB waste storage. Initially, this approach was used by only a few of the larger generators for their own PCB wastes, however, there now are several facilities that accept PCB waste from any source in Ontario. These facilities operate under Cs of A from the MOE. The generators of the PCB waste need to register as described in Section 4.5.

In 2001 and 2002, both Environment Canada and the MOE announced that PCB storage needed to be phased out, and that destruction and disposal technologies had evolved to the point where this was a realistic goal. For its part, the MOE has drafted amendments to Regulation 362 that would require PCB waste currently in storage to be destroyed within three years, and even shorter time frames for storage at sensitive sites such as schools, hospitals, residences, parks, playgrounds and other institutions.

Similarly, Environment Canada has been proposing to replace existing regulations with new regulations that that would pose additional restrictions on the uses of PCBs, accelerate the phase-out of PCB equipment in use at sensitive locations, mandate the labelling and reporting of PCB equipment, prohibit storage from extending beyond one year, phase out all PCB equipment by December 31, 2007 in accordance with goals of the NAFTA action plan for PCBs, and require the destruction of all PCB waste by 2009 (another goal of the NAFTA action plan).

7.1.12   Transportation

The movement of PCB wastes on roads within Ontario is governed by Regulation 347. The requirements specified in that regulation are described in Chapter 4. It also must comply with the requirements of the Ontario DGTA.

Interprovincial and international air, water and rail transportation of equipment and wastes that contain PCBs are governed by TDGA, 1992 and TDGR as described in Section 4.9.

A clear language version of the TDGR came into effect in August, 2002. The clear language version includes several revisions that are pertinent to PCBs.

1. PCBs or articles containing PCBs fall into Class 9, which includes various types of hazardous substances. There no longer are divisions in Class 9.
2. All references to waste manifests have been removed from the new regulations.
3. Shipping labels are required for eight classes of means of containment (see s. 3.5(1)(e) of the TDGR).
4. When an emergency response assistance plan (ERAP) is required, the acronym ERAP, ERP, or PIU must appear on the shipping document.
5. The special PCB label no longer is needed. Instead, the standard Class 9 label will be used. The label is shown in Figure 7.5 and the corresponding placard (for larger containers) is shown in Figure 7.6.
6. The previous TDGR specified the means of containment for electrical equipment and material containing PCBs. Those specifications (which appeared in ss. 7.34 and 8.7 of the previous regulation) have been eliminated.

The transportation of PCB wastes within Canada as well as between Canada and the United States went through major changes between 1995 and 1997. Until 1995, transportation inside Ontario required instructions from an MOE director. The agreement of the municipalities through which the PCBs were to travel as well as the receiving facility were notified and could object. Instructions were difficult to obtain and relatively little movement of PCB waste occurred. By 1995, the destruction facility in Swan Hills, Alberta expressed an interest in receiving PCB wastes from anywhere in Canada. By 1995, the MOE was approving carriers to transport PCB wastes through Ontario (and similar decisions were made in other provinces).

Between 1995 and 1997, another sequence of events caused considerable confusion with respect to the transporting of PCB waste between Canada and the United States. In 1996, the U.S. lifted a ban on importing PCB waste. The Canadian government announced a similar lifting. After several reversals on both sides of the border, shipments stopped.

7.1.13   Destruction

Three technologies are most commonly used for destroying PCBs or PCB waste. The most common is high-temperature incineration. This approach is a widely used technology in Europe and the United States to destroy PCB liquids and PCB wastes. In Canada, there are stationary high-temperature incinerators approved for the destruction of PCB wastes at Swan Hills, Alberta and at Baie Comeau, Quebec. Mobile incinerators have been used for brief periods at Goose Bay, Labrador, and at Smithville, Ontario. When used at a federal facility or operating under federal contract, the unit must comply with the Federal Mobile PCB Treatment and Destruction Regulations (SOR/90-5), which require that destruction systems must have a minimum PCB destruction limit of 99.9999%.

When used in Ontario, a mobile incinerator would need to comply with Regulation 352, which sets out requirements for siting, operation, environmental control, monitoring, bonding requirements, record keeping and record retention.

Regulation 352 also addresses mobile chemical destruction facilities. The most commonly used chemical destruction technology often is referred to as low temperature dechlorination. This technology works best on liquids and PCB content less than 14,000 ppm. This makes it well suited for treating dielectric fluid in transformers.

The third destruction technology often is referred to as gas phase reduction. Patented by a Canadian company, it has been demonstrated in the United States and used in Australia.

Various organizations and agencies maintain lists of companies that operate PCB destruction facilities. One such example can be found in the *Environmental Guide for Federal Real Property Managers*, produced by the Treasury Board of Canada Secretariat at **www.tbs-sct.gc.ca**.

In 1997, the MOE proposed to revise the approval requirements for PCB destruction facilities because it was felt that the current regulations might impede the use of non-incineration technologies. The MOE also indicated that it was considering consolidating the regulatory requirements of Regulation 352 and Regulation 362 into Regulation 347. As of late 2002, this had not occurred.

As noted in Section 7.1.11, the MOE posted a draft regulation in March, 2002 that would require the mandatory destruction of PCB wastes in storage within three years. If such a regulation is passed, the types of amendments first proposed in 1997 would also need to be made.

7.1.14 Disposal

As noted in Section 7.1.4, the disposal of PCB wastes in Ontario and the operation of disposal sites are addressed in Regulation 362. To receive PCB wastes at a disposal site, written instructions are required from an MOE director, or the site must be operated under a C of A that specifies the circumstances under which PCB waste may be accepted.

Once accepted at a site, the PCB waste can not be disposed, decontaminated or otherwise managed or diluted if in the form of a liquid unless the conditions are specified in the C of A or in accordance with written instructions of an MOE director.

The record-keeping requirements of Regulation 362 for waste disposal sites include:

- method and time of PCB delivery to and from the site
- the source of the PCB waste and/or destination and the contact person
- description of the nature and quantity of PCB waste at the site
- location of the waste disposal site
- method of storage of the PCB waste at the site
- notification to the director immediately by telephone, and in writing within three days after PCB waste first comes on the site
- notification to the director within 30 days after any other PCB waste is taken to or from the site

The operator must maintain the records until two years following written notice to the MOE that the operator has ceased to be a holder of PCB wastes.

A PCB waste disposal site is exempt from the Part V requirement that a C of A or provisional C of A is needed for all waste management systems and waste disposal sites, provided the PCB disposal site complies with all the requirements of Regulation 362. These requirements include those discussed above for record keeping, and either written instructions from the director prior to the removal of PCB waste are required, or the waste management system or waste disposal site must have a C of A that states that PCB wastes may be stored, handled, treated, collected, transported, processed or disposed of

- if the waste being removed contains over 50 L of PCB liquid, the removal must be in accordance with written instructions of the director, regardless of its destiny, and
- PCB liquid must not be removed from equipment or a container except to transfer the liquid from a leaking container (after notifying the director of the transfer) or pursuant to instructions of the director

Certain conditions apply if a PCB waste disposal site is offered for sale or lease. The prospective purchaser, tenant or person taking possession of the site must be made aware of the existing legal requirements of the site in addition to notifying an MOE director of the change. The director must be notified 10 days after the sale, lease or change in possession of the location of the site and the nature and quantity of PCB waste.

7.2   Asbestos

7.2.1   Overview

Asbestos is a generic term that applies to naturally occurring hydrated mineral silicates that are separable into flexible, incombustible fibres. The family of asbestos minerals can be subdivided into serpentine and amphibole fibres. Chrysotile is the most common fibrous serpentine and accounts for more than 90% of the world's production of asbestos.

Canada, specifically the Thetford area of Quebec, has been a major producer of asbestos for more than 100 years. Canada was the world's largest producer of chrysotile (white) asbestos, accounting for over 40% of world production.

Asbestos has been used for ceiling and floor tiles, pipe insulation, cement and insulating materials. In addition, asbestos is incorporated into cement construction materials (roofing, shingles and cement pipes), friction materials (brake linings and clutch pads), venting and gaskets, asphalt coats and sealants (Mossman *et al.*, 1990).

Around 1900, asbestos was shown to cause asbestosis, a fibrotic lung disease. In the 1950s and 1960s, it was shown to cause lung and pleural tumours in asbestos miners and workers (Mossman *et al.*, 1990). The inhalation of asbestos fibres has also been shown to produce lung cancer and mesothelioma, a cancer of the lining of the lung and chest cavities. Epidemiological research in the 1960s revealed that insulation workers who had dealt with asbestos for 20 years or more were dying of lung cancer and the complications of asbestosis at alarming rates, particularly those who had smoked (Zurer, 1990).

7.2.2   Regulatory Framework

Much of the regulatory framework for asbestos in Ontario has been passed under the Occupational Health and Safety Act (OHSA). Asbestos was one of the first designated substances to be regulated. The original regulation (O. Reg. 570/82) addressed uses of asbestos-containing material (ACM) except in construction projects, which was to be addressed in a second regulation. Subsequent to the findings of a Royal Commission into the use of asbestos, a regulation for the use of asbestos on construction projects and in buildings and repair operations (O. Reg. 645/85) was filed. Both regulations have been amended on several occasions and re-titled. The current regulations under OHSA that address asbestos are:

- Designated Substance – Asbestos (Regulation 837)
- Designated Substance – Asbestos on Construction Projects and in Buildings and Repair Operations (O. Regulation 838)

Regulation 837 addresses the mining of asbestos, the manufacture of asbestos parts and all occupations where people may inhale or ingest asbestos fibres. Regulation 838 addresses the use of asbestos on construction projects and in buildings and repair operations. The only "environmental" regulation that pertains to asbestos is Regulation 347, which defines asbestos waste and the requirements for managing that waste (see Section 7.2.4).

7.2.3   Removal versus Management

ACM can be present in any type of building but generally was used more often in large buildings including offices, schools and hospitals. While ACM does not spontaneously shed asbestos fibres, physical damage due to wear, decay, renovation or demolition can lead to the release of airborne fibres.

Public pressure, fuelled by notions such as inhaling one fibre of asbestos can cause cancer, has resulted in the removal of ACM from many schools and public buildings even though ACM is found most commonly in boiler rooms and other areas that are relatively inaccessible to building occupants.

ACM removal in large buildings can cost millions of dollars. Removal must be done carefully and thoroughly. The improper removal of previously undamaged or encapsulated asbestos can lead to increases in airborne concentrations of fibres in buildings, sometimes for months afterwards. (Mossman *et al.*, 1990).

In the United States, the 1986 Asbestos Hazard Response Act provides the Environmental Protection Agency (U.S. EPA) with the mandate to require schools to develop plans to manage (not necessarily remove) asbestos. It also requires asbestos-removal consultants and workers to have some minimal training and requires that asbestos be taken out of buildings before they are demolished or renovated (Zurer, 1990).

The U.S. EPA has published numerous documents to help building owners identify and control asbestos hazards. An index of these publications can be found at **www.epa.gov/opptintr/asbestos/**.

If asbestos is suspected of being present the following action should be taken:

1. A sample should be taken as visual inspection of ACM is unreliable.
2. The type of asbestos should be identified. Chrysotile asbestos forms curly fibres while amphibole types of asbestos, including crocidolite and amosite, crystallize as sharp needles.
3. Indoor air monitoring should be conducted.
4. If air quality levels are unacceptable, a detailed assessment should be made of physical damage to ACM.
5. Damaged areas should be repaired and/or encapsulated.
6. If conditions warrant removing the ACM, it should be removed by trained individuals (as per Regulation 838) and the material disposed as per s. 17 of Regulation 347.

7. The ACM should be wet prior to stripping to suppress dust.
8. Individuals involved in the removal must wear appropriate protective clothing including respirator, coveralls and gloves.
9. When respiratory equipment is required, the Ontario Ministry of Labour *Code for Respiratory Equipment for Asbestos* (June 2000) should be consulted.

7.2.4   Management of Asbestos Waste

Regulation 347 defines asbestos waste as solid or liquid waste that results from the removal of asbestos-containing construction or insulation materials or the manufacture of asbestos-containing products and contains asbestos in more than a trivial amount or proportion. No definition is provided for what constitutes "a trivial amount or proportion".

Section 17 of Regulation 347 specifies various aspects of managing asbestos waste. The management of asbestos waste must be carried out in accordance with the provisions outlined in Table 7.5. One of the requirements of s. 17 is that every precaution must be taken to avoid the asbestos waste from becoming airborne.

Generators of asbestos waste are not required to register with the MOE because asbestos waste is identified in s. 1 of Regulation 347 as non-hazardous solid industrial waste. It therefore does not require manifesting and can be disposed at a non-hazardous landfill site, however, the federal TDGA, 1992 and provincial DGTA require that the asbestos waste be manifested for transportation.

Regulation 347 requires that the word "CAUTION" in letters at least 10 cm high be present on both sides of all rigid containers and vehicles used to transport asbestos waste. The words "CONTAINS ASBESTOS FIBRES", "Asbestos May Be Harmful To Your Health", and "Wear Approved Protective Equipment" also need to be present. TDGA, 1992 and DGTA also require that labels and placards be present during shipment.

7.2.5   Acceptable Concentrations in Air

The MOE has established an interim half-hour point-of-impingement limit for total asbestos of 5 fibres per cm^3, and an ambient air quality criterion of 0.04 fibres per cm^3 for fibres greater than 5 μm in length (MOE, 2001a).

7.3   Summary

PCBs are a family of compounds that bioaccumulate in the food chain. Birds, aquatic invertebrates and most species of fish are particularly sensitive to PCBs. The potential health effects from long-term exposures to relatively low concentrations are still not well known. Exposures to relatively high levels in occupational settings clearly show that PCBs can cause a severe form of acne called chloracne. Other health effects include numbness in limbs, weakness and problems with the nervous system. There have been suggestions that PCBs can cause cancer in people, but according to Health Canada, "there is no proof yet of a definite risk between PCBs and cancer".

PCBs came into commercial use in the late 1920s. The major Canadian use was in dielectric fluid for industrial electrical equipment. All non-electrical uses of PCBs were prohibited in Canada in 1977. There remain many pieces of electrical equipment in service that contain PCBs.

The use of PCBs and equipment that contains PCBs are addressed in a comprehensive framework of federal and provincial regulations. Collectively, the legislation defines various types of PCB materials, place several prohibitions on their use and importation and impose requirements on their storage, transportation, destruction and disposal. There also are guidelines and criteria that identify maximum acceptable concentrations in air, water supplies, soils and sediments. In general, materials are considered to be PCB waste when the concentration exceeds 50 ppm.

Research has demonstrated that the inhalation of asbestos fibres may produce asbestosis, lung cancer and mesothelioma.

Asbestos-containing materials have been used widely in public buildings, including schools and hospitals, as well as commercial and industrial establishments. Asbestos in buildings does not spontaneously shed fibres, but physical damage due to wear, decay, renovation or demolition can lead to the release of airborne fibres. It is therefore important that experienced personnel assess the condition of ACM in a building. If the material is damaged, it should be repaired and/or encapsulated, or if conditions warrant, be removed. Trained individuals must do the removal.

In Ontario, the management of asbestos is covered by Regulation 837 (mining, manufacturing of asbestos parts, and general requirements) and Regulation 838 (construction projects and buildings). The management of asbestos waste is specified under Regulation 347. The regulation covers the storage, removal, transportation and disposal of asbestos waste. The transportation of asbestos waste must comply with the requirements of Regulation 347, DGTA and TGDA.

References

Basrur, S.V. June 2001. *Harmonization of PCB Tranfers, Decontamination and Waste Storage Procedures*. Staff Report to the Toronto Board of Health.

Canadian Council of Ministers of the Environment (CCME). August 1986. *The PCB Story*.

Canadian Council of Ministers of the Environment (CCME). September 1989. *Guidelines for the Management of Wastes Containing Polychlorinated Biphenyls (PCBs)*. CCME-TS/WM-TRE008, Manual EPS 9/HA/1 (revised).

Canadian Council of Ministers of the Environment (CCME). 1995. *PCB Transformer Decontamination Standards and Protocols*. 3rd ed. CCME EPC-HW-105E.

Environment Canada. 1988. *Handbook on PCBs in Electrical Equipment*. EN 47-310.

Environment Canada. 1991. *Identification of Lamp Ballasts Containing PCBs*. EPS 2/CC/2.

Environment Canada. July 2001. *Compliance Promotion Bulletin #5, Options for Decommissioning & Disposal of PCB Equipment*. Download from **www.on.ec.gc.ca/pollution/fpd/cpb/3005-e.html**.

Health Canada. November 2001. *It's Your Health. PCBs*. Downloaded from **www.hc-sc.gc.ca/english/iyh/pcb.html**.

Miller, M.M., P.W. Stanley, G.L. Huang, W.Y. Shiu and D. Mackay. 1985. "Relationships between Octanol-Water Partition Coefficient and Aqueous Solubility." *Environ. Sci. and Technol.* 19:6: 522-528.

Ontario Ministry of the Environment (MOE). January 2000a. *Protocol for Sampling and Testing at PCB Storage Sites in Ontario*.

Ontario Ministry of the Environment (MOE). August 2000b. *Ontario Drinking Water Standards*.

Ontario Ministry of the Environment (MOE). September 2001a. *Summary of Point of Impingement Standards, Point of Impingement Guidelines, and Ambient Air Quality Criteria (AAQCs)*. Standards Development Branch.

Ontario Ministry of the Environment (MOE). December 2001b. *Managing Hazardous Waste In Ontario*. Media Backgrounder. Downloaded from **www.ene.gov.on.ca/envision/news/121801mb.htm**.

Ontario Ministry of Environment and Energy (MOEE). July 1994. *Water Management — Goals, Policies, Objectives and Implementation Procedures of the Ministry of Environment and Energy*.

Ontario Ministry of Environment and Energy (MOEE). February 1997. *Guideline for Use at Contaminated Sites in Ontario*. PIBS 3161E01.

Mossman, B.T., J. Bignon, M. Corn, A. Seaton, and J.B.L. Gee. 1990. "Asbestos: Scientific Developments and Implications for Public Policy." *Science* 24 (January).

Zurer, P.S. 1990. "Many Asbestos Removal Projects Cited as Needless, Costly, and Risky." *Construction & Engineering News*. February 5.

Table 7.1

Summary of PCB Acts, Regulations, Guidelines and Sources of Criteria

Environmental Criteria	MOE Provincial Water Quality Objectives
	MOE Drinking Water Objectives
	MOE Air Quality Criteria and Standards
	MOE Soil and Ground Water Quality Criteria
	MOE Sediment Quality Criteria
	CCME Ambient Water Quality Guidelines
	CCME Interim Environmental Quality Criteria for Soil
Environmental Releases	Chlorobiphenyls Regulations, SOR/91-152
	Federal Mobile PCB Treatment and Destruction Regulations, SOR/90-5
	Disposal at Sea Regulations, SOR/2001-275
Importation, Manufacture, and Use	Chlorobiphenyls Regulations, SOR/91-152
Decommissioning Equipment	MOE Notification
	CCME Guidelines
Decontaminating Equipment	CCME Guidelines
Equipment Storage	MOE and Environment Canada Notification
Waste Storage	Ontario Regulation 362
	Chlorobiphenyls Regulations, SOR/91-152
Movement/Transportation	Ontario Regulation 347
	Dangerous Goods Transportation Act
	Transportation of Dangerous Goods Act
Labelling	Dangerous Goods Transportation Act
	Transportation of Dangerous Goods Act

Mobile Destruction	Ontario Regulation 352 Federal Mobile PCB Treatment and Destruction Regulations, SOR/90-5
Disposal	Ontario Regulation 362
Record Keeping	Ontario Regulation 362
Import/Export	PCB Waste Export Regulations, SOR/97-109

Table 7.2

Physical Parameter Values for Selected PCBs

Compound	Molecular Wt. g/mol	Solubility mol/m^3	Log Kow
biphenyl	154.2	4.35x10-2	3.76
2-	188.7	2.68x10-2	4.50
2,6-	223.1	6.23x10-3	4.93
2,4,6-	257.5	8.76x10-4	5.51
2,3,4,5-	292	7.17x10-5	5.72
2,3,4,5,6-	326.4	1.68x10-5	6.3
2,2',4,4',6,6'-	360.9	1.13x10-6	7.55
2,2',3,3',4,4',6-	395.3	5.49x10-6	6.68
2,2',3,3',5,5',6,6'	429.8	9.15x10-7	7.11
2,2',3,3',4,5,5',6,6'-	464.2	3.88x10-8	8.16

Reference: Miller et al., 1985

Table 7.3

Allowable Concentrations of PCB

Water

1 ng/L in ambient water — provincial water quality objective (MOEE, 1994)

1 ng/L in ambient water — federal water quality guideline (CCME,1989)

3 µg/L in drinking water — interim maximum acceptable concentration (MOE, 2000b)

5 µg/L — maximum in liquid effluents (Federal Mobile PCB Treatment and Destruction Regulations, SOR/90-5)

Groundwater

0.2µg/L — criterion (MOEE, 1997)

Air

35 ng/m^3 (annual average) — ambient air quality criterion (CCME, 1989; MOE, 2001a)

150 ng/m^3 (24-h average) — ambient air quality criterion (CCME, 1989; MOE, 2001a)

450 ng/m^3 (0.5-h average) — point-of-impingement limit (MOE,2001a)

1 mg/kg — maximum in gaseous releases (Federal Mobile PCB Treatment and Destruction Regulations, SOR/90-5)

Soil

0.5 ppm for agricultural — criterion (MOEE, 1997) and interim guideline (CCME, 1989)

5 ppm for residential and park land sites — criterion (MOEE, 1997) and interim guideline (CCME, 1989)

25 ppm for industrial and commercial sites — criterion (MOEE, 1997) and interim guideline (CCME, 1989)

Sediment

0.07 µg/g (dry weight) — lowest effect level: provincial guideline (MOEE, 1997)

Waste

50 ppm by weight — Ontario Regulation 362

Release Rate

1 g/d — Chlorobiphenyls Regulations, SOR/91-152

Table 7.4

PCB Storage Requirements

1. Spill containment (have a collection system for rain water if storage is outdoors; can place PCB-adsorbent material in drain lines).
2. There should be a fire alarm system and suitable portable or flood-type fire extinguishers.
3. Emergency response training for personnel.
4. The storage facility should be separate from processing and manufacturing operations.
5. Access to the site should be restricted.
6. An isolated room (alternatively a 2 m high woven mesh fence with lockable gate).
7. Placement of personnel protection equipment and clean-up kits within easy access.
8. Adequate ventilation (air intake and exhaust outlets should be on the exterior walls of buildings).
9. If the storage site is equipped with a mechanical exhaust system that exhausts into a building, the system must be provided with a smoke sensory control to stop the fan and close the damper(s) in the event of a fire.
10. Ventilation switch placed outside of storage room if mechanical ventilation (the system should allow for several minutes of ventilation prior to entry).
11. Floors should be made of steel, concrete or similar durable material.
12. The storage area should have continuous curbing (designed to accommodate the larger of two times the largest piece of equipment or 25% of the total volume of PCB liquid at the location, or for a single unit — 125% of the container's volume).
13. Concrete should be sealed with a PCB-resistant sealant.
14. All floor drains, pumping systems and sumps leading from the storage location should be sealed.
15. The storage facility should be located and engineered so that no PCBs will be released in the event of flood, storm or runoff from fire-fighting.
16. Use storage containers as described in s. 9 of SOR/92-507.
17. Drums or other portable containers should be placed on skids or pallets.
18. Stacking should be limited to two containers high, unless special shelving, bracing, strapping, etc. is provided.
19. If equipment containing liquid PCBs is stored outside, equipment and containment should be covered by a weatherproof roof or barrier that protects the equipment or liquids and the curbing or drip pans under them.
20. Solids, including drained PCB equipment, may be stored outside without being covered by a roof or other secondary covering providing the drained equipment and containers are structurally sound.
21. Bulk containers (for example, large international shipping containers

or approved commercially manufactured metal storage containers or structures) may be used for either primary or secondary containment for outdoor storage.

22. There should be an emergency fire plan that has been approved by the local fire department.

Table 7.5

Requirements for Managing Asbestos Wastes

1. Asbestos waste that leaves a site must be sent to a waste disposal site where the operator has previously agreed to accept the material and has been advised of its arrival time.
2. Unless transported in bulk, asbestos waste must be stored in a rigid, impermeable, sealed container capable of accommodating the weight of the material.
3. Only a waste management system operating under a C of A that specifically authorizes the transportation of asbestos in bulk can be employed to transport bulk asbestos.
4. If a cardboard box is used for storage, the asbestos must be sealed in a polyethylene bag 6 mm thick placed within the box and must be transported within a closed vehicle.
5. The external surfaces of the container and the vehicle or vessel used for transportation must be free of asbestos waste.
6. Asbestos waste shall not be transported in a compaction type waste haulage vehicle.
7. Both sides of the vehicle transporting the asbestos wastes and every container (see item 2) must display in large, easily read letters that contrast in colour with the background, the word "CAUTION" in letters not less than 10 cm high. The words "CONTAINS ASBESTOS FIBRES", "Asbestos May Be Harmful To Your Health" and "Wear Approved Protective Equipment" also need to be present.
8. The driver must be trained in the management of asbestos waste.
9. Asbestos waste shall not be transported with any other cargo in the same vehicle.
10. The vehicle must be equipped with emergency spill equipment.
11. Asbestos waste may be deposited at a landfilling site only while the depositing is being supervised by the operator of the site or a person designated by him for the purpose and the person supervising is not also operating machinery or the truck involved.
12. If deposited in a landfill, the location within the site must have been adapted for this purpose and at least 125 cm of garbage or cover material must be placed forthwith over the deposited asbestos waste.
13. Protective clothing and respiratory equipment must be worn.

Note: This is a slightly abbreviated list of requirements that appear in Regulation 347, s. 17.

Figure 7.1
PCB Label for Large Equipment

ATTENTION
PCB   BPC
CONTAINS   CONTIENT DES
POLYCHLORINATED BIPHENYLS   BIPHÉNYLES POLYCHLORÉS

A TOXIC SUBSTANCE SCHED-   SUBSTANCE TOXIQUE MENTION-
ULED UNDER THE CANADIAN   NÉE DANS L'ANNEXE DE LA
ENVIRONMENTAL PROTECTION   LOI CANADIENNE SUR LA PRO-
ACT. IN CASE OF ACCIDENT,   TECTION DE L'ENVIRON-
SPILL OR FOR DISPOSAL   NEMENT. EN CAS D'ACCIDENT,
INFORMATION, CONTACT THE   OU DE DÉVERSEMENT, OU
NEAREST OFFICE OF ENVIRON-   POUR SAVOIR COMMENT
MENTAL PROTECTION, ENVI-   L'ÉLIMINER, CONTACTER LE
RONMENT CANADA.   BUREAU DE LA PROTECTION
DE L'ENVIRONNEMENT, MINI-
STÈRE DE L'ENVIRONNEMENT
LE PLUS PRÈS.

OR 26900

Figure 7.2
PCB Label for Small Equipment

Figure 7.3
PCB General Warning Label

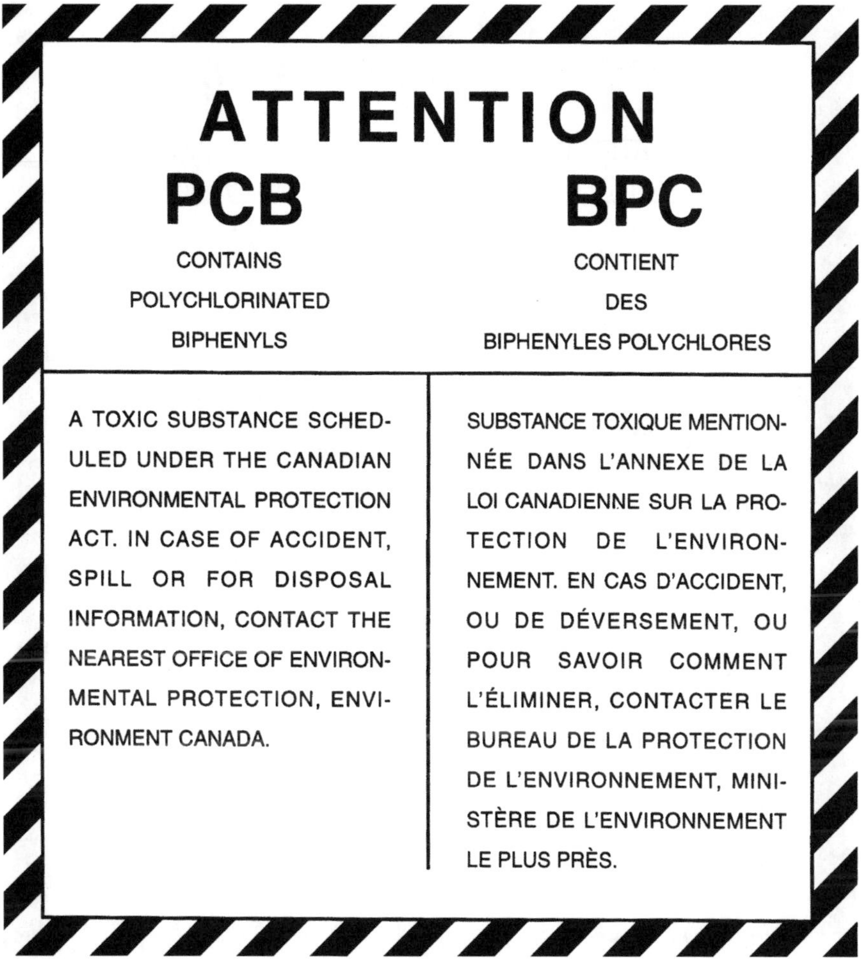

ATTENTION

PCB	BPC
CONTAINS	CONTIENT
POLYCHLORINATED	DES
BIPHENYLS	BIPHENYLES POLYCHLORES
A TOXIC SUBSTANCE SCHEDULED UNDER THE CANADIAN ENVIRONMENTAL PROTECTION ACT. IN CASE OF ACCIDENT, SPILL OR FOR DISPOSAL INFORMATION, CONTACT THE NEAREST OFFICE OF ENVIRONMENTAL PROTECTION, ENVIRONMENT CANADA.	SUBSTANCE TOXIQUE MENTIONNÉE DANS L'ANNEXE DE LA LOI CANADIENNE SUR LA PROTECTION DE L'ENVIRONNEMENT. EN CAS D'ACCIDENT, OU DE DÉVERSEMENT, OU POUR SAVOIR COMMENT L'ÉLIMINER, CONTACTER LE BUREAU DE LA PROTECTION DE L'ENVIRONNEMENT, MINISTÈRE DE L'ENVIRONNEMENT LE PLUS PRÈS.

Figure 7.4

PCB Warning Label For Contaminated Equipment

ATTENTION

CONTAMINATED WITH PCBs
(POLYCHLORINATED BIPHENYLS)

CONTAMINÉ PAR BPCs
(BIPHÉNYLES POLYCHLORÉS)

THE CONTENTS OF THIS EQUIPMENT ARE CONTAMINATED WITH PCBs, A TOXIC SUBSTANCE SCHEDULED AND REGULATED UNDER THE CANADIAN ENVIRONMENTAL PROTECTION ACT. IN CASE OF ACCIDENT, SPILL OR FOR DISPOSAL INFORMATION, CONTACT THE NEAREST OFFICE OF ENVIRONMENTAL PROTECTION, ENVIRONMENT CANADA.

LE CONTENU DE CET APPAREIL EST CONTAMINÉ PAR DES BPCs, UNE SUBSTANCE TOXIQUE ANNEXÉE ET RÉGLEMENTÉE EN VERTU DE LA LOI CANADIENNE SUR LA PROTECTION DE L'ENVIRONNEMENT. EN CAS D'ACCIDENT, OU DE DÉVERSEMENT, OU POUR SAVOIR COMMENT L'ÉLIMINER, CONTACTER LE BUREAU LE PLUS PROCHE DE LA PROTECTION DE L'ENVIRONNEMENT, MINISTÈRE DE L'ENVIRONNEMENT.

PCB CONCENTRATION (parts per million)
CONCENTRATION DE BPC (parties par million) _____

DATE ANALYZED
DATE D'ANALYSE _____

COMPANY NAME
NOM DE LA COMPAGNIE _____

AUTHORIZED COMPANY OFFICIAL
AGENT OFFICIEL AUTORISÉ_____

Figure 7.5
TDGA, 1992 Class 9 Label

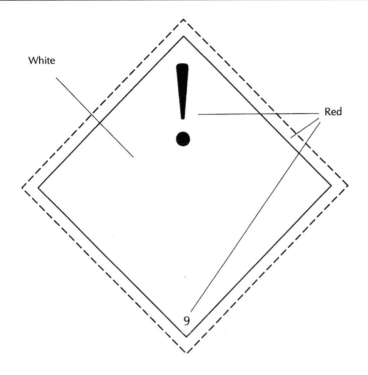

Figure 7.6
TDGA, 1992 Class 9 Placard

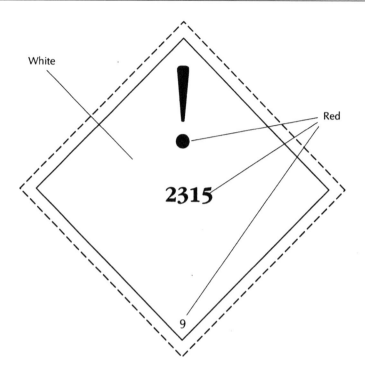

CHAPTER

8

Enforcement

8.1 Background

In Ontario, and elsewhere in Canada, environmental agencies have tended to direct more effort to abatement activities (those focused on bringing about and maintaining compliance with regulations and approvals) than on enforcement (investigating and prosecuting non-compliance). While statutes and regulations long have contained sections that created positions with titles such as "inspector" or "compliance officer", charges seldom were laid, convictions were few and fines were small. Between 1972 (when the Ministry of Environment was created) and 1980, the Ministry obtained 69 convictions, or approximately seven to eight convictions a year (Emond, 1985).

In 1985, Ontario became the first province to create a branch with the prime function of enforcement. With the Investigations and Enforcement Branch in place, the numbers of charges increased dramatically. By 1992, over 2,000 charges were laid, there were close to 400 convictions (individuals and companies combined), and fines totalled $3.6 million.

Shortly thereafter, enforcement activities began to decline almost as sharply as they had risen just a few years earlier. While optimists suggested that this was because many offenders had seen what was happening and decided to come into compliance, pessimists pointed to cuts in MOE resources. Another theory was that the MOE decided to ease up on prosecutions when it determined that the average fine being levied was far less than the average cost of obtaining a conviction.

The events at Walkerton in 2000 and the subsequent public inquiry in 2001 put this declining trend in enforcement under critical scrutiny. Before 2000 was over, the MOE announced the formation of its SWAT team, a mobile abatement and enforcement unit. And by early 2001, a review of how the MOE could become more effective emphasized "strong, effective, tough inspection, investigation, and enforcement are essential" (ERG, 2001). And so it would seem that we have entered another period where enforcement will be given a higher profile at the MOE.

While this ebb and flow in enforcement is difficult to anticipate, those who ignore compliance issues put themselves in jeopardy. As described in this chapter, if violations are observed or suspected, regulatory agencies can be expected to pursue corrective actions and, if necessary, seek convictions.

245

8.2   Overview of Enforcement Activities

8.2.1   Federal Agencies

The main federal statutes include sections that identify those empowered to perform enforcement activities and describe the types of activities they can undertake.

In the Canadian Environmental Protection Act, 1999 (CEPA, 1999), Part 10 is focussed on enforcement. Those who can undertake enforcement activities are designated as "enforcement officers" and "analysts".

An enforcement officer can inspect facilities, investigate suspected violations and direct that corrective measures be taken. The powers of enforcement officers include entry, search, seizure and detention of items related to enforcement. They also have the powers of a peace officer. Those powers can be limited when the Minister designates an individual as an enforcement officer.

Analysts have more limited powers. They can enter facilities, open containers or receptacles, take samples, conduct tests, take measurements, and require that documents or information be provided to them. They can only exercise these powers when accompanying an enforcement officer.

In the Transportation of Dangerous Goods Act, 1992 (TDGA, 1992), those who can undertake enforcement activities are designated as "inspectors". The Fisheries Act uses the terms "inspector" and "analyst". Although there are differences in the powers assigned to these individuals, their roles and activities mirror those outlined in CEPA, 1999.

Environment Canada plays an important role in enforcement of all three statutes and is the main source of enforcement for CEPA, 1999 and the Fisheries Act.

The enforcement of federal statutes was the focus of a House of Commons standing committee in the second half of the 1990s. In its 1998 report, the standing committee agreed that ensuring compliance through voluntary rather than coercive approaches has its place, but that Environment Canada may have been relying too much on voluntary approaches. It also was apparent that enforcement activities were declining since approximately 1995, likely in response to reductions in staff and budgets (Standing Committee on Environment and Sustainable Development, 1998).

When CEPA, 1999 came into force, enforcement was one area that was strengthened. The powers of enforcement officers were expanded to those noted above, and officers were authorized to issue certain orders to bring an immediate halt to any infraction.

Environment Canada subsequently developed and published a compliance and enforcement policy for CEPA, 1999 (Environment Canada, 2001). The general principles of the policy are: compliance is mandatory; enforcement should be fair, predictable, and consistent; enforcement should emphasize preventing damage to the environment; every suspected violation should be examined; and the reporting of suspected violations is encouraged. A similar policy has been drafted for the Fisheries Act.

While enforcement officers, inspectors and analysts may uncover and investigate violations, deciding whether or not a violation warrants prosecution rests with the Attorney General (the Minister of Justice) and the Crown prosecutors acting on behalf of the Attorney General.

Additional information about enforcement of CEPA, 1999 can be obtained at the Environment Canada "Green Lane" Web site at **www.ec.gc.ca**. That site provides links to the CEPA Environmental Registry and a section entitled "Environmental Law Enforcement".

8.2.2   Provincial Agencies

The Environmental Protection Act (EPA), Ontario Water Resources Act (OWRA) and Pesticides Act designate those who can undertake enforcement activities as "provincial officers".

In broad terms, provincial officers administer the statutes by attempting to identify, contain, clean up and prevent emissions or spills from being repeated. Similar to federal enforcement officers, provincial officers have the powers of entry, search, seizure and detention of items as part of an abatement function only. It can be difficult to determine if provincial officers are restricting their activities to those related to abatement, or whether, in fact, they are obtaining evidence for a prosecution.

When a provincial officer (or any Ministry staff member) comes upon a situation that appears to be a violation, the *Compliance Guideline F-2* (MOEE, 1995) requires that an occurrence report be prepared and submitted to the Investigations and Enforcement Branch (IEB). The provincial officer may recommend a follow up investigation or suggest that abatement measures will be adequate. IEB staff members are investigators who visit facilities only to determine whether or not a law has been broken, to collect evidence of an offence and to lay charges.

In September, 2000, the environmental SWAT team, a mobile enforcement unit, was created to investigate specific industrial sectors in Ontario, arriving unannounced to conduct inspections. As of 2001, the SWAT team had inspected electro/metal platers, hazardous waste transfer and processing facilities, industrial waste haulers, pesticide applicators, recycling in multi-unit residential buildings, and septic waste haulers.

Violators brought to the attention of the IEB through occurrence reports or by its own investigations are reviewed by IEB and written briefs are forwarded to the Legal Services Branch (LSB) for those violations that may warrant prosecution. The LSB will review the brief and decide whether to recommend prosecution. LSB lawyers are seconded from the Ministry of the Attorney General. The main role of the LSB is to recommend which cases should go forward for prosecution and to conduct those prosecutions under provincial environmental legislation.

Overall, Ministry policy on enforcement has been expressed as four principles. In short, these are: all persons are entitled to equal protection; prosecution will be the result of informed judgment; when enforcement is not pursued the reasons shall be recorded; and enforcement will be administered in an even handed, non-discriminatory and fair manner that advances and protects the public interest (MOEE, 1995).

The Technical Standards and Safety Act, 2000 addresses various matters including boilers, pressure vessels and the handling of liquid fuels such as gasoline, diesel, and heating oil. The Act, associated regulations, codes and guidance documents are administered by the Technical Standards and Safety Authority Inc. (TSSA), a privatized organization that formerly was part of the Ministry of Consumer and Corporate Relations, now the Ministry of Consumer and Business Services. The Act designates inspectors, who can issue compliance orders to bring about conformity with the Act's regulatory regime. Directors may issue safety orders, which are equivalent to "tag out" orders. For intransigent situations, the TSSA may prosecute contraventions.

8.2.3   Municipalities

The Municipal Act allows municipalities to establish by-laws and to designate employees or agents to enforce by-laws. Possible violations of by-laws may come to the attention of a municipality as a result of its efforts (such as routine sewer monitoring), complaints from the public, or the review of information it requests.

Municipalities and these designated individuals have the powers of entry to check on compliance with by-laws and to investigate possible violations. It is an offence to hinder or obstruct any person exercising or performing a duty under the Municipal Act or any by-law under the Act. Similar provisions appear in the Municipal Act, 2001 which will take effect in January, 2003.

Municipalities also enforce the Ontario Building Code and the Ontario Fire Code, both of which contain provisions related to environmental matters. These provisions include recycling activities, tanks used to store chemicals and the assessment of environmental suitability (largely soil and groundwater quality issues) during the site development process.

8.3   Inspections and Investigations

8.3.1   Timing and Notice

Regulatory agencies (including the MOE) differentiate between "inspections" and "investigations". Inspection most often is used to describe an abatement activity (to verify compliance), while investigation is an enforcement activity (to gather evidence of a violation). If, during the course of an inspection, the officer or inspector sees or suspects a violation, the inspection can turn into an investigation. It can be difficult for all parties to determine exactly when this important change happens.

The following discussion is most directly relevant to inspections by provincial officers and so that term is used throughout, but the discussion would not change substantially if focussed around the inspectors, officers or analysts designated in various federal and provincial statutes.

A provincial officer may enter a business premise at any reasonable time to check for compliance. What constitutes a reasonable time may vary with the circumstances. It is generally accepted that routine inspection should be conducted during normal business hours.

The MOE may give a company notice that an inspection will take place. If the suggested time is not convenient, the company may request an alternate time. It is not unreasonable to request that an inspection be delayed to allow the company time to consult a lawyer. Note, however, that most Certificates of Approval (Cs of A) contain a provision requiring access for inspectors at any time.

If prearrangements have not been made and a provincial officer arrives to inspect a situation, the officer should be treated courteously and the concerns addressed without delay. Some companies, fearing that incriminating evidence or practices may be uncovered, may refuse entry or assume an unco-operative attitude. This type of conduct may give rise to a charge of obstructing a provincial officer in the execution of the officer's duty.

If a provincial officer arrives unannounced, the officer should be asked to show credentials or badges, etc., to indicate the department represented and then be asked questions concerning:

- the purpose of the inspection and/or the nature of the concern
- whether there is a specific date or series of dates that are of concern
- whether a specific location or source is under scrutiny
- the section of which legislation or the approval called into question

If the responses indicate that the inspection is related to an investigation into a possible violation, copies of the responses should be sent to a lawyer, and it should be decided if a co-operative approach will be taken to the investigation.

If an officer is refused entrance or finds the premises locked or abandoned, the officer can seek a warrant from a justice of the peace or a judicial order from a court.

8.3.2   Information Gathering

During an inspection, a provincial officer may take copies of documents or photographs, conduct tests, or inspect equipment as long as the requests are reasonable.

A list should be kept of all documents or photocopies provided to the provincial officer. Documents that are provided should be numbered to avoid later confusion.

All testing and sampling procedures should be reviewed to ensure that they were done properly and allow for a fair conclusion. It may be appropriate to take a duplicate sample and have the sample analyzed by an independent laboratory.

It is preferable to designate one company representative to show the provincial officer the facility and to co-ordinate responses to requests or questions. The representative should be familiar with the legal implications of the inspection and should have a working knowledge of the facility. In addition, the representative should take careful notes of what is seen and photographed, who is interviewed, what is said, and the sampling procedures and locations.

8.3.3 Search and Seizure

Provincial officers (and federal enforcement officers) have broad powers to search for, examine, seize and detain records, products, and equipment.

In addition to the authority granted by legislation, provincial officers may, in certain circumstances, be required to obtain a court order or a search warrant.

Case law on the Canadian Charter of Rights and Freedoms strongly suggests that provincial officers should not search premises to seize or obtain evidence of an offence without a search warrant or court order (Jetten and Smith, 1989). In 2002, the Supreme Court of Canada ruled that once a provincial officer has reasonable grounds to believe that an offence has been committed, an investigation for information that is potential evidence constitutes an unreasonable search and seizure (a breach of a company's or individual's Charter rights) unless:

- the party subject to the investigation consents to it and co-operates in it
- a pollution offence or an offence involving unlawful handling or disposal of hazardous or liquid industrial waste has occurred or is occurring, the place of the investigation is likely to afford evidence of the offence and exigent circumstances make it impractical for the provincial officer to obtain judicial authorization before proceeding
- there is judicial authorization (judicial order or search warrant) obtained for the investigation

Since one type of provincial officer — an IEB investigator — has as their sole purpose the collection of evidence for prosecuting an offence, the attendance of an investigator will always mean an investigation, which requires one of the three above circumstances to be lawful.

If an unreasonable search and seizure occurs, an application can be made for a Charter remedy, which could include exclusion of the evidence and, if the evidence is critical to the outcome, the ordering of a new trial. Any subsequent search for evidence must be conducted according to the Charter. This indicates that the broad inspection powers of a provincial officer are available only for abatement purposes and not enforcement purposes (Willms and Shier, 2002). Expressed another way, the broad powers are available for inspections but not for investigations without proper authorization, consent or exigent circumstances.

8.3.4   Obligations of Those Being Inspected

Some legislation contains sections that, in effect, impose obligations on individuals to co-operate when an investigation is launched. Section 184 of the EPA prohibits an individual from hindering, obstructing or providing false information to a provincial officer in the lawful performance of his or her duties. In addition, an individual may not refuse any reasonable request for information required for the purposes of the Act or regulation.

Similar sections that prevent obstructing or hindering an inspection can be found in CEPA, 1999, TDGA, 1992, and the Municipal Act. Some federal statutes including CEPA, 1999 and the Fisheries Act further require that inspectors be given all reasonable assistance so that they can carry out their duties and functions.

The courts have not resolved the tension between such mandatory compulsions to furnish information with certain rights (such as the right to remain silent) under the Canadian Charter of Rights and Freedoms.

Reporting by an individual may incriminate that person. This can be avoided by having a person not involved in the event making such reports.

8.3.5   Confidentiality

One approach to the unwanted disclosure of documents is to use the concept of solicitor-client privilege. In most cases, information passing between a lawyer and client is privileged if it is secured for the purpose of obtaining legal advice, created in contemplation of litigation, or is otherwise kept confidential.

If the question of confidentiality does arise during an investigation, especially over potentially sensitive documents, legal counsel should be sought.

8.4   Interviews

During an inspection or investigation, a federal or provincial officer may ask to interview people. A company should try to prearrange the procedures used for interviews. Before the interview, companies should brief individuals on their individual rights and what is required of them. In cases where this is not possible, or for some other reason the prearrangement has not been made, individuals should be made aware of their individual rights at the beginning of the interview.

Individuals should approach interviews with the attitude that statements made may be used in a court of law, whether or not the individual being interviewed is charged. Some guidelines to remember during an interview include (Rovet, 1988):

1. Stick to the facts; do not speculate.
2. Make sure you understand the question, or ask for clarification or a repeat of the question.
3. Take your time answering the questions.
4. Do not wander off the topic.
5. Do not volunteer information.
6. If you do not know an answer, say so.
7. Take notes during the interview.

If the purpose of the questioning is to obtain evidence of an offence for the purpose of prosecuting the individual, that person may refuse to answer. A provincial officer cannot force a person into self-incrimination. If the officer has reasonable or probable grounds to believe that the individual being interviewed has committed an offence, the officer must recognize the rights against self-incrimination and give a warning (McKenney, 1989).

During the interview, an individual is required to give only the information that is pertinent to the investigation. Most provincial officers have been trained as interviewers and may try to get an individual to be open. This may reveal information that in some cases is not pertinent to the investigation.

A company can be incriminated and convicted through statements made by an employee. Damaging statements given by employees must be made voluntarily without threat, promise or inducement.

A provincial officer may write a narrative of the answers while interviewing a person and ask the person to sign it as his or her own statement. The statement may be paraphrased in words selected by the provincial officer to highlight issues the officer feels are important. Provincial statutes or regulations do not require individuals to sign or initial anything. The choice of whether to sign or not sign the document is up to the individual being interviewed. Signing may be interpreted as condoning the method used to record the statement, and as confirming the accuracy of the statement.

A provincial officer may suggest the use of a tape recorder for interviewing purposes. There is no statutory authority or obligation that would require an individual to participate in tape-recorded interviews and, therefore, it is up to the individual to object to the use of tape recorders. A person should not feel embarrassed or be afraid to refuse to participate in recorded interviews.

Whenever possible, a company representative not related to the incident but familiar with the particular Act under which the investigation is being conducted, should be present at all meetings and interviews during the investigations. If this has not been prearranged, the person being interviewed can request the right to have someone present. The provincial officer has the legal right to disallow a third party to be present during the interviews unless the third party is the person's legal counsel. It may be legitimate for corporate counsel to be present where the seniority or responsibility of the individual being interviewed is such that the individual's statements speak for the company.

8.5 Responses to Violations

8.5.1 Federal Agencies

Enforcement officials, whether federal, provincial, or municipal, are expected to respond to violations according to the sufficiency of the evidence and predetermined response criteria.

When a violation of the CEPA, 1999 is discovered, enforcement officials consider the nature of the violation, the desired response by the violator and consistency in enforcement when deciding what action to take.

The nature of the violation covers the seriousness of the harm or potential harm, the intent of the alleged violation, whether the occurrence has been repeated, and whether there were attempts to conceal information or otherwise subvert the objectives and requirements of the Act.

The desired result of any response is compliance with the Act within the shortest possible time and with no further occurrence of violation. Officials may consider: the violator's history of compliance and willingness to cooperate with enforcement officials; evidence of corrective action already taken; and the existence of enforcement actions by other federal or provincial authorities as a result of the same activity.

To achieve consistency in enforcement, officials will consider how similar situations were handled when deciding what enforcement action is to be taken.

Environment Canada has various types of responses available to deal with violations of CEPA, 1999 and its regulations (Environment Canada, 2001).

A warning may be issued if it is believed that a violation is continuing or has occurred, and the degree of harm or potential harm to the environment, human life or health appears to be minimal. A warning may be issued if the violator has a record of compliance and/or has made reasonable efforts to remedy or mitigate the consequence of the offence. Warnings are given in writing.

When there is a minimal threat to the environment or human life or health, a ticket may be issued. Offences which may be ticketed include failure to provide information on a new substance, or failure to provide, within the allowable time, a pre-shipment notice. For any offence that is designated as ticketable, inspectors will always issue a ticket except when a warning is considered appropriate.

The list of offences that are ticketable and the associated ticket amounts are presented in the Contraventions Act. Examples of ticketable CEPA, 1999 offences are the failure to provide information or the failure to provide information within the stipulated time limit. As of 2002, tickets can be for amounts up to $500. The Fisheries Act has similar provisions for issuing tickets.

The accused can plead guilty and pay the ticket, plead guilty with an explanation and appear in court to request a lesser fine, or submit a plea of not guilty and go to court. Failure to choose an option within the stated time limit means that the ticket cannot be challenged. A new ticket can be issued for each day an offence continues.

In response to a release of a substance in contravention of the CEPA, 1999, an enforcement officer may give directions to remedy a dangerous condition or reduce any danger to the environment or human life or health that results from the release of a substance.

Several types of orders can be issued, but the one that is most likely to be used is the environmental protection compliance order. Enforcement officers can issue these orders to deal with any offence under CEPA, 1999 when a warning, ticket or direction has not had the desired effect. The order will require a return to compliance but impose no financial penalty.

For more serious situations, the Minister has the authority to seek injunctions to stop or prevent violations. If an individual or company is not complying with an injunction, the Minister can seek a contempt of court ruling.

Enforcement officers can lay a charge and seek a prosecution for every violation under CEPA, 1999, except where warnings, tickets, directions or orders are considered to be more appropriate. Prosecution will always be pursued in the following situations: a violation involves death or bodily harm, there is serious harm or risk to the environment or human health, false information is given or information is concealed, an enforcement officer or analyst is obstructed in carrying out his or her duties, information that an offence has occurred is concealed, or less than all reasonable measures are being taken to comply.

8.5.2   MOE Responses

Like the federal agency responses described in Section 8.5.1, the MOE has several responses to violations at its disposal. These are described below in order of increasing severity, culminating in prosecution. More detailed descriptions can be found in MOE *Compliance Guideline F-2* (MOE, 1995).

Occurrence Report

The MOE will identify non-compliance situations (violations) by various means including inspections, responding to spills, addressing complaints, responding to Environmental Bill of Rights requests, reviewing submitted information, and pursuing abatement programs. When a provincial officer is of the opinion that a violation has occurred, the suspected violation and a recommendation on the need for enforcement will be recorded in an occurrence report. All such reports are forwarded to the IEB.

Tickets

For some minor offences, a provincial officer can issue a ticket (more properly called a Provincial Offences Act Part I offence notice). The maximum fine currently is $500. Examples of the types of violations that may warrant a ticket are listed in Table 8.5. The issuing of a ticket will be documented in an occurrence report. For other minor offences, a Part I summons can be issued. Unlike a ticket, there is no provision to plead guilty to a summons and pay a fine out of court.

Voluntary Abatement

Another option that a provincial officer has is to request voluntary abatement. All such requests will be confirmed in writing (usually in an inspection report, letter or in a notice from a provincial officer that a violation has been observed) unless the action was completed forthwith, in which case the only documentation is the occurrence report. Since March, 2002, relying on voluntary abatement requires written justification from abatement staff to their supervisors. In response to a voluntary program that has been undertaken, the MOE may issue a program approval, signed by a director and titled as such.

Control Documents

When mandatory abatement is considered appropriate, MOE will issue one or more control documents or authorizing documents. Control documents are authorized by statutes. They are binding and can be enforced by prosecution when issued by a director. Some examples of control documents are:

- control order
- stop order
- remedial order
- preventative measures order
- field order

The Ministry will provide advance notice of the intention to issue a control document and post a draft version on the Environmental Bill of Rights (EBR) registry for public comment.

In general, a control document will require that corrective actions be completed in a timely fashion. Various types of orders can be issued, and only a few are highlighted here. When a discharge or emission is found to be in non-compliance, several sections of the EPA allow for the issuing of a control order. Such an order may specify directions or the manner of discharge, additional studies of the discharge and implementation of procedures.

1. Stop order – Section 8 (and others) of the EPA allow for the issuing of a stop order to immediately cease polluting activities. There must be an immediate danger to human life or health or property for a stop order to be issued.

2. Remedial order – Section 17 of the EPA allows for the issuance of a remedial order to any individual or company who caused or permitted a discharge or spill of a contaminant compelling the individual or company to take necessary steps to repair the injury or damage. For spills, s. 97 allows an order to restore the environment to be issued.

3. Preventative measures order – Section 18 of the EPA allows for the issuing of a preventative measures order if it is believed that the discharge of a contaminant likely will cause an adverse environmental effect even if no discharge has occurred. This type of order may require specific facilities or equipment or the procedures to be followed. It may also require monitoring or the study of the effects of the contaminant.

4. Field order/provincial officer order – Another type of control document is a provincial officer order, also known as a field order. Field orders can be issued under several sections of the EPA, the OWRA, and the Pesticides Act. A field order can direct a person to prevent or repair damage or injury, to implement specified procedures, to install or modify pollution control or prevention devices, to monitor and record environmental conditions or emissions, to remove waste, or to bring conditions into compliance.

Within five days of receipt, a written request can be made for review of the field order by the assigned director. If the director confirms the field order or does not respond within another five days, a second appeal can be made within a further 15 days to the Environmental Review Tribunal (formerly the Environmental Appeal Board).

Authorizing documents include Cs of A, licences and permits issued under various sections of the EPA, OWRA, or Pesticides Act. Like control documents, they are binding and can be enforced by prosecution.

During the development of control and authorizing documents, except for emergency approvals, the public is notified via the EBR registry.

An offender may be required to provide financial assurance to ensure that funds are available in the event that the offender fails to comply with an authorizing document. MOE policy is to not require financial assurances from public sector proponents.

Occurrence reports are forwarded to the IEB. The IEB then uses factors related to enforcement and informed judgement to decide whether enforcement action should be taken. When it is determined that action is necessary, that action usually is an investigation by IEB staff.

Once the IEB has undertaken an investigation, the IEB officer in charge must decide whether or not to recommend prosecution. If they decide prosecution is warranted, they prepare a written brief that is reviewed by the IEB director before being forwarded to the LSB. The LSB then makes a recommendation to prosecute or not.

As noted above, informed judgement is used at various stages of enforcement to determine appropriate responses or next steps by the MOE. This encompasses many factors, including: the risk a violation poses to human health and the environment; whether the violation is deliberate; whether the violation is the result of negligence; whether the violation has been repeated or is ongoing; whether the offender has disregarded warnings; and the offender's compliance record and attitude (MOEE, 1995).

8.5.3 TSSA Responses

The Technical Standards and Safety Act, 2000 designates "directors" and "inspectors" at the Technical Standards and Safety Authority Inc. (TSSA) who can take actions to bring about conformity with the regulatory regime of the Act.

A director can refuse to grant, suspend or revoke authorizations to applicants who fail to meet requirements, where there is reason to suspect the activities will not comply with the requirements, the applicant lacks the required resources or skill, or the applicant has not been honest. A director may issue safety orders that require any thing to which the Act applies to be shut down or used only in accordance with the order. A director may apply to a judge of the Superior Court of Ontario for an order directing compliance.

An inspector can issue an inspection order directing the contravener to come into compliance, or "seal" any thing to which the Act applies that may be a threat to public safety.

For intransigent situations, the TSSA may prosecute contraventions.

8.5.4 Municipal Responses

Like their federal and provincial counterparts, municipalities can respond to environmental violations with various informal and formal actions. Informal actions include informal notice (for example, a telephone conversation), meeting, warning letter, and/or meeting to show cause. At the latter, the municipality will present the facts concerning the non-compliance and ask the violator to show cause as to why the municipality should not proceed with one or more formal enforcement actions. Formal actions include the issuance of a compliance program, an order of prohibition, a charge of contempt for failing to comply with an order of prohibition, or prosecution.

Where a by-law passed under the authority of the Municipal Act or under any other general or special statute is contravened and a conviction entered, the court may make an order prohibiting the continuation or repetition of the offence by the person convicted. Similar provisions appear in the Municipal Act, 2001 which takes effect in January, 2003.

Table 8.8 presents a typical example of an enforcement strategy that a municipality may follow when ensuring compliance with the local sewer-use by-law.

8.6   Penalties and Liabilities

8.6.1   Canadian Environmental Protection Act, 1999

CEPA, 1999 includes sections that stipulate the penalties for convictions and criteria to guide the courts when imposing those penalties (although the courts are not compelled to follow those guidelines).

Table 8.1 presents a summary of penalties under CEPA, 1999. The Act provides for maximum fines of $1 million and up to five years in jail. It also provides a civil cause of action to any person who has suffered loss or damage as a result of conduct contrary to the Act or the regulations, including the right to seek injunctive relief.

Upon conviction, the enforcement officer responsible for the case may request that the courts include one or more orders provided under CEPA, 1999. The orders requested can prohibit the offender from any activity that might repeat the offence, require the offender to take corrective action, require the offender to implement a pollution prevention plan or environmental emergency plan (see Section 12.2), or encourage creative sentencing such as the performing of community service (see Section 8.6.11).

CEPA, 1999 empowers the Crown to recover costs by civil suit under certain circumstances. It is possible to recover costs even if there is no prosecution, the prosecution did not result in conviction, or prosecution brought conviction but not an order to recover costs.

In an action to recover costs, the defendant can be an individual, or a company or organization that owned or had charge of a substance immediately before its initial release into the environment.

8.6.2   Transport of Dangerous Goods Act

Penalties for violations under the TDGA, 1992 are presented in Table 8.2. The Act provides for maximum fines of $100,000 a day for a summary conviction and imprisonment for up to two years if the conviction is on indictment.

Upon conviction, the court may impose orders similar to those for CEPA, 1999 (see Section 8.6.1). The total value of penalties for a single offence must not exceed $1 million.

8.6.3   Fisheries Act

To be prosecuted under the Fisheries Act, the Crown does not have to prove that fish were actually harmed, but only that a "deleterious" substance was deposited into waters frequented by fish, or that the fish habitat was harmed. Regulations made under the Act allow for specific materials to be discharged under certain circumstances. Such regulations have been developed for chloro-alkali effluents, liquid mercury effluents, petroleum refinery effluents, and pulp and paper effluents, among others.

Penalties prescribed under the Fisheries Act are presented in Table 8.3. The penalties cover the addition of a deleterious substance to waters frequented by fish, the harmful alteration of a fish habitat and throwing material overboard.

When the court is satisfied that a convicted offender has realized monetary benefit by committing an offence, the court may order the offender to pay an additional fine equal to the amount of those monetary benefits.

8.6.4   Ontario Environmental Protection Act

Table 8.4 presents a summary of EPA penalties. The maximum fine can be levied for each day a violation occurs. In addition to the fines, violators can be required to pay an amount equal to the monetary benefit obtained due to the violation. There are provisions for imprisonment (terms of up to five years) for violations of specific sections or subsections of the Act.

There are penalties for individuals and corporations. Section 192 of the EPA deems the actions and omissions of an officer, official, employee or agent of a corporation in the course of his or her employment, or in the exercise of his or her powers or the performance of his or her duties, as the actions or omissions of the corporation also.

When corporations are involved, it is up to the discretion of the Crown to decide whether individuals will be charged along with the corporation. Often the Crown is not interested in convicting individuals except those with considerable corporate authority, or when individuals are acting outside the scope of their authority.

As noted in Section 8.5.2, a provincial officer can issue a ticket if a violation poses minimal or no threat to the environment or human life or health. Table 8.5 presents examples of ticketable offences. The fines themselves may not be large but the conviction will affect a company's compliance record — a record that may be used against the company should it be charged and/or convicted of a more serious offence. As of late 2002, tickets carry a maximum fine of $500. The details of issuing tickets are provided in the Provincial Offences Act.

8.6.5　Ontario Water Resources Act

The penalties for the contravention of any part of the OWRA are presented in Table 8.6. In addition to fines, some convictions may result in imprisonment for not more than one year. The penalties are similar in severity to those possible under EPA.

As noted in Section 8.5.2, a provincial officer can issue a ticket if a violation poses minimal or no threat to the environment or human life or health. The fines are not large (not more than $500 as of late 2002) but the conviction will affect a company's compliance record. The details of issuing tickets are provided in the Provincial Offences Act.

8.6.6　Toughest Environmental Penalties Act

In 2000, the Government of Ontario introduced the Toughest Environmental Penalties Act. The Act increased the maximum fine for a corporation's first conviction of a major offence from $1 million to $6 million per day. Fines for an individual's first conviction of a major offence increased from $100,000 to $4 million per day. Maximum jail terms for a person convicted of a major offence increased from two years to five years less a day. These increases are evident in the penalties shown in Table 8.4 and Table 8.6.

8.6.7　Technical Standards and Safety Act, 2000

The penalties for the contravention of the Technical Standards and Safety Act, 2000 are presented in Table 8.7. In addition to fines, convictions may result in imprisonment for not more than one year. The penalties are similar in severity to those possible under the EPA and OWRA. The Technical Standards and Safety Act, 2000 includes both a fine for a corporation that commits an offence and penalties for every director or officer who has failed to take all reasonable care.

8.6.8　Proposed Administrative Monetary Penalties

In January, 2002, the Ontario government announced the proposed Administrative Penalties Regulations that would allow the MOE to impose monetary penalties for a broad range of clear and simple infractions under the EPA, OWRA, and Pesticides Act where the expense of prosecuting the offending party cannot be justified.

The legislative provisions for administrative monetary penalties (AMPs) establish a maximum penalty cap of $10,000 per day per contravention. The proposed regulations set out the violations subject to monetary penalties, the amounts that could be imposed, how a monetary penalty must be calculated and the rules to be followed for issuing AMPs. The amount of the AMP will be based on various factors including recent convictions, prosecutions and AMPs.

It will be up to the MOE to choose either an AMP or prosecution, but not both. The informed judgment matrix in the *Compliance Guideline F-2* (MOEE, 1995) likely indicates how this choice will be made. AMPs are not the equivalent of prosecution and a due diligence defence will not apply except that it can be used to reduce the amount of the penalty. Those who receive an AMP will have the right to appeal to the Environmental Review Tribunal. As of late 2002, these regulations are not in force.

It also is anticipated that the TSSA will be imposing AMPs by the end of 2002 for violations related to the distribution, storage and handling of fuels.

8.6.9 Other Provincial Initiatives to Encourage Compliance

The 2001 review of best practices for managing the environment emphasizes that the challenge of effective environmental management is broader than any one agency and that there is a need to co-ordinate efforts for achieving compliance (ERG, 2001).

In April, 2002, the MOE announced a pilot project to increase compliance through co-operative efforts with the metal finishing sector and the auto body repair sector. At the same time, the MOE posted a proposal to implement a co-operative agreement with the Automotive Parts Manufacturers' Association to improve environmental performance above and beyond what is currently required. The latter program is an attempt to put into action several of the recommendations from the 2001 review of best practices. These types of initiatives could be expanded to other sectors.

8.6.10 Municipal Act

Table 8.9 presents the maximum penalties for improper discharges to a municipal sewer system. The penalties range from a maximum of $5,000 a day for a first offence to $50,000 a day for a subsequent offence. The same penalties appear in the Municipal Act, 2001 which will take effect in January, 2003.

The Municipal Act allows a municipality to obtain an order prohibiting the continuation or repetition of a contaminant discharge to a sewer by a person convicted of an offence. If the offender knowingly acts in contravention of the order of prohibition, an application can further be brought for an order citing the offender in contempt of court and committing the offender to jail. Municipalities can lay charges and prosecute offenders. The Act also allows a municipality to recover costs or to add penalties to taxes. Similar provisions appear in the Municipal Act, 2001.

8.6.11   Creative Sentencing

Creative sentences are different from, or in addition to, jail terms or fines. Until 2000, these sentences were gaining favour in environmental law as the courts in Ontario looked for additional ways to make convicted offenders pay their debt to society by directly helping the environment they harmed through restoration, repayment, education and training. Examples of cases in the early 1990s that involved creative sentencing are listed below.

1. A large chemical company repaid the MOE $26,000 for outside laboratory costs it incurred as a result of a spill. These monies were in addition to the $125,000 fine for the incident.
2. A waste disposal company was ordered to forfeit $70,000 in profits made from an offending operation; the operation was in repeated violation of conditions in the company's C of A. This type of creative sentencing is known as profit-stripping. This payment was in addition to a fine of $70,000 against the company and fines of $10,000 and $7,500 levied against two company directors.
3. A large forest products company was ordered to make a $25,000 contribution to a graduate scholarship in environmental and forestry management at a university following a conviction for failing to report a spill. The company was also fined $5,000.
4. A development company was ordered to put $25,000 in fines into a fund to be used by the Ministry of Natural Resources to restore fish habitats damaged by the actions of the company.

In 2000, a defendant who was offered a creative sentence protested to Ontario's premier. Pending a review by the Ministry of the Attorney General, creative sentencing has stopped in Ontario.

8.6.12   Liability of Corporate Officers and Directors

Prior to 1986, the liability of officers and directors was limited to persons who could be said to be the directing minds and wills of the corporation. This artificial distinction between those who make policy and those who implement it no longer exists as illustrated by the provisions found in federal and provincial jurisdictions.

In CEPA, 1999, s. 280 states that where a corporation commits an offence, any officer, director or agent of the corporation who directed, authorized, assented to, acquiesced in or participated in the commission of the offence is a party to and guilty of the offence, and is liable, whether or not the corporation has been prosecuted or convicted. The same wording appears in s. 78 of the Fisheries Act.

Under the TDGA, 1992, any officer, director or agent of a corporation who directed, authorized or consented to the commission of an offence is guilty of the offence and liable to conviction whether or not the corporation is prosecuted.

In the EPA, s. 194(1) clearly states that every director or officer of a corporation that engages in an activity that may result in the discharge of a contaminant into the natural environment contrary to the Act has a duty to take all reasonable care to prevent the corporation from causing or permitting such unlawful discharge. Furthermore, it is an offence for anyone who fails to carry out that duty and s. 194(3) permits a conviction of an officer or director despite the fact that the corporation has not been prosecuted or convicted for the offence.

In the OWRA, s. 116 indicates that every director or officer of a corporation that engages in an activity that may result in the discharge of any material into or in any waters or on any shore or bank or that may impair the quality of the water has a duty to take all reasonable care to prevent the corporation from causing or permitting such unlawful discharge.

In the Technical Standards and Safety Act, 2000, s. 37(2) indicates that every director or officer of a corporation has a duty to take all reasonable care to prevent the corporation from committing an offence. In addition to penalties for the corporation there also are penalties for directors (see Table 8.7).

Court decisions provide further interpretations of how these statutory provisions are translated into the duties of corporate officers and directors. For example in *R. v. Sault Ste. Marie (City)*, the court determined that officers must report back periodically to the board on the operation of the company's environmental system and officers must report substantial non-compliance matters to the board in a timely manner.

In *Queen v. Bata Industries Ltd.*, the minimum duties on the officers and directors was expanded substantially:

1. The board of directors is ultimately responsible for environmental compliance.
2. If the board chooses to delegate its responsibilities, it must ensure that a system of adequate and effective supervision is in place, that the Board is kept regularly informed on environmental matters, and that the persons to whom the responsibility has been delegated are properly educated in environmental matters and have the resources to deal with environmental problems.

3.   The board is entitled to place reasonable reliance on delegates, and on the reports such delegates prepare.
4.   An environmental policy dealing with pollution prevention (including waste management and disposal) is to be approved by the board and the board must be assured that the policy is being complied with. One way to establish compliance is through a comprehensive environmental review or audit.
5.   The board should be aware of the standards in its own industry as well as those in other industries that may deal with similar environmental concerns.
6.   Directors should ensure that minutes of the board's meetings accurately reflect the consideration of environmental matters.
7.   The board should be informed of and react to environmental concerns that affect the company as quickly as possible.

8.7   Defences and Mitigating Factors

8.7.1   Criminal Offences

For offences under the Criminal Code or other statutes with significant penal sanctions, such as environmental laws, the burden of proof to be overcome by the prosecution is proof beyond a reasonable doubt, strict proof as to the identification of accused and proof of causal connection to the offence (Petrie, 1989).

The prosecution must show that the source of the pollution was a specific company or individual, or at least demonstrate that there is no reasonable likelihood of another source of the pollution. This can be demonstrated by the use of photographs showing the plume of smoke dispersing from the stack on the roof of the plant, by testimony of witnesses who saw effluent pouring from the drain pipe into a stream, by scientific methods of "fingerprinting" the chemical and matching that fingerprint to a product known to be used by the accused, or by eliminating other likely sources (Petrie, 1989).

Once the accused has been positively identified, a causal connection to the environmental damage must be made. Neighbours may testify that they could smell, taste or feel the pollutant settling on their properties. Experts may testify as to the probable health effects of the pollution. Alternatively, it may be sufficient to prove that the regulatory standards or performance standards in a C of A were exceeded. Types of technical defences, although rare, may include:

- improper or missing chain of custody evidence
- lapse of the limitation period (usually two years)
- officially induced error

- breach of the Canadian Charter of Rights and Freedoms (such as unreasonable search and seizure, involuntary admission, unreasonable delay in coming to trial)
- act of God

8.7.2   Quasi-Criminal Offences

Absolute liability offences must be specified as such in the statute. Absolute liability means that, once the Crown proves that the accused has committed the offence, the accused cannot raise a defence of due diligence. The defendant, subject to the Canadian Charter of Rights and Freedoms, may be free of fault and yet liable under an absolute liability offence (Petrie, 1989).

Most environmental statutes create strict liability offences. An example is in s. 283 of CEPA, 1999, which states that "[n]o person shall be found guilty of an offence under this Act . . . if . . . the person exercised all due diligence to prevent its commission."

To establish a defence based on due diligence, the onus of proof is on the defendant to establish that reasonable care was taken. What constitutes reasonable care evolves with time. In the 1999 case of *Queen v. Courtaulds Fibres Canada*, the judge ruled that:

> Reasonable care and due diligence do not mean superhuman efforts. They mean a high standard of awareness and decisive, prompt and continuing action. . . . The decision of how long a company must prove that it was duly diligent before it can become a valid defence, cannot be clearly answered by considering only how long the company was so engaged. The state of the facility, the age of the facility, the problems to be addressed, and the scope of the actions taken to deal with them, as well as the time the company was engaged in remedial action, must all be weighed and balanced.

The proposed AMP regulations illustrate in a generic way what the MOE considers (in 2002) to be "all reasonable steps" when evaluating the circumstances of contraventions to which an AMP may be issued:

1. There is an adequate system in place that would otherwise have prevented the contravention.
2. There is compliance with an industry standard that is otherwise adequate to protect the environment.
3. No other feasible alternatives are available.
4. It was not reasonable to foresee the contravention.
5. The party and its employees have sufficient skill and knowledge to engage in activities related to the contravention.
6. The appropriate degree of control was exercised over employees and agents at the time of the contravention.
7. Steps are taken promptly to rectify a contravention.
8. Steps are taken to prevent a continuation or reoccurrence of a contravention.

9. Steps are taken to reduce or eliminate damage caused by the contravention.
10. The person or organization responsible for the contravention co-operated with public authorities.

In cases where reasonable care does not afford a full defence of due diligence, the efforts may still be used as a mitigating factor in determining sentence. In the case of the proposed AMP regulations, the amount of an AMP can be reduced by up to 75% if the MOE director is satisfied that all reasonable steps were taken to prevent the contravention.

Section 99(3) of the EPA lists several possible defences for spills:

- due diligence
- an act of war, civil war, insurrection, terrorism or an act of hostility by the government of a foreign country
- a natural phenomenon of an exceptional, inevitable and irresistible character
- an act or omission with intent to cause harm by a person other than a person for whose wrongful act or omission the owner of the pollutant or the person having control of the pollutant is by law responsible

8.8   Role of Common Law

8.8.1   Origins

Section 8.5 (Responses to Violations) and Section 8.6 (Penalties and Liabilities) of this chapter cover environmental statutory law. In recent years, especially in the United States, there has been a significant increase in private lawsuits for pollution-related activities such as clean-up, contaminated property transfers and elevated exposures to nearby residents from various types of activities. However, the statutory context in the United States (notably the Comprehensive Environmental Response Compensation and Liability Act) is substantially different from that in Canada. In essence, fault does not have to be proven in the United States and all related parties must bear a share of the responsibility and financial burden.

Common law has its origins in the morals, traditions and business practices of England. The law has evolved over the centuries through the decisions of judges interpreting these norms and legislation that codified society's standards. Torts are part of common law. They are civil wrongs other than breach of contract. Courts may remedy a wrong by awarding damages or injunctive relief (prohibitive or mandatory orders) to compensate the injured property and prevent further wrongs. Tort causes of action may include negligence, nuisance, strict liability, trespass and riparian rights.

8.8.2   Negligence

Negligence occurs when a person's conduct falls below the standard of care regarded as reasonable among his peers. To prove negligence, one must show that the standard of reasonable care was breached, that the defendant owed a duty of care to the injured party, and that the defendant should have foreseen the damage that resulted.

8.8.3   Nuisance

Nuisance is the unreasonable interference with the use, comfort or enjoyment of property, or causing damage to another's property. Noise, odours, vibrations and actual physical intrusions of deleterious substances can create a cause of action in nuisance.

The law recognizes both private and public nuisances. Private nuisance involves the interference with an interest in land, such as the occupier's interest in the use and enjoyment of the land. Public nuisance is the interference with the public in its exercise of public rights. A nuisance against the public normally was pursued with the consent of the Attorney General, but the EBR registry has alleviated the need for this. An individual does not have to be negligent to be found liable in nuisance.

Nuisances can be continued or adopted by the subsequent owner of property who, with the knowledge or presumed knowledge of the nuisance, allows it to continue after a reasonable period of time available to stop the nuisance.

8.8.4   Strict Liability

A long-standing example of the role of strict liability in environmental matters is the 1868 decision of the British House of Lords in the case of *Rylands v. Fletcher*. The decision supported the proposition that a party who brings onto his land a non-natural use is answerable for the damages that the thing causes if it escapes or is discharged from that party's land. A water reservoir constructed on land that contained an abandoned mine shaft ruptured through the shaft flooding the neighbouring coal mine. The degree of care exercised by the defendant was not an issue. Liability was a direct consequence of proof of the physical damage resulting from the non-natural use of the land. Liability can be imposed even if there is no negligence.

8.8.5   Trespass

Trespass is a direct, unauthorized interference with private property. Liability for trespass can occur where there is no actual damage suffered by the person in possession of the property. The discharge of pollution onto another's property is a form of trespass.

The owner of land that borders on a watercourse has a riparian right to the continued flow of the water in its natural quantity and quality subject to the ordinary, reasonable use of owners upstream along the watercourse. The right to sue for damages and injunctive relief is created not only with diminished quantity or quality of water; one can also sue if surface water flow is altered so that another's property is flooded. The OWRA and its permit to take water provisions give statutory variations to this principle.

8.9   Ontario Environmental Bill of Rights

The Ontario Environmental Bill of Rights (EBR) was passed in 1994 to provide increased government accountability for the environment. Under the EBR, selected Ministries are required to:

- develop and implement "statements of environmental values" that outline how environmental considerations are integrated into decision-making process
- determine which of their laws, regulations and policies have an impact on the environment

The Act established the EBR environmental registry. The registry can be accessed by internet or telephone modem at **www.ene.gov.on.ca/envision/ebr**.

The EBR has provisions to increase protection to employees who blow the whistle on polluting employers. Also, the EBR is intended to increase public participation in environmental issues, by mechanisms that include:

- an electronic registry where any member of the public has access to information
- increased public participation in environmental decision-making by government (public notice is given in an electronic registry of proposed Acts, policies, regulations and other legal instruments that are environmentally significant to allow for public comment)
- improved access to the courts for Ontario residents to protect the environment

The Environmental Commissioner's Office was introduced in 1994 to serve as a conduit through which the public may lobby the government for change or introduce an investigation regarding an environmental matter. The Environmental Commissioner assists Ministries in screening public complaints to determine if a full investigation is warranted. The public may also issue, through the Environmental Commissioner, a request for a review of a Ministry's Acts, regulations, approvals or policies.

8.10   Summary

Several federal and provincial statutes empower enforcement staff to inspect sites, take copies of documents, collect samples, take tests, interview people, check equipment and containers, and take photographs. Some environmental legislation imposes obligations on individuals to co-operate with an investigation.

Environmental agencies can respond to violations in various ways. These range from simple warnings, to issuing tickets, to orders, to charges. If prosecuted and found guilty, substantial fines, imprisonment, or both can result. If an offender is found to have profited or benefited by committing a violation, it can be ordered to forfeit the profits.

References

Emond, D.P. 1985. "Environmental Law and Policy: A Retrospective Examination of the Canadian Experience," in I. Bernier and A. Lajoie, *Consumer Protection, Environmental Law, and Corporate Power*. University of Toronto Press.

Environment Canada. March 2001. *Compliance and Enforcement Policy for the Canadian Environmental Protection Act, 1999 (CEPA, 1999)*.

Executive Resource Group (ERG). January 2001. *Managing the Environment, A Review of Best Practices*.

Jetten, B. and B. Smith. November 1989. "When the Environmental Police Call, It Could Become a Corporate Nightmare." *Environmental Science and Engineering*.

McKenney, M. 1989. "Environmental Enforcement in Ontario." Presented at the Environmental Auditing Workshop, 10 and 11 October, University of Toronto.

Ontario Ministry of Environment and Energy (MOEE). June 1995. *Compliance Guideline F-2 (formerly Policy 05-02)*.

Petrie, P.D. 1989. "What Management and Staff Should Know about Environmental Laws to Minimize Personal and Corporate Liability." Presented at the Environmental Auditing Workshop, 10 and 11 October, University of Toronto.

Rovet, E. 1988. *Canadian Business Guide to Environmental Law*. Vancouver: International Self-Counsel Press Ltd.

Standing Committee on Environment and Sustainable Development. May 1998. *Enforcing Canada's Pollution Laws: The Public Interest Must Come First!*

Willms & Shier, Environmental Lawyers. 2002. "MOEE Search and Seizure Subject to Charter". Environmental Law – Municipal/Corporate Report. Spring.

Table 8.1

Penalties under CEPA, 1999

Type of Offence	Type of Conviction	Maximum Penalty
Contravention of the Act, the regulations, an order, direction or agreement	Indictment	$1 million or imprisonment for a term of not more than three years, or both
	Summary	$300,000 or imprisonment for a term of not more than six months, or both
Knowingly providing false or misleading information	Indictment	$1 million or imprisonment for a term of not more than three years, or both
	Summary	$300,000 or imprisonment for a term of not more than six months, or both
Negligently providing false or misleading information	Indictment	$500,000 or imprisonment for a term of not more than three years, or both
	Summary	$200,000 or imprisonment for a term of not more than six months, or both
Intentionally damaging the environment or causing a risk of death or harm to others	Indictment	a fine (not specified) or imprisonment for a term of not more than five years, or both and potential punishment under the Criminal Code

Note
Further details can be found in ss. 272-274 of CEPA, 1999.

Table 8.2
Penalties under TDGA, 1992

Type of Offence	Type of Conviction	Maximum Penalty
Contravention or non-compliance with TDGA, 1992	Indictment	imprisonment for a term of not more than two years
	Summary	$50,000 for a first offence $100,000 for a subsequent offence

Table 8.3
Penalties under the Fisheries Act

Type of Offence	Type of Conviction	Maximum Penalty
Harmful alteration or fish habitat	Indictment	$1 million for the first offence $1 million or imprisonment for a term of not more than three years or both for a subsequent offence
	Summary	$300,000 for a first offence $300,000 or imprisonment for a term of not more than six months or both for a subsequent offence
Throwing overboard prohibited substances or depositing a deleterious substance in water frequented by fish	Indictment	$1 million for the first offence $1 million or imprisonment for a term of not more than three years or both for a subsequent offence
	Summary	$300,000 for a first offence $300,000 or imprisonment for a term of not more than six months or both for a subsequent offence
Failure to provide requested information or make a required report, or failure to take reasonable measures required by the Act or failure to comply with a direction	Summary	$200,000 for a first offence $200,000 or imprisonment for a term of not more than six months or both for a subsequent offence

All other offences under the Act	Indictment	$500,000 for the first offence $500,000 or imprisonment for a term of not more than two years or both for a subsequent offence
	Summary	$100,000 for a first offence $100,000 or imprisonment for a term of not more than one year or both for a subsequent offence

Table 8.4
Penalties under the Ontario EPA

		Maximum Daily Penalty	
Offence	**Occurrence**	**Individual**	**Corporation**
Contravene any part of the Act or regulations	First Subsequent	$20,000 $50,000 + up to one year's imprisonment or both	$100,000 $200,000
Contravene the Act in a way that poses a risk of an adverse effect OR fail to comply with an order or C of A OR obstruct a provincial officer	First Second	$50,000 $100,000 + up to one year's imprisonment or both	$250,000 $500,000
Contravene the Act in a way that causes an adverse effect OR haul liquid industrial waste or hazardous waste in a way that may cause an adverse effect OR fail to comply with a stop order	First Second	$4,000,000 $6,000,000 + up to five year's imprisonment or both	$6,000,000 $10,000,000

Notes

For some violations, additional terms can include:
- penalties equal to the monetary benefit accrued due to the violation
- the court can order action to prevent or eliminate effects on the environment
- the court can order any other conditions it feels appropriate
- changes to conditions can be made as the court sees fit
- compliance even if in jail
- suspension of one or more existing or pending licences if fines are not paid

Table 8.5

Examples of Ticketable Offences under EPA

Regulation   **Activity**

Regulation 347 (Waste Management)
Operate a landfilling site:
- allow use by unauthorized persons
- allow access while attendant not on duty

Operate waste management system:
- vehicle valves not locked with driver absent
- fail to clearly mark vehicle
- fail to keep C of A in vehicle

Asbestos handling:
- permit asbestos waste to leave location in inadequate containers
- asbestos waste not covered with tarpaulin or net in unenclosed vehicle

Transporting waste:
- carrier - fail to promptly transport subject waste to receiving facility
- carrier - fail to section B of manifest
- carrier - fail to give manifest to generator at time of transfer
- generator - fail to complete section A of manifest
- generator - fail to retain copy 2 of manifest for two years

Regulation 346 (Air Pollution)
- cause or permit visible emission – obstruct passage of light more than 20%
- burn a fuel or waste for which a combustion unit is not designed
- emit contaminants beyond property limits from prescribed activities

Regulation 362 (PCB Management)
- fail to keep records of all PCB waste held
- fail to report required information to MOE in writing within three days

Table 8.6

Penalties under the OWRA

		Maximum Daily Penalty	
Offence	**Occurrence**	**Individual**	**Corporation**
Contravene any part of the Act or regulations	First Subsequent	$20,000 $50,000 + up to one year's imprisonment or both	$100,000 $200,000
Contravene the Act in a way that poses to impair the quality of any waters OR obstruct a provincial officer	First Second	$50,000 $100,000 + up to one year's imprisonment or both	$250,000 $500,000
Contravene the Act in a way that impairs the quality of any water OR contravene regulations that relate to water treatment or distribution systems	First Second	$4,000,000 $6,000,000 + up to five year's imprisonment or both	$6,000,000 $10,000,000

Notes

For some violations, additional terms can include:
- penalties equal to the monetary benefit accrued due to the violation
- the court can order action to provide an alternate water supply
- the court can order any other conditions it feels appropriate
- changes to conditions can be made as the court sees fit
- compliance even if in jail unless imprisonment makes it impossible to comply

Table 8.7

Penalties under the Technical Standards and Safety Act, 2000

	Maximum Daily Penalty	
Offence	**Individual**	**Corporation**
Contravene any part of the Act or regulations OR knowingly providing false information OR fails to comply with an order OR obstructs an inspector	$50,000 + up to one year's imprisonment or both	$1,000,000
Every director or officer of a corporation who fails to take all reasonable care	$50,000 + up to one year's imprisonment or both	

Table 8.8

Examples of Municipal Enforcement Strategy using the Model Sewer-use By-law

Sample Violations	Responses
Informal Actions	
Minor exceedance of discharge limit—one-time violation	Informal meeting
Minor exceedance of discharge limit on an infrequent basis	Warning letter
Failure to notify about a one-time minor spill into sewage works	Show cause meeting
Major exceedance of discharge limit—one-time violation	Show cause meeting
Formal Actions	
Exceeding of discharge limits on a regular basis	Prosecution
Failure to notify municipality of a major spill into the sewage works	Prosecution
Exceeding discharge limit with known damage to sewage treatment plant or failure to comply with compliance program	Order of prohibition
Exceeding discharge limits with known damage to sewage treatment plant on a regular basis	Order of prohibition
Failure to comply with order of prohibition to stop discharging sewage exceeding by-law limits and causing sewage treatment plant damage	Contempt of court

Table 8.9

Penalties for Improper Discharges to a Municipal Sewer System

	Individual	**Corporation**
First Offence	$5,000/day	$25,000/day
Subsequent Offence	$10,000/day	$50,000/day

Note

These same penalties appear in s. 92 of the Municipal Act, 2001, which will take effect in January, 2003.

CHAPTER 9

Environmental Fate

9.1 Why Be Concerned about Environmental Fate?

Anyone who deals with environmental issues can benefit from having a basic understanding of the factors that influence the fate of chemicals in the environment. An understanding of environmental fate can be applied to tasks as diverse as assessing treatment options, developing monitoring programs and estimating the potential risks during the production, use and disposal of a chemical.

Current and proposed legislation, including the Canadian Environmental Protection Act (CEPA, 1999), consider the impact of chemical releases on the environment. CEPA, 1999 requires a "cradle-to-grave" approach for managing chemicals.

A basic understanding of the environmental fate of a chemical can be used to assess the effectiveness of treatment options. For example, consider two water treatment options — an activated carbon column and air stripping — and two compounds of concern — benzo[a]pyrene and trichloroethylene (TCE). Because benzo[a]pyrene has a relatively high octanol-water partition coefficient and strongly adsorbs to carbon, the carbon column should be effective. Conversely, air stripping would not be expected to be cost-effective because of the compound's low vapour pressure. For TCE, which has a high vapour pressure and low affinity for carbon, air stripping is the preferred option.

To develop cost-effective monitoring systems, companies and environmental agencies must be able to predict the compartment of the environment where a chemical is most likely to be found. The compartments that usually are considered include air, water, soil, sediment and biota. For example, a decision must be made whether to sample sediment or water for PCBs near an old dump site. Because of its high affinity for sediment, the majority of the PCBs likely will be found in the sediment. Water column samples would be expected to show no detectable or very low concentrations.

Understanding the environmental behaviour and fate of a chemical is an essential component of a risk assessment. The behaviour of a chemical will strongly influence the extent to which exposure routes or pathways will occur. For example, if a sudden release occurs of a highly volatile compound, it may be necessary to evacuate the area downwind. If the chemical is soluble and could reach a nearby river or water supply, then appropriate interceptors (such as dykes or trenches) may be needed.

The federal Pesticide Control Products Act requires that exhaustive tests be conducted, both in the laboratory and field, prior to the introduction of a new pesticide to the marketplace. These tests are used to assess the environmental fate of a pesticide and determine if it can adversely affect the environment. CEPA, 1999 requires a similar approach for all new chemicals being produced or imported into Canada. One of the main objectives of CEPA, 1999 is to avoid the introduction of chemicals that may cause an adverse effect on the environment in the short and long term.

9.2   Key Chemical Properties

Every substance possesses a unique set of physico-chemical properties. While many such properties exist, the following five are important when determining environmental behaviour and fate.

1. **Vapour pressure** is a measure of the volatility of a chemical in its pure state. Vapour pressure indicates the maximum concentration that a chemical may achieve in the air compartment.

2. **Water solubility** is the maximum concentration of a chemical that dissolves in pure water. Highly soluble chemicals are easily and quickly distributed by the hydrologic cycle. These chemicals tend not to adsorb to soil or sediments and to bioconcentrate only slightly in organisms. Most tend to be biodegradable by micro-organisms in soil, surface water and sewage treatment plants.

3. **Density** can indicate how a chemical will move in the environment. For example, chemicals more dense than water will sink in aquatic environments or, if released into the subsurface, will move downward until encountering a relatively impermeable layer of soil or rock. Similarly, a release of dense vapours will follow the contours of the land and may gather in low-lying areas.

4. **Melting point** and **boiling point** determine whether a chemical is a gas, liquid or solid according to the temperature of the environment of concern.

5. **Molecular weight** (or molar mass) is important when stoichiometric relationships can be used to estimate various processes such as dissolution, precipitation and hydrolysis.

Useful compilations of key environmental parameters for many chemicals include Mackay *et al.* (2000) and Howard (1989, 1990, 1991, 1993, 1997). If values for environmental parameters cannot be found in references such as these, then parameter estimation methods described in Mackay and Boethling (2000) may be used.

9.3   Key Characteristics of Environments

When a chemical is introduced or released into the environment, that environment has numerous innate characteristics that will influence the behaviour of the chemical. Various examples are listed in Table 9.1.

Key characteristics of the air compartment pertain to the transport, diffusion and/or convection of a chemical in that compartment. The atmospheric stability class indicates whether a gaseous emission may readily rise and disperse or be trapped by an inversion condition.

In the surface water compartment, the behaviour of a chemical will be influenced by the flow velocity, flow rate and residence time (or turnover rate) of a water body. Fast-moving streams may promote uniform mixing within a short distance, whereas large lakes can stratify into several layers. The pH and cation/anion balance are useful in determining the precipitation and dissolution of ionic species.

The accumulation of a chemical in the sediment compartment is influenced by the rates of sediment deposition and resuspension. It is also influenced by the adsorptive capability, organic carbon content and porosity of the sediment.

In the soil compartment, the movement of a chemical that has been applied to soil or deposited on the surface can be influenced strongly by hydrological factors (runoff and rainfall), soil type (porosity and adsorption capability) and chemical properties of the soil, such as pH and organic carbon or organic matter content.

The movement of chemicals in the ground water compartment is influenced by soil type and flow conditions (hydraulic conductivity and gradient). It is also influenced by the recharge rate to the aquifer (the amount of water flowing into the aquifer).

9.4   Partition Coefficients

Partition coefficients describe the tendency of a chemical in the environment to transfer between environmental compartments such as from water to air or soil to air. Table 9.1 presents various equilibrium partition coefficients. Useful compilations of partition coefficients such as Henry's Law Constant and octanol-water partition coefficient for many chemicals include: Mackay *et al.* (2000); and Howard (1989, 1990, 1991, 1993, 1997). When values for partition coefficients cannot be found in references such as these, then estimation methods described in Mackay and Boethling (2000) may be used.

For non-equilibrium conditions, the partition coefficients described in Table 9.1 may be used in combination with other parameters such as kinetic rate constants, although the relationships can become mathematically complex. Usually the assumption of equilibrium conditions is valid for a first approximation of the fate of a chemical, especially if several months or years have passed since the chemical was introduced to the environment.

9.4.1   Henry's Law Constant

Henry's Law Constant or the air-water partition coefficient relates the vapour pressure of a chemical to its solubility in the water phase:

Equation 9.1

$$H = P \div S$$

where H = Henry's Law Constant ($Pa \bullet m^3/g$ or $Pa \bullet m^3/mol$)

P = vapour pressure (Pa)

S = solubility (g/m^3 or mol/m^3)

Values of Henry's Law Constant range from greater than 5 $Pa \bullet m^3/g$ for a volatile chemical with a low solubility to less than 10^{-6} $Pa \bullet m^3/g$ for a low-volatility chemical with a high solubility.

The air-water partition coefficient is defined as:

Equation 9.2

$$K_{aw} = H \div RT$$

where K_{aw} = air-water partition coefficient

R = 8.314 $Pa \bullet m^3/mol \bullet K$

T = temperature in K

9.4.2   Octanol-Water Partition Coefficient

The octanol-water partition coefficient is defined as the ratio of the concentration in octanol to the concentration in water. At low concentrations, many organic compounds are miscible in octanol so solubility cannot be measured. As a result, this coefficient often is expressed as:

Equation 9.3

$$K_{ow} = S_o \div S$$

where K_{ow} = octanol-water partition coefficient

S_o = solubility in octanol (i.e., an alcohol with eight carbons) (g/m^3)

S = solubility (g/m^3)

Scientists have found that octanol is a good surrogate for the lipid materials present in the fat portion of fish and organic matter in soils and sediments. Hence, K_{ow} is used to estimate the amount of an organic chemical that can be bioaccumulated by biota. The coefficient is also used in correlations to predict adsorption partition coefficients for organic chemicals in soils and sediments.

K_{ow} values range from 10^{-1} to 10^7 and are usually expressed as logarithms. A chemical with a log K_{ow} greater than 3 is considered to have a high affinity for lipids in biological media and the organic carbon portion of soil and sediment, while a chemical with a log K_{ow} value less than 1 is considered to have a higher affinity for water.

9.4.3   Organic Carbon Partition Coefficient

The organic carbon partition coefficient, or K_{oc} value, is a measure of the relative potential of organic chemicals to sorb to organic carbon. K_{oc} is closely related to K_{ow} (Karickhoff, 1981):

Equation 9.4

$$K_{oc} = 0.41 \times K_{ow}$$

where K_{oc} = organic carbon partition coefficient (L/kg)
        K_{ow} = octanol-water partition coefficient

The constant, 0.41, has units of L/kg.

Other empirical relationships have been developed for estimating K_{oc} using other properties, including solubility and bioconcentration factors. For example:

Equation 9.5

$$\log K_{oc} = a \times \log (S, K_{ow}, \text{ or BCF}) + b$$

where S   = solubility
        K_{ow} = octanol-water partition coefficient
        BCF   = bioconcentration factor
        a,b   = constants

Other correlations are available in the literature.

9.4.4   Sorption Distribution Coefficients

Sorption distribution coefficients are used for both organic and inorganic chemicals:

Equation 9.6

$$C_s = C_w \times K_d$$

where C_s = soil concentration
        C_w = aqueous phase concentration
        K_d = sorption distribution coefficient

For organic chemicals, the K_d value can be determined from:

Equation 9.7

$$K_d = K_{oc} \times f_{oc}$$

where K_d = adsorption distribution coefficient for an organic compound

K_{oc} = organic sorption partition coefficient

f_{oc} = fraction of organic carbon in the soil (g/g)

For inorganic parameters, no simple correlation exists for sorption, and hence literature K_d values must be sought. The K_d value selected should reflect the inorganic chemical as well as the type of soil and the pH of the soil.

In addition to sorption, solubility-controlled dissolution should be considered when assessing inorganic substances. To estimate the dissolution rate, additional information may be required, such as soil water pH, ionic balance, concentrations of other ionic species in ground water and solid phases present.

9.4.5　Bioconcentration Factor

The bioconcentration factor, or BCF, is used to estimate the concentration of the chemical in biotic phases (such as fish). For some chemicals, especially organic compounds, this concentration may be orders of magnitude higher than in the water phase. Several correlations have been developed for BCF. The following simple correlation is one of the more widely used ones for organic chemicals (Mackay, 1982a):

Equation 9.8

$$BCF = L \times K_{ow}$$

where BCF = bioconcentration factor

L = volume fraction lipid content of fish (typically about 0.05)

K_{ow} = octanol-water partition coefficient

Some compounds partition into fish very slowly. Hydrophobic chemicals have low solubilities and therefore a large volume of water must pass through the gills of the fish to accomplish the necessary transfer. For relatively large molecules, there may be additional resistance to bioconcentration due to transport resistance in the cell membranes (Mackay and Hughes, 1984).

Equation 9.8 does not take into account biomagnification effects that may be caused by fish feeding on other aquatic organisms, or the ability of fish to metabolically convert a substance. Rates of metabolism for fish are poorly documented. If fish are organisms of interest, consideration also must be given to the migratory nature of the fish.

9.4.6　Plant Uptake Factors

Plant uptake factors are used to estimate the transfer of substances from soil, water and air to plants. There are three ways for plants to take up a chemical in the environment: foliar deposition, root uptake and uptake of vapours via leaves. The first two mechanisms are considered to dominate most scenarios and various equations have been formulated to estimate uptake. For certain compounds (those which are not soluble, persistent and sparingly volatile), recent studies suggest that the uptake of vapours may be the major long-term uptake route.

One equation developed to estimate the uptake of compounds due to foliar deposition is (Hetrick and McDowell-Boyer, 1984):

Equation 9.9

$$K_{pf} = R/(Y\ W)\ [1 - e^{(-W\ t)}\ /(W\ t)\ D_p]$$

where K_{pf} = uptake factor due to foliar deposition (on a dry weight basis)
    R = initial fraction of material intercepted
    Y = vegetative productivity or yield
    W = weathering constant
    t = crop growth period
    D_p = deposition rate of particulate matter

A relatively simple equation for estimating root uptake of organic compounds in soil is (California Department of Toxic Substances Control, 2000):

Equation 9.10

$$K_{ps}\ (\text{roots})\quad = 273\ K_{ow}^{\ -0.622}$$

where K_{ps} (roots)   = root-soil partition coefficient for an organic (on a dry weight basis)

Uptake of organic vapours via leaves can be estimated as (Hiatt, 1999):

Equation 9.11

$$K_{pa} = 0.01\ K_{oa}$$

where K_{pa} = the plant-air partition coefficient
    K_{oa} = the octanol/air coefficient and can be estimated as
        $K_{ow}\ /\ K_{aw}$

Other equations as well as variations of Equations 9.9 and 9.10 have been developed; however, in general, the understanding of plant uptake is still limited. Examples of these modelling efforts can be found in Trapp *et al.* (1990), Paterson *et al.* (1991, 1994), Trapp and Matthies (1995), and Hung and Mackay (1997).

9.5   Reaction Rates and Half-Lives

9.5.1   Reaction Rates

An organic compound in the environment can be subjected to many types of reactions, most of which involve the degradation of the compound and therefore influence its concentration in the environment. The overall effect of the reaction rates often is referred to as a compound's persistence or environmental stability. Various reaction rates are identified in Table 9.1. Useful compilations of reaction rates and half-lives for many chemicals are included in Howard *et al.* (1991).

Hydrolysis is a chemical transformation process in which an organic molecule reacts with water, forming a new carbon-oxygen bond and often cleaving a carbon bond in the original molecule. Typically, the net reaction is the direct displacement of a component of the organic molecule by a hydroxyl ion (Harris, 1982). For example, acetic acid and ethanol are formed by the hydrolysis of ethyl acetate.

Photolysis is the degradation of a chemical due to exposure to light. Degradation typically results via the rupture of covalent bonds. The process is most relevant to chemicals in the atmosphere but also occurs in surface water and on the surface of soils.

Biodegradation involves the breakdown of organic or inorganic materials by organisms, most often micro-organisms. Bacterial metabolism alone accounts for 65% of the total metabolism of a soil community because of high bacterial biomass and metabolic rates (Scow, 1982). Biodegradation may occur in the presence of oxygen (aerobic conditions) or absence of oxygen (anaerobic conditions).

Chemical oxidation can involve the addition of oxygen, removal of hydrogen or the removal of electrons. A chemical that is responsible for oxidizing another is called an oxidizing agent or an oxidant. Examples of oxidizing agents include ozone (O_3) and potassium permanganate ($KMnO_4$).

9.5.2   Overall Rate Constants

In addition to specific reaction rate constants, total compartment rate constants are sometimes used to express the overall rate of loss of a chemical from a compartment. For example, the total soil loss rate constant for an organic chemical may include losses from volatilization, leaching, degradation and runoff.

9.5.3   Half-Lives

An alternative to the use of reaction rates or soil loss rate constants is the half-life, which can be calculated, once reaction rate constants (k values) are known, according to the equation:

Equation 9.12
$$T_{1/2} = \ln 2/k = 0.693/k$$

For each environmental compartment, a persistence half-life can be estimated and used to assess the chemical's fate. The use of half-lives when expressing persistence is sometimes easier to interpret. For example, a half-life of 10 days for toluene in soils following a spill may be easier to understand than a rate constant of 0.069 days^{-1}.

9.6 Using Environmental Fate to Evaluate Releases

9.6.1 Overview

The release of a chemical into the environment, even in a small quantity, may be disruptive or lead to adverse effects. A chemical may be transported by several pathways, reaching unsuspecting or non-target organisms (including humans) that may experience subtle and delayed effects (Mackay, 1982b). Such long-range transport is demonstrated by the presence of trace organic compounds at measurable concentrations in snow in remote regions of Canada (Gregor and Gummer, 1989).

Wherever chemicals have been released to the environment or a release is being considered, an understanding of environmental fate can be used to ensure that the relevant transport and transformation processes are assessed. The underlying objective of assessing environmental fate is to assist in estimating the exposures that humans and biota may experience and to determine how environmental concentrations of the chemical likely will change with time. This is particularly relevant in light of the environmental policy being advocated by growing numbers of regulatory agencies towards the virtual elimination of persistent and bioaccumulative toxic chemicals (also referred to as persistent organic pollutants or POPs, see Section 2.5).

9.6.2 Stack and Fugitive Emissions to the Atmosphere

The emission rate and behaviour of a chemical from a stack may be strongly influenced by the type of pollution control equipment used and its operating parameters, production rates and stack exit gas velocity. In recent years, there has been a shift towards longer residence times (more than two seconds) and higher temperatures (greater than 1000 °C) during the combustion of organic chemicals to ensure complete destruction.

In addition to stack emissions, fugitive emissions can be major sources of releases to the air. Fugitive emissions may result from leaks, storage tanks, vents, open windows near process areas and waste disposal and treatment site emissions.

Once a chemical is emitted to the atmosphere, it can be transported by turbulent mixing and convection to the surrounding area or even to distant locations. The distance a chemical can travel is a function of several parameters, including the release height, location of adjacent buildings, atmospheric stability, wind speed, deposition and prevailing long-range transport phenomena.

The processes that may act upon a chemical in the air compartment are illustrated in Figure 9.1. From the air compartment a chemical may be deposited on water, land or plants.

Transformation in the form of photochemical degradation or reaction with other compounds is also important for compounds in the atmosphere. An example of the latter is the formation of acidic precipitation from the interaction of water vapour with emissions of sulphur and nitrogen compounds.

9.6.3   Spills onto Water

Whenever a chemical is stored, used or transported near an aquatic environment, there exists the risk that a spill may occur and the chemical may reach a river, stream, lake, etc. Releases also can result from inadequate effluent treatment at a manufacturing plant, spillage during manufacturing and distribution, losses during transportation or leakage from disposal sites.

In 1989, the *Exxon Valdez* released approximately 42 million L of oil into the marine environment. When oil is spilled at sea, various processes act on it, most of which are influenced by the oil's properties. Early behaviour is dominated by the spreading tendency of the oil. The spreading process is very complex, involving unsteady-state transient behaviour of an oil of changing composition as material evaporates on a mobile water surface. The usual behaviour is for the oil to form slicks less than a millimetre thick surrounded by sheens only a few micrometres thick (Mackay *et al.*, 1983).

The oil will also drift and potentially form a viscous water-in-oil emulsion (mousse) that is difficult to pump. Eventually, wind and waves may shear the mousse into pieces referred to as "pancakes," which in turn break, form tar balls, and may wash ashore.

Evaporation is often an important source of loss for chemicals spilled in aquatic environments. The evaporation rate is a function of the chemical's vapour pressure, temperature and characteristics of the air above the chemical (primarily wind speed). In the case of an oil spill, the most soluble components are also the most volatile and hence will evaporate readily (Mackay *et al.*, 1983).

Other processes that will act on the chemical are illustrated in Figure 9.2. The chemical may dissolve into the water column resulting in the potential uptake by biota, sorption onto suspended solids or interaction with sediment. The material may undergo oxidation, hydrolysis and biodegradation in the water column and in the sediment. Oil drops will form by either breaking waves or a local surface convergence that may occur on steep waves. The smaller drops may be conveyed by eddy diffusion currents to depths in the water and become essentially permanently incorporated in the water column (Mackay *et al.*, 1983).

Spilled chemicals that have densities greater than water will sink to the bottom and may interact with sediment. These "sinkers" may become a source of contamination over an extended period of time as components gradually dissolve from the material and enter the water column.

9.6.4   Leaks from Underground Storage Tanks

Liquids released from an underground storage tank into unsaturated soil are subjected to numerous processes that influence its fate. Some can become sorbed onto soil particles. Some can become vapours. Some becomes trapped in soil void spaces. Some can be degraded by chemical or biological processes. The degrees to which these occur are influenced by factors such as the type of soil, the vertical distance from the release point to the surface and the vertical distance from the release point to the water table.

In general, gravity will draw the released liquid downward. The rate of migration will be faster in coarse textured soils than in fine textured soil. The liquid will continue to move downward as long as the amount released is more than the amounts removed by the other processes noted above. Where that occurs, and the liquid is present in the subsurface in essentially the same form as it was released, the liquid component often is referred to as non-aqueous phase liquid or NAPL. It may also be referred to as free product, or free phase liquid.

If the liquid is less dense then water, and it reaches the water table, it will accumulate on the surface of the water as a film, layer or pool. The NAPL can be carried laterally by the movement of the ground water. Some liquid may become dissolved and transported with the water. Some may become vapours and be dispersed in the unsaturated zone. Some may be degraded by chemical or biological processes. Typically, all of these conditions are occurring simultaneously.

As the water table moves up and down, it will carry the layer of floating liquid with it, creating a "smear zone" in the saturated soil. Whether the liquid travels vertically or horizontally, it may leave small, relatively immobile pockets of liquid behind.

If the liquid is more dense than water, it will not accumulate on the water table, but will sink or continue to move downward until it encounters a relatively impermeable barrier such as a layer of fine soil or rock. Depending on the amount of liquid that reaches such a barrier, blebs, blobs, puddles or pools of the liquid may accumulate. Liquids that can behave in this way often are referred to as dense, non-aqueous phase liquids or DNAPLs.

When DNAPL forms on an impermeable barrier, it then will migrate in whatever directions gravity can take it. This may even include opposite to the ground water flow direction. As ground water passes over or around the DNAPL, small amounts of the liquid will become dissolved and be carried with the ground water. In addition, some of the liquid may become vapours and be dispersed in the unsaturated zone. Some may be degraded by chemical or biological processes. Typically, all of these conditions are occurring simultaneously.

If the released liquid is relatively volatile, the vapours that volatilize into the unsaturated zone can migrate into outdoor air, into underground utility corridors or migrate into structures, particularly those with basements or sumps. The vapours can be inhaled by people who spend time at those locations. In confined spaces, there have been cases where vapours accumulated to explosive levels.

While releases of all types of liquids can prove to be challenging in terms of understanding their fate in the subsurface, there is more experience with liquids that are less dense than water. Good examples include gasoline and other petroleum products. The fates of liquids that are more dense than water pose more variables and are proving to be more difficult to understand, although important advances are occurring (Pankow and Cherry, 1996).

9.7　Environmental Fate Models

Numerous models have been developed to assess the environmental fate of contaminants. Many of the models focus on a specific environmental compartment or fate process. For example, Table 9.2 presents a sampling of models that address vapour transport from soils, chemical movement in both the unsaturated and saturated zones of soil and chemical fate in surface water environments.

The majority of the unsaturated zone models address the influences of runoff, infiltration and volatilization on chemicals applied to soil. The MINTEQA2 model (U.S. EPA, 1991; HydroGeoLogic Inc. and Allied Geoscience Consultants, 1999) predicts the aqueous speciation, adsorption, precipitation and dissolution of inorganic substances.

The saturated zone models presented in Table 9.2 simulate the movement of a chemical in the saturated zone under the influence of convection, dispersion, adsorption and reaction processes.

The surface water quality models address the tendency of a substance to become distributed among the various compartments of aquatic environments. These can include the water column, suspended material, biota, sediment and overlying air.

There are also models that can be used to assess environmental fate in a broader or more general context. Some of these types of models are included in Table 9.2.

References

Bennett D.H., T.E. McKone, M. Matthies and W.E. Kastenberg. 1998. "General formulation of characteristic travel distance for semi-volatile organic chemicals in a multimedia environment". *Environ. Sci. Technol.* 32: 4023-4030.

Bennett D.H., T.E. McKone and W.E. Kastenberg. 1999. "General formulation of characteristic time for persistent chemicals in a multimedia environment". *Environ. Sci. Technol.* 33: 503-509.

Beyer, A., D. Mackay, M. Matthies, F. Wania and E. Webster. 2000. "Assessing long-range transport potential of persistent organic pollutants". *Environ. Sci. Technol.* 34: 699-703.

California Department of Toxic Substances Control. 2000. "CalTOX Frequently Asked Scientific Questions". **www.dtsc.ca.gov/ScienceTechnology/faq.html**.

Gregor, D.J. and W.D. Gummer. 1989. "Evidence of Atmospheric Transport and Deposition of Organochlorine Pesticides and Polychlorinated Biphenyls in Canadian Arctic Snow". *Envir. Sci. and Tech.* 23 (No. 5): 561-565.

Groundwater Services, Inc. 1999. *RBCA Tool Kit - Atlantic Canada - Tier 2 RBCA Tool Kit.*

Harris, J.C. 1982. "Rate of Hydrolysis". In *Handbook of Chemical Property Estimation Methods.* W.J. Lyman, W.F. Reehl and D.H. Rosenblatt, eds. New York: McGraw-Hill Book Co.

Hetrick, D.M. and L.M. McDowell-Boyer. 1984. *User's Manual for TOX-SCREEN: A Multimedia Screening-Level Program for Assessing the Potential Fate of Chemicals Released to the Environment.* Prepared by the Oak Ridge National Laboratory for the United States Environmental Protection Agency Office of Toxic Substances, EPA Report 560/5-83-024.

Hiatt, M.N. 1999. "Leaves as an Indicator of Exposure to Airborne Volatile Organic Compounds". *Environ. Sci. Technol.* 33: 4126-4133.

Howard, P.H. 1989. *Handbook of Environmental Fate and Exposure Data for Organic Chemicals.* Volume I. Large Production and Priority Pollutants. Lewis Publishers. Boca Raton, Florida.

Howard, P.H. 1990. *Handbook of Environmental Fate and Exposure Data for Organic Chemicals.* Volume II. Solvents. Lewis Publishers. Boca Raton, Florida.

Howard, P.H. 1991. *Handbook of Environmental Fate and Exposure Data for Organic Chemicals.* Volume III. Pesticides. Lewis Publishers. Boca Raton, Florida.

Howard, P.H. 1993. *Handbook of Environmental Fate and Exposure Data for Organic Chemicals.* Volume IV. Solvents 2. Lewis Publishers. Boca Raton, Florida.

Howard, P.H. 1997. *Handbook of Environmental Fate and Exposure Data for Organic Chemicals.* Volume V. Solvents 3. Lewis Publishers. Boca Raton, Florida.

Howard, P.H., R.S. Boethling, W.F. Jarvis, W.M. Meylan and E.M. Michalenko. 1991. *Handbook of Environmental Degradation Rates.* Lewis Publishers. Boca Raton, Florida.

Hung, H. and D. Mackay. 1997. "A novel and simple model of the uptake of organic chemicals by vegetation from air and soil". *Chemosphere* 35: 959-977.

HydroGeoLogic Inc. and Allied Geoscience Consultants, Inc. 1999. *MINTEQA2 and PRODEFA2, A Geochemical Assessment Model for Environmental Systems: User Manual Supplement for Version 4.0.* Prepared for United States Environmental Protection Agency.

Javandel, I.C., C. Doughty and C.F. Tsang. 1984. *Groundwater Transport: Handbook of Mathematical Models.* Water Resources Monograph Series 10, American Geophysical Union Washington D.C.

Jury, W.A., W.F. Spencer and W.J. Farmer. 1983. "Behavior Assessment Model for Trace Organics in Soil: I Model Description". J. Environ. Qual. 12: 558.

Karickhoff, S.W. 1981. "Semiempirical Estimation of Sorption of Hydrophobic Pollutants on Natural Sediments and Soils". *Chemosphere* 10: 833-849.

Kincaid, G.T. and P.J. Mitchell. 1986. *Review of Multiphase Flow and Pollutant Transport Models for the Hanford Site.* Pacific Northwest Laboratory, PNL-6048.

Mackay, D. 1982a. "Correlation of Bioconcentration Factors". Environ. Sci. Technol. 16: 274-278.

Mackay, D. 1982b. "Nature and Origin of Micropollutants". Wat. Sci. Tech. 14: 5-14.

Mackay, D. 2001. *Multimedia Environmental Models: The Fugacity Approach.* 2nd ed. Lewis Publishers, Boca Raton, Florida.

Mackay, D. and R.S. Boethling. (eds.) 2000. *Handbook of Property Estimation Methods for Chemicals: Environmental Health and Sciences.* CRC Press.

Mackay, D. and A.I. Hughes. 1984. "Three-Parameter Equation Describing the Uptake of Organic Compounds by Fish." Environ. Sci. and Technol. 18:6: 439-444.

Mackay, D., S. Paterson, B. Cheung and W.B. Neely. 1985. "Evaluating the Environmental Behavior of Chemicals with a Level III Fugacity Model". *Chemosphere* 14 (No. 3/4): 335-374.

Mackay D., S. Paterson, G. Kicsi, C.E. Cowan, A. Di Guardo and D.M. Kane. 1996a. "Assessment of Chemical Fate in the Environment Using Evaluative, Regional and Local-Scale Models: Illustrative Application to Chlorobenzene and Linear Alkylbenzene Sulfonates". Environ. Toxicol. Chem. 15: 1638-1648.

Mackay D., S. Paterson, G. Kicsi, A. DiGuardo and C.E. Cowan. 1996b. "Assessing the fate of new and existing chemicals: A five stage process". Environ. Toxicol. Chem. 15: 1618-1626.

Mackay D., S. Paterson, A. DiGuardo and C.E. Cowan. 1996c. "Evaluating the environmental fate of a variety of types of chemicals using the EQC model". Environ. Toxicol. Chem. 15: 1627-1637.

Mackay D., S. Paterson, G. Kicsi, C.E. Cowan, A. DiGuardo and D.M. Kane. 1996d. "Assessment of chemical fate in the environment using evaluative, regional and local-scale models: Illustrative application to chlorobenzene and linear alkyl-benzene sulfonates". Environ. Toxicol. Chem. 15: 1638-1648.

Mackay, D., W.Y. Shiu and K.C. Ma. 2000. *Physical-Chemical Properties and Environmental Fate Handbook*. CRCnetBase 2000, Boca Raton, Florida. [This is a compilation of five volumes of physical-chemical properties and environmental fate information published by the authors from 1991 through 1997 by the same publisher.]

Mackay D., W. Stiver and P.A. Tebeau. 1983. "Testing of Crude Oils and Petroleum Products for Environmental Purposes". Proceedings 1983 Oil Spill Conference, San Antonio, TX. American Petroleum Institute, Washington, D.C.: 331-337.

McKone T.E. 1993. *CalTox, a multi-media total-exposure model for hazardous waste sites part II the dynamic multi-media transport and transformation model*. A report prepared for the State of California, Department of Toxic Substances Control, by the Lawrence Livermore National Laboratory. No. UCRL-CR-111456PtII. Livermore, CA.

Pankow, J.F. and J.A.Cherry. 1996. *Dense Chlorinated Solvents and Other DNAPLs in Groundwater*. Waterloo Press.

Paterson, S., D. Mackay, E. Bacci and D. Calamari. 1991. "Correlation of the equilibria and kinetics of leaf-air exchange of hydrophobic organic chemicals by leaves". *Environ. Sci. Technol.* 25: 866-871.

Paterson S., D. Mackay and C. McFarlane. 1994. "A model of organic chemical uptake by plants from soil and the atmosphere". *Environ. Sci. Technol.* 28: 2259-2266.

Scheringer, M. 1996. "Persistence and spatial range as endpoints of an exposure-based assessment of organic chemicals". *Environ. Sci. Technol.* 30: 1652-1659.

Scheringer, M. 1997. "Characterization of the environmental distribution behavior of organic chemicals by means of persistence and spatial range". *Environ. Sci. Technol.* 31: 2891-2897.

Scow, K.M. 1982. "Rate of Biodegradation". In *Handbook of Chemical Property Estimation Methods*. W.J. Lyman, W.F. Reehl and D.H. Rosenblatt, eds. New York: McGraw-Hill Book Co.

Short, T.E. 1985. "Mathematical Modeling of Land Treatment Processes". Invited presentation to the National Specialty Conference on Land Treatment, University of Texas, Austin TX, 16 to 18 April.

Trapp S. and M. Matthies. 1995. "Generic one-compartment model for uptake of organic chemicals by foliar vegetation". *Environ. Sci. Technol.* 29:2333-2338.

Trapp, S., M. Matthies, I. Scheunert and E.M. Topp. 1990. "Modelling the bioconcentration of organic chemicals in plants". *Environ. Sci. Technol.* 24: 1246-1255.

United States Environmental Protection Agency (U.S. EPA). 1991. *MINTEQA2 and PRODEFA2, A Geochemical Assessment Model for Environmental Systems: Version 3.0 User's Manual.* EPA/600/3-91/021.

United States Environmental Protection Agency (U.S. EPA). 2000. *Exposure Analysis Modeling System (EXAMS): User Manual and System Documentation*. EPA/600/R-00/081.

United States Environmental Protection Agency (U.S. EPA). 2001. *Center for Exposure Assessment Modeling - PRZM3*. **http://www.epa.gov/ceampubl/**.

Wania F., D. Mackay, Y.-F. Li, T. F. Bidleman and A. Strand. 1999a. "Global chemical fate of α-hexachlorocyclohexane. 1. Evaluation of a global distribution model". *Environ. Toxicol. Chem.* 18: 1390-1399.

Wania F. and D. Mackay. 1999b. "Global chemical fate of α-hexachlorocyclohexane. 2. Use of a global distribution model for mass balancing, source apportionment, and trend prediction". *Environ. Toxicol. Chem.* 18:1400-1407.

Table 9.1

Key Properties and Parameters that Influence Environmental Fate

---

Chemical Parameters
- vapour pressure
- density
- molecular weight
- solubility
- melting point and boiling point

Reaction Rates
- hydrolysis
- biodegradation
- reaction with other chemicals
- photolysis
- chemical oxidation

Partition Coefficients
- Henry's Law constant (H)
- bioconcentration factor (BCF)
- adsorption coefficient (K_d)
- octanol-water partition coefficient (K_{ow})
- organic carbon partition coefficient (K_{oc})
- plant uptake factor (K_p)

Physical Properties of the Environment

Air
- wind velocity and direction
- physical topography
- temperature and rainfall
- atmospheric stability class

Water
- pH and temperature
- concentration of suspended sediment
- water velocity and flow
- residence time of water
- cation/anion balance

Sediment
- oxygen concentration
- resuspension rate
- porosity
- organic carbon content
- deposition rate

Soil
- pH and temperature
- runoff
- density and porosity
- surface vegetation
- rainfall
- organic carbon content
- soil water content
- solid phases present (*e.g.*, carbonate)

Ground water
- organic carbon content
- hydraulic conductivity
- dispersion
- infiltration rate
- aquifer soil type (density, porosity)
- gradient
- pH and cation/anion balance

Table 9.2
Selected Environmental Fate Models

Volatilization from Soil

Mackay, 2001 - Soil model allows a simple assessment of the relative potential for degrading reaction, evaporation and leaching of a pesticide applied to a surface; simplified version of the Jury *et al.* (1983) model (**www.trentu.ca/academic/aminss/envmodel/Soil.html**).

Unsaturated Zone

United States Environmental Protection Agency (U.S. EPA), 1991; HydroGeoLogic Inc. and Allied Geoscience Consultants, 1999 - Metal Speciation Equilibrium Model (MINTEQA2) is a thermodynamic equilibrium model for aqueous speciation, adsorption and precipitation dissolution of solid phases (**www.epa.gov/ceampubl/minteq/index.htm**).

United States Environmental Protection Agency (U.S. EPA), 2000 - Pesticide Root Zone Model (PRZM3) is a one-dimensional model that estimates runoff, erosion, plant uptake, leaching, decay, foliar wash-off and volatilization (**www.epa.gov/ceampubl/gwater/przm3/index.html**).

Short, 1985 - Regulatory and Investigative Treatment Zone Model (RITZ) estimates movement of chemicals following land treatment of oily wastes (**www.epa.gov/ada/csmos/models/ritz.html**).

Saturated Zone

Yeh and Ward, 1981[*] - Finite Element Model of Waste Transport (FEMWASTE) simulates the transport of dissolved constituents in two-dimensional ground water (unsaturated/saturated zones) (**www-rsicc. ornl.gov/codes/ccc/ccc4/ccc-451.html**).

Prickett, Naymik and Lonnquist, 1981[*] - RANDOM WALK + can be used to evaluate one- or two-dimensional flow and solute transport (**www.mines.edu/research/igwmc/software/igwmcsoft**).

Voss, 1984[**] - SUTRA + simulates two-dimensional, transient, or unsteady state, saturated or unsaturated, transport of energy or chemically reactive single species (**http://water.usgs.gov/software/sutra.html**).

Surface Water

United States Environmental Protection Agency (U.S. EPA), 2000 - EXAMS (Exposure Analysis Modeling System) estimates chemical fate in various types of aquatic environments including lakes, rivers, and estuaries; has steady-state and non-steady-state options (**www.epa.gov/ceampubl/swater/exams/index.htm**).

Multi-media Models

Mackay *et al.*, 1985 - Fugacity Level III is a steady-state, non-equilibrium chemical fate model; fugacity is a thermodynamic concept that represents the tendency of a chemical to escape from one environmental compartment to another (**http://www.trentu.ca/academic/aminss/ envmodel/VBL3.html**).

Mackay *et al.*, 1996a - ChemCAN is a Level III model of 24 regions of Canada which was written for Health Canada. It estimates average concentrations in air, fresh surface water, fish, sediments, soils, vegetation, and marine near-shore waters (**http://www.trentu.ca/academic/ aminss/envmodel/ChemCAN.html**).

Mackay *et al.*, 1996b,c,d - EQC uses physical-chemical properties to quantify a chemical's behaviour in an evaluative environment using Level I, II or III models. Levels I and II assume thermodynamic equilibrium is achieved; Level II includes advective and reaction processes. Level III is a non-equilibrium, steady state assessment (**http://www.trentu.ca/academic/aminss/envmodel/EQC.html**).

McKone, 1993 - CalTox is a Level III model developed in California that also calculates exposure to humans (**http://www.dtsc.ca.gov**).

Groundwater Services, Inc., 1999 - RBCA Toolkit for Atlantic Canada is an Excel-based model that incorporates the approach and calculation methods described in the American Society for Testing and Materials (ASTM) Standard Guide for Risk-Based Corrective Action (ASTM ES 38) (**http://www.environmental-center.com/software/gsi/canada.htm**).

Long-Range Transport Models

Beyer *et al.*, 2000 - TaPL3 model is an evaluative tool used in the detailed assessment of chemicals for persistence and potential for long-range transport in a mobile medium, either air or water, in a Level III (steady state environment) (**http://www.trentu.ca/academic/ aminss/envmodel/TaPL3.html**).

Wania *et al.*, 1999a,b - GloboPOP describes the global fate of persistent organic pollutants on a time scale of decades (**http://www.scar.utoronto.ca/~wania**).

Scheringer *et al.*, 1996, 1997 - Chemrange and SCHE are circular multi-box models for calculating persistence and long-range transport **http://ltcmail. ethz.ch/hungerb/research/product/chemrange.html, http://ltcmail.ethz.ch/ scheri**).

Bennett *et al.*, 1998, 1999 - BENNX is a model used in the assessment of chemicals for persistence and potential for long-range transport.

Notes

*   Reference: Javendel *et al.*, 1984
**   Reference: Kincaid and Mitchell, 1986

Figure 9.1

Atmospheric Transport and Transformation Processes

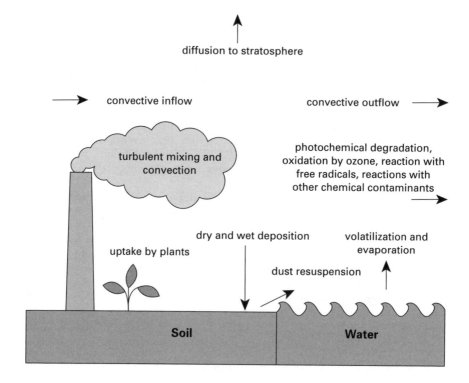

Figure 9.2

Transport and Transformation of Chemicals in Water

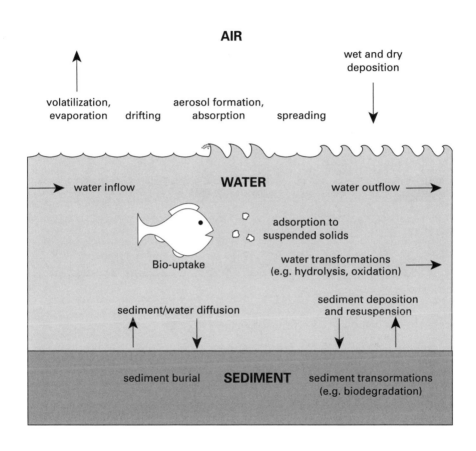

Figure 9.3

Leakage from an Underground Storage Tank

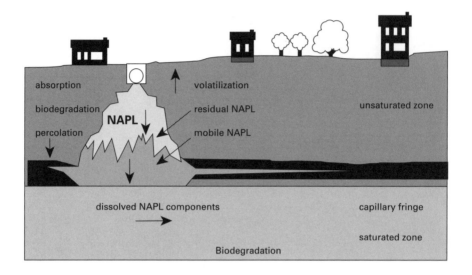

CHAPTER 10

Environmental Audits

10.1   What Is an Environmental Audit?

An environmental audit is a systematic and verifiable process of objectively obtaining and evaluating information (sometimes referred to as evidence) to determine whether or not an organization is conforming to the environmental requirements that may be set by regulatory agencies, other standard-setting agencies or even the organization itself.

In addition to assessing compliance, organizations may use environmental audits to identify ways to improve compliance, prioritize compliance issues and demonstrate reasonable care.

Important advances have occurred in environmental auditing in Canada over the last few years, largely as a result of organizations such as the Canadian Standards Association (CSA), the Canadian Environmental Auditing Association (CEAA), and the International Organisation for Standardisation (ISO). The programs of these organizations provide principles and practices that ensure auditor objectivity and professional competence and care, and ensure that well-defined and systematic procedures are being used.

Several guidelines or standards for environmental auditing have been published that should be consulted prior to conducting environmental audits in Ontario. Some examples are:

1. CSA has published several guidelines that pertain to environmental auditing. A contributing member to ISO initiatives in this area, CSA has adopted several ISO standards and published them as CSA standards. More information can be obtained at **www.csa.ca**.

2. ISO is the main source of environmental auditing standards for many countries. Many of its key guidelines such as 14001 (environmental management systems), 14010 (general principles for environmental auditing), 14011 (procedures for environmental auditing), and 14012 (qualification criteria for auditors) are cited in this chapter. ISO currently is working on Guideline 19011 which will replace 14010, 14011, 14012 and 10011 (guideline for auditing quality systems). That guideline is scheduled for release at the end of 2002.

3. The European Community (EC) Eco-Audit Scheme is directed at satisfying an EC regulation that requires the public reporting of environmental performance information.

The ISO 14001 standard requires an organization to have an Environmental Management System (EMS) program and procedures that clearly define the scope and frequency of audits, responsibilities for managing and conducting audits, communication of audit results, auditor competence, and information on how the audits will be conducted. The audit program should be based on the environmental importance of the activity concerned and the results of previous audits (ISO, 1996a).

The EC Eco-Audit regulation requires organizations to provide environmental performance information to the public via an environmental statement. This statement has to be validated by an environmental verifier. Participating companies are required to implement an Environmental Protection System (EPS) and undertake environmental audits. The EPS must include effective management control of practices relevant to the environment. The environmental audits are to involve a comprehensive analysis of the environmental issues and impacts and performance relating to a site. Following the audit, an environmental statement regarding the performance of a company's environmental activities must be made public. The audit methodology and procedures must be undertaken in accordance with the guidelines set out in ISO 10011 and the principles and requirements of the EC regulation.

The certification of environmental auditors has become more formalized over the past few years. The CEAA, founded in 1994, is dedicated to the continual improvement and development of the practice of environmental auditing in Canada. As part of its mandate, CEAA developed certification programs for Canadian environmental auditors. The first auditors were certified in 1996.

Certification and re-certification (necessary every three years) are based on an assessment of criteria in the areas of education, work experience, formal training and audit experience. Additional information can be obtained at **www.ceaa-acve.ca**.

10.2 Pre-Audit Decisions and Considerations

10.2.1 Types of Audits

Compliance Audit

While environmental audits are performed for various reasons, they can be grouped into the two broad categories of compliance audits and EMS audits.

A compliance audit is defined as the systematic and verified (by documentation) process of objectively obtaining and evaluating audit evidence to determine whether an organization's operations and activities comply with the requirements set out in environmental legislation, and then communicating the results of this process to the client. The regulatory requirements can include federal and provincial acts, regulations, standards, criteria, permits, orders and guidelines, as well as municipal by-laws.

Failure to comply with regulatory requirements may result in control or stop orders, revocation of permits, court injunctions, fines, and even criminal prosecution of company personnel, including corporate officials. The definition of failure to comply has recently been broadened to include failure to notify regulatory agencies of events or conditions. Regulatory reporting requirements have increased in recent years. For example, the Environmental Protection Act (EPA) and the Ontario Water Resources Act (OWRA) contain extensive reporting requirements. Failure to notify agencies forthwith of an environmental incident (such as a spill or unscheduled release) has become one of the most frequently cited compliance violations.

A compliance audit also may consider self-imposed requirements. For example, a company may want to establish its own environmental objectives or may strive to comply with a code of practice established by an industrial association.

Environmental Management System Audit

An EMS audit is defined as the systematic and verifiable (by documentation) process of objectively obtaining and evaluating audit evidence to determine whether an organization's EMS conforms to the environmental management system audit criteria, and communicating the results of this process to the client (ISO, 1996c). Such an audit consists of a critical review of a facility's EMS and so is broader in scope than a compliance audit. ISO 14011 clearly defines the goals of an EMS audit and the procedures to be followed.

Neither definition specifically includes environmental site assessments (ESAs), although ESAs often are described as a type of investigation that falls under the broader definition of environmental audit. (See Section 5.4 for more information about ESAs.) Nevertheless, ISO 14011 does stipulate that an audit should examine the basis of relevant sampling programs and the procedures for ensuring effective quality control of sampling and measurement processes. These are ESA activities.

10.2.2   Setting Objectives, Scope and Boundary Conditions

The objectives, scope and boundary conditions of an audit will depend upon who is requesting the audit and reasons why the audit is to be undertaken. Uncertainties in the scope of the audit may result in large variances in quotations to perform the work, or the writing of an audit report that does not respond to the objectives of the client.

Objectives will normally make reference to the types of audit criteria to be employed.

For a compliance audit, the following examples of criteria apply:

- determine conformance of auditee's activities, products or services to criteria derived from environmental legislation
- determine conformance of auditee's activities, products or services to audit criteria derived from company's internal standards or objectives on environmental performance, or from appropriate industrial codes of practices

For an EMS audit, the following criteria are taken from s. 4.1 of ISO 14011:

- determine conformance of auditee's EMS to EMS audit criteria
- determine whether auditee's EMS has been properly implemented and maintained
- identify areas of potential improvement in the auditee's EMS
- assess the ability of the internal management review process to ensure the continuing suitability and effectiveness of the EMS
- evaluate EMS of an organization where there is a desire to establish a contractual relationship

The scope of the audit should define the activities required to meet the audit objectives and summarize the boundary conditions. It is important that the initial audit scope and any changes be agreed upon by the auditor and auditee.

Boundary conditions that may be included in the scope of the audit include the following:

- physical location and organizational activities
- time period (that is, current and/or past operations)
- assessment of off-site impacts (for example, air dispersion modelling)
- on-site subcontractors/lessees
- interaction with government agencies
- define the extent of environmental legislation to be considered – for example, only items that have force of law such as Acts, regulations and standards, etc. or also items that do not have force of law such as government policies and guidelines
- involvement of third parties
- inclusion/exclusion of recommendations in the audit report

The current approach to auditing as indicated by ISO 14011 is that an EMS audit includes a report on findings and conclusions with references to supporting evidence. Recommendations are not included. Most compliance audit reports include an action plan, however, this needs to be discussed with the client and defined by the scope of the work or in the request for proposal.

It is important that the relevant boundary conditions be clearly documented.

While the scope, objectives and procedures employed in environmental audits vary widely, there also are some common, prerequisite elements that need to be in place to demonstrate the benefits that an environmental audit can provide. Regardless of whether the audit is to be performed by outside auditors or by in-house staff, the following conditions should be in place:

- the audit is supported by the company's senior management
- the audit team is given the authority, resources and training to conduct a thorough and proper job
- the auditors receive good co-operation from staff at the facility to be audited
- the company's management is prepared to respond to any deficiencies identified during the audit

10.2.3   Audit Confidentiality

Two issues related to confidentiality can arise during an environmental audit. The first is related to confidentiality between an auditor and the company. Sensitive information collected during an audit may affect a company's competitiveness and reputation if passed on to competitors, members of the financial community or the press. Contract conditions can be used to restrict the release of information collected during the audit.

Unless required by law, auditors should not disclose information or documents obtained during the audit, or the final report, to any third party, without the express approval of the client and, where appropriate, the approval of the auditee (ISO, 1996c).

The second issue concerns the accessibility of audit reports by government agencies or through court-directed disclosure. Both industry and regulators continue to struggle with this issue.

Environment Canada recognizes that environmental audits are an effective management tool for assessing environmental compliance, and intends to promote their use. To encourage the practice of environmental auditing, enforcement officers and analysts will not request environmental audit reports during routine inspections to verify compliance with the Canadian Environmental Protection Act, 1999 (CEPA, 1999) (Environment Canada, 2001).

Access to environmental audit reports may be required when enforcement officers have reasonable grounds to believe that:

- an offence has been committed;
- the audit findings will be relevant to a particular violation, necessary to its investigation, and required as evidence; or
- the information being sought cannot be obtained from other sources through the exercise of the enforcement officer's powers.

Environment Canada has stated that an environmental audit must not be used to shelter monitoring, compliance or other information that would otherwise be accessible to an enforcement officer or analyst under CEPA, 1999. In addition, the demand for access to an environmental audit during an investigation will be made under the authority of a search warrant. An exception can be made when the delay necessary in obtaining a warrant would likely result in danger to the environment or human health, or the loss or destruction of evidence (Environment Canada, 2001).

The MOE document *Policy and Guideline on Access to Environmental Evaluations* (MOEE, 1995) states that the MOE may request voluntary disclosure of evaluations (including environmental audits) during investigations. If a request is denied, a judicial inspection order or a search warrant is required. The government may only obtain environmental evaluations without an order or warrant during emergencies involving a serious risk to human health or the environment. The MOE has indicated that it will not, as a matter of course, request evaluations and will only seek access under a specified set of conditions. The policy (also referred to as Guideline H-9) is available at **www.ene.gov.on.ca.envision/gp/hp**.

Prior to initiating an audit, a client should discuss the issue of confidentiality with legal counsel. Two schools of thought on the confidentiality of environmental audit revolve around the concepts of attorney-client privilege and due diligence.

Solicitor-Client Privilege

Several companies use "solicitor-client" privilege when conducting an environmental audit and restricting access to the audit report. As noted in Section 8.3.5, in most cases, information that passes between a lawyer and client is privileged if it is secured for the purpose of obtaining legal advice, created in contemplation of litigation, or is otherwise kept confidential. An outside consultant and/or corporate staff can assist the lawyer in the preparation of an audit report without jeopardizing the solicitor-client privilege.

Counsel responsible for an audit should develop procedures for maintaining confidentiality and ensure that the procedures are followed by all participants. One disadvantage of this approach is that it may deny access to the report by individuals who will implement the corrective actions. This may be partially addressed by releasing a separate, corrective action plan that focuses on the necessary improvements without describing the potential or actual non-compliance.

Due Diligence

A second option is the due diligence approach that promotes the flow of information from an auditing program. The argument put forward is that the improved communications on potential issues outweigh any negative concerns associated with disclosure. Advocates of this approach argue that a defence of due diligence or reasonable care based on an active audit program, which critically and impartially identifies and corrects environmental deficiencies, may be the best defence.

10.3   Pre-visit Activities

10.3.1   Selecting the Audit Team

It is essential that auditors be objective, independent and competent. Independence can be defined as being free from bias and conflict of interest. If in-house audit-team members are used, they should not be accountable to those directly responsible for the subject-matter being audited (ISO, 1996b). If audits are to be performed on a regular basis, it may be appropriate to select and train several individuals from different facilities to audit their counterparts.

Although it is not mandatory in Canada, at least one of the audit team members should be a certified auditor (see Section 10.1). If the audit is a component of a company's ISO 14001 program or a registration/maintenance audit by a registrar (that is, a third party that confirms the company's conformance to ISO 14001 standard) all audit team members must be certified.

As stipulated in both the CEAA and ISO auditor documentation, auditors should use the care, diligence, skill and judgment expected of any auditor in similar circumstances and follow quality assurance procedures (see Section 10.3.3).

The use of external auditors, especially for a first audit, can be beneficial. External auditors can assist in training internal audit staff. It may also be appropriate to have external consultants perform periodic audits to assess the adequacy of the internal audit program depending on the degree of potential non-compliance and regulatory changes.

Audit team members should be selected based upon the scope and type of audit to be undertaken. Audit team members should include professionals with appropriate specialist skills. These skills may include:

- environmental affairs and policy
- regulatory affairs expertise
- management systems expertise
- specific environmental and engineering expertise
- scientific expertise such as biologists, chemists, toxicologists or hydrogeologists
- familiarity with operations

Collectively, the team must understand the operations of the facility or company from a technical perspective and be familiar with the audit criteria to be used. Note that for ISO 14001 certification, an organization must commit, in its policy, to comply with relevant environmental legislation and put in place a procedure to identify and have access to said legislation.

The responsibilities of each member of the team must be clearly established. One of the members of the audit team must be appointed as lead auditor. Ideally, the lead auditor will be certified as per CEAA (see Section 10.1), and be familiar with the activities performed at the facility.

When organizing the audit team, the availability of team members for the required time period must be unequivocal. The amount of time required to audit a facility will depend on the objective of the audit, the size of the facility and its complexity, as well as the number of people on the audit team. A two-person team may need two to four days to audit a small- or medium-sized chemical plant, while a four-member team may require two to five days for a larger facility.

Audits initiated by regulatory agencies tend to be performed by government staff or hired consultants. In Ontario, the MOE on occasion has designated consultants as provincial officers during the course of the audit, thereby providing the auditors greater access to records. Steps that can be taken when a facility is the subject of an external audit are discussed in Section 10.6.

10.3.2   Identification of Criteria

Criteria must be employed to assess the degree of compliance, performance or liability. Depending on the type of audit to be undertaken, the criteria may include government legislation, company policies, practices, standards and procedures and/or industrial codes of practices against which the auditor compares collected audit evidence about the subject matter.

Compliance audit – To establish criteria for a compliance audit, the auditor will rely on federal and provincial statutes, regulations, and standards, as well as municipal by-laws. If the scope of the audit extends beyond legislation, other sources of criteria will include policies, guidelines and objectives. Care should be taken to ensure that the current versions of legislation are used, since regulatory requirements frequently change and numerous changes to Ontario legislation have been proposed or are being developed.

EMS audit – The criteria for an EMS audit should be based upon the ISO 14001 standard and the organization's EMS documentation since the audit will seek to determine whether the EMS has been designed to address all of the requirements of the ISO 14001 standard, and has been properly implemented and maintained.

10.3.3   Develop an Audit Plan

A plan should be developed for the collection of evidence using substantive procedures relevant to the objectives of the audit. The plan should contain the following (ISO, 1996c):

- audit objectives, scope and criteria
- identification of auditee's organizational and functional units to be audited
- identification of functions and/or individuals within the auditee's organization having significant direct responsibilities regarding the auditee's EMS
- identification of those elements of the auditee's EMS that are of high audit priority
- procedures for auditing the auditee's EMS elements as appropriate for the auditee's organization
- working and reporting languages of the audit
- identification of reference documents
- expected time and duration for major audit activities
- dates and places where the audit is to be conducted
- identification of audit-team members
- schedule of meetings to be held with the auditee's management
- confidentiality requirements
- report content and format, and expected date of issue and distribution of the audit report
- document retention requirements

It is important that the client reviews and approves the plan prior to conducting the audit. Any objections should be dealt with prior to conducting the audit.

Travel arrangements may need to be made along with arranging for permission to inspect the facility. The time of the initial meeting with plant personnel to discuss the audit and a tour of the facility should be arranged in advance. The availability of plant personnel must also be established prior to the audit. At many types of facilities, personal protective equipment (PPE) or special security passes are needed. PPE may range from hard hats, safety shoes and safety glasses to respirators and hazard suits. Passes may be required to ensure security and also safety in the event of an accident. Once again, arrangements must be made prior to the site visit.

10.3.4   Working Documents

Working documents may include forms for documenting supporting audit evidence and audit findings, procedures and checklists for evaluating EMS elements. These documents should be maintained at least until the completion of the audit and those that may involve confidential or proprietary information should be suitably safeguarded by the audit-team members (ISO, 1996c).

Audit Procedures

The audit procedure should include tests to assess specific items on the checklists. For example, the number of manifests to be randomly selected and reviewed for compliance with Regulation 347 (Waste Management) should be identified. The use of specific statistically sound test methods can allow the audit to be conducted in a reasonable amount of time while minimizing the audit risk. The following are ranges of sizes of samples that may be appropriate for assessing waste manifests: <25 manifests – assess all, <50 manifests – assess 25, <100 manifests – assess 30, <250 manifests – assess 40.

It may also be appropriate to assess the audit risk (that is, the probability of reaching incorrect conclusions before an audit is undertaken). The level of risk is determined by the type of audit, the objectives of the audit, and the materiality, extent and nature of the collected data and evidence. An example of an audit risk is the aforementioned practice of randomly selecting manifests to assess compliance with Regulation 347. The potential exists that non-compliance issues may be overlooked since not all manifests are reviewed during the audit. Once the potential risks are identified, the audit plan and level of investigation can then be formulated so that the risk is reduced to an agreed-upon level.

Audit Checklist

Questions in a compliance audit checklist should be worded in such a way that they can be answered "yes" or "no". This allows the audit team to quickly assess and summarize non-compliance issues. In the case of an EMS audit, the question may be worded to "How do you ensure that . . . ?".

The checklist also should require all responses be referenced. A protocol should describe how references are to be recorded (for example, numbered and presented at the back of the checklists) and whether copies of all pertinent parts of documents (letters, reports and pictures) indicating non-compliance are to be copied and attached to the audit report.

10.3.5   Data Request and Review

Ideally, the audit team should have been provided with and reviewed various types of information before actually visiting a facility. The types of information requested in advance will be strongly influenced by the type of audit as well as by scope and boundary conditions.

Compliance Audit

Foremost among the information that should be provided in advance of a compliance audit is a comprehensive site plan. Depending upon the scope of the audit (for example, all discharges versus just air emissions), the plan should show:

- locations of all buildings
- storage and on-site transportation facilities
- waste storage/treatment/disposal facilities
- waste discharge or release points
- environmental monitoring equipment
- sewer connections
- site drainage
- neighbouring properties
- environmentally sensitive areas
- similar pertinent information

Environmentally sensitive areas can include drinking water supplies, streams, wetlands, parks, areas frequented by birds or animals, and any areas used by endangered species.

A process and emissions block diagram showing the physical or chemical processes conducted on the site, the handling of raw materials brought onto the site, and all points where waste products (solids, liquids or vapours) leave the facility also should be provided to the team, or the team should produce one. A typical, simple block diagram is presented in Figure 10.1. A block diagram is essential since all the emissions or waste streams leaving the facility need to be identified.

All documentation relating to previous environmental inspection reports must be collected. Notices of complaints, violations and prosecutions should also form part of the pre-audit materials. The audit team should be provided with hazardous materials reports that identify the locations of materials such as asbestos insulation, equipment that contains PCBs, radioactive materials or solvents.

Depending on the agreed-upon scope of the audit and boundary conditions, it may be appropriate to contact the MOE to obtain copies of available Certificates of Approval (Cs of A), waste registration information, complaints and orders, and to discuss any concerns they may have with the facility. A similar request can be made of the municipality to identify potential problems with discharges to sewer system and noise levels.

EMS Audit

All documentation related to the scope of the audit should be provided in advance. These may include documentation related to the following:

- environmental policy
- environmental aspects and impacts
- legal and other requirements
- objectives and targets
- environmental management programme
- roles, responsibility and authorities
- training needs and programs
- communications
- descriptions of the core elements of the EMS
- document control and operational control
- emergency preparedness and response
- monitoring and measurement
- non-conformance and corrective and preventive action
- record retention
- EMS audits
- management review

ISO 14001 and 14011 standards provide additional detail on required documentation.

10.4   On-Site Visit Activities

10.4.1   Opening Meeting and Orientation Tour

A meeting between the audit team and appropriate representatives of the facility being audited should be the first on-site activity. The meeting should allow the following (ISO, 1996c):

- introduce members of the audit team
- review and discuss the scope, objectives and audit plan
- discuss the audit schedule, including the time and date of the closing meeting
- provide a short summary of the methods and procedures to be used to conduct the audit
- establish the official communication links between the audit team and the auditee
- confirm that the resources and facilities needed by the audit team are available
- promote active participation by the auditee
- review relevant site safety and emergency procedures for the audit team

Following this meeting, it may be appropriate to perform an orientation tour to provide an overview of the facility and its operations. The tour is relatively brief and auditors should note areas of concern where further follow-up is required. A site plan should be taken with the audit team as part of the orientation.

The activities described in the following sections are methods for collecting evidence to complete the checklists or working papers described in the audit plan. Information collected during any one of these activities should be tested to an appropriate extent by acquiring corroborating information.

10.4.2   Record Review

A major component of almost any type of audit is the review of environmental records. Depending on the type and scope of the audit, the review may include the following:

- minutes of meetings
- complaints or requests for EMS-related information
- progress toward achieving objectives and targets
- spill and incident reports
- waste manifests and inventories
- observations of all discharges or releases to the environment
- the performance logs of environmental control equipment
- calibration reports for measuring devices
- written environmental procedures and policies
- results of environmental monitoring efforts
- Cs of A or orders issued by regulatory agencies
- reports of unusual or unscheduled events

The review of environmental records can be very time-consuming. The practices and theory from financial accounting can be used to advantage in some instances. For example, the review of shipping documentation for dangerous goods can be time-consuming if every document must be reviewed. Instead, a representative sample of the documentation can be retrieved from the files and reviewed for compliance. As discussed in Section 10.3.4, the number of manifests to be randomly selected should be identified in the audit plan.

When reviewing approvals or permits, care must be taken to check for any major additions, alterations or modifications that have occurred since the original issuance. These may require either an amendment to the current approval or the obtaining of a new approval.

10.4.3   Site Investigation

Depending upon the type of audit, the audit team may conduct a thorough site investigation of all operations, including, but not limited to, processing and storage areas/facilities; waste storage, treatment and disposal locations; wastewater treatment and discharges; and air pollution control equipment and release points. The actual information to be collected will be outlined in the audit checklists.

All team members should make sure that they have the appropriate safety equipment to enter all parts of a facility (for example, hard hat, safety glasses, ear protection, etc.).

10.4.4   Interviews with Plant Personnel

As part of the collection of audit evidence, interviews should be conducted with pertinent staff to assist in the formulation of findings related to the audit criteria. The particular personnel to be interviewed and types of questions will depend upon the audit criteria and type of audit.

It is important to use an interview form or checklist when asking questions to ensure that all areas are covered and to confirm or identify discrepancies with information collected during the facility tour and record review. Areas of non-conformance identified during the interviews should be verified by acquiring supporting information from independent sources, such as observations, records and results of existing measurements. Non-verifiable statements should be identified as such (ISO, 1996c).

10.4.5   Closing Meeting

A closing meeting should be held upon the completion of the evidence collection phase and prior to the preparation of the audit report. At this meeting, the preliminary findings should be presented to the auditee's management and those responsible for the functions being audited. It is important that the findings be presented so that the auditee understands and acknowledges the basis of the audit findings. Disagreements should be resolved if possible before the audit report is issued (ISO, 1996c).

10.5   Post-Site Visit Activities

10.5.1   Preparation of Audit Report

The report should be prepared under the direction of the lead auditor, who is responsible for its accuracy and completeness. The report should be dated and signed by the lead auditor and contain the audit findings and/or a summary thereof with reference to supporting evidence (ISO, 1996c).

According to the ISO 14001 standard, the audit report may also include:

- identification of the organization audited and of the client
- agreed objectives, scope and plan of the audit

- agreed upon criteria, including list of reference documents against which the audit was conducted
- period covered by the audit and the date(s) the audit was conducted
- identification of the auditee's representatives participating in the audit
- identification of audit-team members
- statement of the confidential nature of the contents
- distribution list for the audit report
- summary of audit process including any obstacles encountered
- audit conclusions

Note that the above list does not include recommendations. The inclusion of recommendations, typically a component of a compliance audit, would need to be discussed and agreed upon in the audit plan.

If the audit was performed under solicitor-client privilege, the draft and final report must be submitted to legal counsel directly. Legal counsel can then assess potential non-compliance issues and provide legal advice to the client.

If the report has not been prepared under solicitor-client privilege, it should be marked "preliminary" or "draft" until plant management and legal counsel have had an opportunity to review it and make written comments. The distribution of the report usually will be determined by plant, divisional or corporate management, and/or corporate environmental staff with some input from legal counsel.

10.5.2   Post-Audit Activities

Senior management must be made aware of all findings from the audit. ISO 14001 standard requires a review of the EMS in the light of the EMS audit results.

Ideally, a summary table presenting the findings from the audit should be provided to senior management along with comments from legal counsel if areas of non-compliance have been encountered. Cost estimates, responsibilities and a proposed schedule to implement corrective action for areas of non-compliance may be included if part of the agreed scope of work. Alternatively, these may be conveyed in a separate document.

The preparation and implementation of an action plan is essential for an effective audit program, and is mandatory under the ISO 14001 standard. In essence, it closes the loop and ensures that deficiencies are corrected in a timely manner. It represents the due diligence element of the program.

Little good is accomplished if the paper trail of positive responses does not result in specific action. In some cases, inaction on the findings can be damaging. Procedural controls cost little to implement, and large capital projects can be implemented in phases.

10.6 Reacting to an External Audit

If an environmental audit is being performed as a result of external actions (that is, it has originated with a government agency or a potential purchaser), the corporate environmental staff and the legal department should be made aware of the upcoming audit, and the scope of the audit should be determined as soon as possible.

Pertinent regulations, Cs of A, orders, permits, compliance history, previous audits and environmental reports should be assembled and reviewed prior to the visit of the external auditors with a critical eye to identifying potential areas of vulnerability.

It is important that the auditors not be given misleading or false information. If a point is raised that has not been considered or the answer is not known, seek clarification of the question and the type of answer being sought. It may be appropriate to respond to the auditor at a later time when more information becomes available. Do not speculate for the sake of providing an answer.

A qualified individual should accompany the auditors throughout the audit, assisting where possible. It is a good idea to carry a notebook and write down important questions, as well as the responses provided.

Record all documents that are provided to the external auditors. Use cover letters and other formalities when providing any written information to the auditors.

10.7 Summary

Environmental audits are an essential component of an environmental management system. Environmental audits can assist a company in the identification of compliance or management issues and thereby prevent incidents of pollution or, in the event of an incident, assist in establishing the defence of due diligence.

Recent publications by the ISO, CSA and CEAA reinforce the need to perform environmental audits using recognized principles and practices. Generally, these principles and practices require that auditors who are trained and independent of the activities being audited develop audit protocols to provide verifiable assertions about the degree of compliance and the communication of the results to the client. It is anticipated that in time these principles and practices will be become generally accepted — analogous to those employed in financial audits.

References

Environment Canada. March 2001. *Compliance and Enforcement Policy for the Canadian Environmental Protection Act, 1999 (CEPA, 1999).*

International Organisation for Standardisation (ISO). 1996a. *Environmental Management Systems – Specifications with Guidance for Use.* ISO/DIS 14001.

International Organisation for Standardisation (ISO). 1996b. *Guidelines for Environmental Auditing – General Principles.* ISO/DIS 14010.

International Organisation for Standardisation (ISO). 1996c. *Guidelines for Environmental Auditing – Audit Procedures – Auditing of Environmental Management Systems.* ISO/DIS 14011.

Ontario Ministry of Environment and Energy (MOEE). 1995. *Policy and Guideline on Access to Environmental Evaluations.* PIBS 3199e.

Figure 10.1
Typical Block Diagram of Emissions, Discharges and Waste Streams

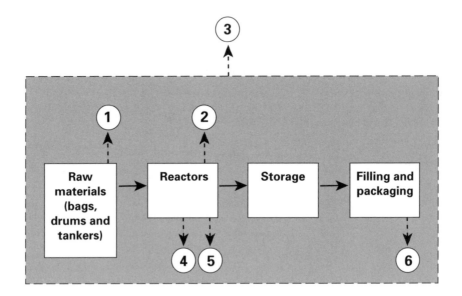

Process flow ⟶

**Emission/
discharge
waste streams** - - - - - - ➤

1.   Duct from raw material storage
2.   Stack for reactor discharge
3.   Fugitive emission to atmosphere
4.   Water discharge to sanitary sewer
5.   Solid waste
6.   Spoiled product for off-site disposal

11

Risk Assessment and Management

11.1   Overview

Risk often is defined as the probability or likelihood that an adverse outcome will be caused by an action or condition. In the context of environmental issues, the actions and conditions of concern usually are associated with the exposure to a contaminant. The Ontario Environmental Protection Act (EPA) definition of contaminant includes a chemical or physical agent such as noise, heat or radiation.

Risk assessment is the process of determining the adverse outcomes that can occur. This can include evaluating the types of potential effects (also referred to as hazards) and estimating the probability of an effect occurring and the numbers of people or other organisms that might be exposed and/or suffer the consequences of exposure.

Risk management is the process of evaluating possible courses of action and selecting among them with the objective of minimizing risks. As such, risk assessment is the vital precursor to risk management. Figure 11.1 shows one interpretation of the steps that make up risk assessment and risk management and how these two processes are related.

During the last two decades, there has been a growing trend to incorporate risk assessment and management into various decision-making processes including the setting of environmental regulations, criteria and policies. These efforts have often proven to be cumbersome and onerous. There are differences of opinion as to the roles that risk assessment and risk management should play. The methods used to assess risk have not been standardized. As a result, a certain amount of skepticism and caution has been directed toward the risk assessment and management processes.

In part, these shortcomings may be the unavoidable outcome of expecting too much too soon from a relatively new aid for decision-making. Risk assessment and risk management often are expected to bring order to large pools of information and resolve conflicting interests and uncertainties en route to identifying appropriate options. Some of the factors with which a risk manager often must contend include situation-specific economic, social, political and legal considerations. The complexity of such situations can be overwhelming.

Despite the difficulties that have been experienced (and no doubt will continue to be experienced), risk assessment and management are likely to play increasingly important roles in decision-making processes. Factors that will continue to support the need for risk assessment and management include:

- the realization that there are insufficient resources to address all concerns (and therefore priorities need to be set)
- the need to compare the relative importance of an ever-increasing number of environmental issues
- government agencies that are asked to use risk as a basis or supporting rationale for regulatory decisions more than ever before
- the increasing number of issues, such as acidic precipitation, global warming or exposure to carcinogens, which are complex and require careful assessment of risks

11.2 Risk Assessment

11.2.1 Purpose and Elements

Risk assessment is directed toward determining the types and probabilities of adverse effects occurring as a result of some activity or condition. Consider the example of a proposed stack that will emit vapours that may be inhaled by downwind residents and cause adverse health effects. To assess the potential risks, it first is necessary to determine the characteristics of the release (how much, when, and at what exit velocity and temperature). It then is necessary to determine how the release will behave in the atmosphere (how much dilution will occur, will the substance degrade while airborne). Various characteristics of the person(s) who could inhale the vapours need to determined (where are they located, how much air do they breathe, will they breathe the vapours all day, every day). Of obvious importance is the toxicological nature of the substance (what kinds of adverse effect can it cause, how much does a person need to inhale to experience an adverse effect). Finally, the estimated doses need to be described in terms of the risk or the probability of adverse effects occurring.

Many descriptions of these elements or steps have been prepared. One of the earlier renditions of the steps and associated terminology was suggested by the U.S. National Academy of Science (NAS, 1983):

- hazard identification
- exposure assessment
- dose-response assessment
- risk characterization

These four steps subsequently have been repeated by numerous agencies, sometimes with subtle variations. For example, the first step may be referred to as "problem formulation" (as is the case in MOEE, 1997) or "data collection and evaluation" (as is the case in U.S. EPA, 1989). The order of the second and third steps may be reversed. Dose-response assessment sometimes is referred to as toxicity assessment (as is the case in U.S. EPA, 1989).

11.2.2   Hazard Identification/Problem Formulation

The first step in a risk assessment is to identify the potential hazards, which in environmental management usually involves identifying where and how chemicals are being released (or have been released) into the environment. Examples of release points include:

- air emissions from stacks and vents, open storage tanks, pressure relief valves, waste ponds and lagoons, fugitive sources, land farms, open process tanks and vessels, tank and drum loading/unloading operations
- liquid discharges from point sources to surface water, discharges to municipal sewers, releases to ground water, leaky storage tanks and ponds
- solid waste shipments to off-site disposal, transfer to on-site disposal, on-site disposal/treatment facilities

At U.S. Superfund sites where soil or ground water is contaminated, the U.S. Environmental Protection Agency has described this step as data collection and evaluation since it typically involves the collection and analysis of environmental samples (soil, water, biota, air, etc.) and the identifying of potential chemicals of concern (U.S. EPA, 1989).

The types of activities that are likely to identify hazards include environmental inspections, site assessments (see Chapter 5), audits (see Chapter 10), or gathering information to comply with regulatory requirements such as completing a waste inventory for a sewer-use by-law (see Chapter 4) or an air emissions inventory for the National Pollutant Release Inventory (see Chapter 2).

11.2.3   Exposure Assessment

When a receptor (usually a person but the term can also be used to refer to an animal or plant) comes into contact with a chemical that has been released into the environment, exposure occurs. For example, a person may be exposed to a chemical in the air by inhaling it. Exposure assessment involves determining the environmental fate of the chemicals released from each release point (see Chapter 9) and the ways in which an individual can be exposed to those chemicals. Exposure pathways can include:

- ingestion of water
- ingestion of soil
- inhalation of vapours
- inhalation of particulate matter
- ingestion of fish
- ingestion of milk and dairy products
- ingestion of produce
- ingestion of dust
- dermal absorption

In many situations, one or two pathways are the major contributors to total exposures. For example, the ingestion of meat and milk accounts for 99% of human exposure to 2,3,7,8-TCDD, one of the more toxic forms of dioxin (Travis and Hattemer-Frey, 1987). The relative importance of a pathway is a function of the physicochemical properties of the substance, characteristics of the receptor, the receptor's behaviour in the environment where contact occurs and characteristics of the environment. Determining which substances and pathways are of greatest concern is a major step in determining the form that remedial measures may need to take.

For compounds with numerical criteria established by regulatory agencies, the estimated or measured concentration of a substance at the point where a receptor makes contact can be compared to those regulatory criteria as a measure of relative hazard if the criteria have been set to avoid a hazard. Some numerical criteria and guidelines are not based on potential hazards but on other factors such as aesthetics or analytical detection limits.

The dose is the amount or mass of chemical taken up by the receptor via the pathways noted above. Dose and exposure often are used incorrectly as synonyms. For example, a receptor may be exposed to a substance present at a concentration of 3 mg/L in water. Upon drinking 2 L of the water, the receptor's intake is 6 mg of the substance. If the receptor weighs 30 kg, their dose is 0.2 mg/kg of body weight (bw). If they drink 2 L of the water every day, the dose is 0.2 mg/kg-bw/d. Dose is a function of several parameters:

- quantity of material taken into the receptor's body
- characteristics of the receptor, for example body weight
- concentration of chemical in the material taken in by the receptor
- frequency of occurrence of exposure
- bioavailability factor(s) specific to the chemical and exposure route

The bioavailability factor refers to the percentage of material taken into the body that actually enters the body tissue or blood. It is the dose at the target tissue that determines the probability of a health effect occurring, not the level of external exposure. Some researchers have suggested that the term "dose" be replaced by "potential dose" if bioavailability is ignored or assumed to be 100% (as in the case described above), and that the term "internal dose" be used once bioavailability is taken into account (Lioy,

1990). It has also been suggested that the fraction of the internal dose that actually leads to a specific effect or physiologic change be called the "biologically effective dose".

It is important to state all of the assumptions employed to derive doses, as the results can vary widely depending upon the assumptions made. There are at least two basic philosophies that have been used to estimate dose:

1. Calculate a maximum or upper extreme dose using conservative assumptions (those deliberately made to avoid underestimating risks) to arrive at a so-called upper bound estimate of the risk.

2. Make assumptions about receptors and pathways that are consistent with what the evidence and the best current scientific information suggest are probable or typical (as opposed to extreme). At the end of the process, apply a final safety factor to ensure public protection.

The first method puts the onus on the risk assessor to make decisions about conservative safety factors as opposed to leaving these assumptions to the judgment of the risk manager. Taken to extremes, the first "conservative" approach can lead to gross overestimations of exposures or portray receptors with unrealistic behaviours or characteristics. The second method is slowly gaining favour in Ontario, although the onus is always on assessors to justify all assumptions.

In evaluations that use the first approach, it often is assumed that the receptor is a hypothetical individual who spends all of his time (24 hours a day, 365 days a year) for 70 years (the receptor's lifetime), at the location of highest contaminant concentration attributable to the source of the substance. In addition, it is often assumed that the source continuously emits the chemical for the receptor's lifetime. Information about the behaviour of people indicates that it would be more realistic to assume that receptors spend one-third of their time indoors at the site, even less outdoors and a substantial portion of their time off-site.

In the conservative approach, the maximum level of exposure is assumed to occur for the duration of the exposure. This requires a constant or infinite source of the substance, which seldom is the case. For example, soil containing a volatile organic compound may be the source of vapours that a receptor can inhale. The vapour emissions of a compound, especially one that volatilizes rapidly, will deplete the amount of the compound in the soil, especially from the top soil layer. In turn, this will lead to decreases in the exposures and doses with time.

In the conservative approach, it often is assumed that 100% of a substance is bioavailable for each pathway (that is, all the chemical is absorbed or metabolized by body tissue or the bloodstream). In many cases, this assumption is made because there is insufficient evidence to support the use of a different value, however, there is a growing amount of data indicating that bioavailability often is substantially less than 100%. This likely is the case for the bioavailability of many organic substances that are ingested in the diet and the bioavailability of substances associated with inhaled particles.

To facilitate the calculation of exposures and doses, mathematical models can be used. Such models should take into account the emission or discharge rate of a chemical, its behaviour or fate in the environment, and the conditions under which receptors come into contact with the chemicals. Many such models exist. Some are highly specialized and may be limited to specific types of release points (for example, point sources such as factories), particular risk agents or specific types of environmental settings, such as watersheds or lakes.

By their very nature, these models are simple representations of real systems and contain parameters that are either imperfectly known or inherently variable by nature. Uncertainties also arise when these models are used to predict effects for conditions or over periods of time different from those for which the models and model parameters were developed. As a result, the outcomes generated by models, or for that matter any other exposure/risk assessment approach, contain a degree of uncertainty.

Until recently, uncertainty has been addressed in models indirectly. The traditional approach has been to run a model with conservative values of the parameters, typically those near the extremes of the expected ranges of the probability distributions for parameters. The result of such an approach is generally a very conservative estimate of the effects.

An alternative approach is to use some form of probabilistic analysis. This approach attempts to take uncertainty into account by specifying probability distributions rather than single values for parameters that are model inputs. The model is run many times with new values selected for the input parameters on each run. The parameter values used in each run are randomly selected from the appropriate distributions. The model output is a probabilistic distribution of possible outcomes.

Output data from the uncertainty analysis can be used to assess the relative importance of the various model parameters by assessing their relative contribution to overall uncertainty in model predictions and determining how sensitive the predicted values are to small changes in model input parameters.

The benefits of this approach are that more realistic estimates of exposure and risk can be made, and the reliance on worst-case assumptions is reduced. As of late 2002, probabilistic modelling is not commonly used in Ontario or is used on relatively small scales (that is, a few parameters in a risk assessment might be assessed as distributions rather than as single values).

11.2.4   Dose-Response Assessment/Toxicity Assessment

The relationship between dose and response is based on an analysis of relevant data that may include epidemiological, clinical, environmental, animal toxicological, biochemical, structure-activity and exposure data.

The exposure levels of interest associated with most environmental issues are often orders of magnitude less than those that cause obvious adverse effects. As a result, it usually is necessary to extrapolate from the available dose-response data down into regions of very small doses that are experimentally impossible to investigate for statistical reasons (that is, natural variation exceeds the effect).

The nature of the dose-response information can be used to divide chemicals into two groups: non-threshold toxicants and threshold toxicants. Non-threshold toxicants are chemicals thought to pose a possibility of adverse effects at any level of exposure. This includes carcinogens. In Canada, only genotoxic carcinogens (those that interact directly with DNA) are considered to be non-threshold toxicants for purposes of risk assessment. It has become common practice to assume that the relationship between dose and response for non-threshold toxicants is linear, however, there is considerable scientific debate regarding the linearity of dose-response relationships, whether or not there may be thresholds below which adverse effects do not occur, and whether the linear models currently used are unduly or inadequately conservative.

Figure 11.2 is a simplistic plot of dose-response relationships for non-threshold toxicants. The results of dose-response studies are in the experimentally accessible region. Passing through those data points are a few of the types of curves used to extrapolate to the inaccessible regions that correspond to most environmental exposures. Most of the models used for non-threshold toxicants show that some response is expected for any exposure.

Threshold toxicants are substances for which the information indicates that there is a dose below which measurable adverse effects should not occur even if there is a lifetime of exposure. The threshold below which no observable response occurs is defined as the no-observable-adverse-effect-level (NOAEL).

Figure 11.2 does not show the dose-response relationships for threshold toxicants. By definition, one or more of the data points for a threshold toxicant would show no response (that is, the data point(s) would "sit" on the exposure or dose axis) and the dose-response curve would have a shape similar to that of a hockey stick with the blade pointed down and intersecting the exposure or dose axis at some point right of the response axis.

11.2.5   Risk Characterization

For non-threshold toxicants, the questions most often asked are "If the dose is known, what risk does it pose?" or "If a certain risk should not be exceeded, what is the corresponding dose?" As a result, non-threshold toxicants often are described in terms of a risk specific dose (RSD). For example, assume that the graph used for arsenic shows that a dose of 0.0006 µg/kg-bw/d corresponds to the risk of one-in-a-million that the exposed person will develop cancer over that person's lifetime. Such a graph might be described by indicating that the RSD at the one-in-a-million risk level is 0.0006 µg/kg-bw/d.

Another way to describe the relationship between dose and response is to use the slope of the dose-response line. The slope factor (SF), is equal to the probability of response (risk) divided by dose. The SF for a non-threshold toxicant can be multiplied by a known or estimated dose to obtain the risk.

Equation 11.1

risk = SF x dose

If the dose mentioned in Equation 11.1 corresponds to a person's average daily dose over a lifetime (often abbreviated as LADD), then the calculated risk is that individual's excess lifetime cancer risk (often abbreviated as IELCR).

Other terms used to express dose-response relationships for non-threshold toxicants include:

- TC_{05} or the tumorigenic concentration that will result in a 5% increase in the incidence of cancer in the exposed population (other percentages can be used)
- ED_{10} or the effective dose that will affect 10% of the exposed population (other percentages can be used)
- UR or unit risk, a value that can be multiplied by the concentration of the chemical to estimate risk (for example the inhalation unit risk or IUR value can be multiplied by the concentration of a chemical in air to estimate the risk of inhaling that air constantly over a lifetime)

If the chemical of interest is a threshold toxicant, then measurable adverse effects should not occur below a specified dose. Such a dose often is referred to as a reference dose or RfD. An RfD value is calculated by dividing the lowest NOAEL by an extrapolation or safety factor (values of 10, 100, 1000 or more are used). The magnitude of the extrapolation factor depends on the confidence that can be placed in the allowable data, the relevance of the data to humans, the severity and type of effects observed, and the types of test systems studied. Unlike RSD values, there are no risk levels associated with RfD values. For some threshold toxicants, RfD values have been determined for both chronic and subchronic exposure periods. It is important to use those RfDs that correspond to the exposure scenario of interest.

Summing Risks for Multiple Exposure Pathways and Chemicals

For both types of toxicants, doses are estimated initially for each unique combination of chemical, receptor and exposure pathway. In many instances, a receptor is assumed to be exposed to a chemical via multiple pathways. The doses from the pathways usually are combined into a total dose for the specific chemical and receptor.

For a non-threshold toxicant, the total dose is used to estimate the total risk.

For a threshold toxicant, the total dose is compared to the RfD. This often is done by dividing the dose by the RfD to produce the hazard quotient (HQ). Care is needed when summing the doses from several pathways. If more than one pathway is built upon worst case assumptions, combining such pathways may lead to impossible scenarios. For example, the amount of time a receptor is assumed to spend at various locations during a day obviously should not exceed 24 hours.

In many instances, a receptor is assumed to be exposed to more than one chemical. If more than one non-threshold toxicant is of interest, the risk estimates can be summed. If more than one threshold toxicant is of interest, the HQ values can be summed if the chemicals produce similar types of adverse effects. An HQ sum often is referred to as a hazard index (HI). If the chemicals produce different types of adverse effects, then the risks are not added together.

Once a dose has been calculated and the corresponding risk, HQ or HI has been estimated for a given situation and set of conditions, the next step often is to comment on the acceptability of the result.

Commenting on the Acceptability of Risks

Establishing acceptability is a complex process influenced by economic, social, political and technical factors as well as scientific factors and therefore is not part of the risk characterization but is part of risk management.

For non-threshold toxicants, decisions by regulatory agencies most often have defined acceptable risks to be in the range of one-in-ten-thousand (1×10^{-4}) to one-in-ten-million (1×10^{-7}). For example, the MOE often recommends an upper limit on acceptable risk of one-in-a-million (1×10^{-6}). How acceptable risks should be defined continues to be a contentious issue.

For threshold toxicants, if the total HI exceeds unity (1) then the risks usually are judged to be unacceptable. HI values between 0.1 and 1.0 indicate a situation of potential concern, although not necessarily unacceptable. Some regulatory agencies prefer that HI values be less than 1. For example, the MOE recommends that HI values not exceed 0.2 for each relevant exposure pathway when using risk assessment to set site-specific soil and groundwater criteria.

Another approach that some agencies use to judge acceptability is to compare exposures or doses to ambient or background conditions. Nuclear scientists have led the way in using natural background exposures as a baseline from which to assess acceptability. The background approach is also one option for setting soil and groundwater clean-up criteria noted in the MOE *Guideline for Use at Contaminated Sites in Ontario* (see Chapter 5). This approach cannot be used when the chemical of interest is man-made and therefore should not be present in "uncontaminated" environments or when there is uncertainty as to how background conditions are to be defined. For example, is it appropriate to sample rural or remote environments and then use that background to assess urban areas, particularly for chemicals widely dispersed into the urban environment for many years, such as those emitted by vehicles or fossil fuel burning facilities that inevitably are present in urban areas?

Chemical Mixtures

Individuals seldom are exposed to a single chemical; most exposures are to complex mixtures of assorted compounds. Assessing human exposure to such mixtures is a formidable task. First, it is generally impractical or impossible to measure the concentration of all constituents contained in a complex mixture. Secondly, even if all constituents were known, the toxicities of all the constituents may not be known. Thirdly, even if the toxicities of individual components are known, the toxicity of a mixture can be substantially different from the composite of the toxicities of its constituents.

Chemicals can interact synergistically or antagonistically to form a mixture that is more or less toxic than expected. Many such interactions are reported in the toxicological literature. For example, cigarette smoke and asbestos are known to interact synergistically, thereby increasing the incidence of lung cancer in asbestos workers. Benzene and toluene interact antagonistically to decrease damage resulting from benzene exposure. While direct exposure of experimental animals to complex mixtures in the laboratory is one way to overcome these limitations of information, it is difficult to duplicate complex environmental conditions in a laboratory setting. The exact composition of environmental mixtures can be unknown and may change over time.

11.3   Risk Management

11.3.1   Need for Risk Management

While there are uncertainties associated with virtually every aspect of assessing risks, there also are ways to compensate for the uncertainties or at least highlight them. Risk management, on the other hand, is not so restricted in scope and often must try to balance factors that are not easily compared or quantified.

Risk management is generally portrayed as an interactive process of identifying the options available for abating unacceptable risks, evaluating the cost-effectiveness of those options and identifying the preferred course of action for achieving the desired risk reductions.

The management process may need to consider both actual and perceived risks. The preferable outcome of risk management is the abatement of an actual risk in a cost-effective manner. The reduction of risks for reduction's sake alone has little long-term benefit and may be detrimental if it consumes resources that could be used more effectively to reduce other sources of risk.

As noted in Section 11.2.5, regulatory agencies that use or support the use of risk assessment and risk management must decide how to define acceptable risks and recognize that there is little merit or economic justification in striving for lower levels of risk. For example, adopting analytical detection limits is not considered a viable method of defining acceptable exposure, dose or risk levels. The search for hazardous micropollutants has on occasion been compared to a "witch hunt" in which the hunters will never be fully satisfied because there is no clear proof of innocence (Mackay, 1982).

11.3.2   Factors that Often Need To Be Considered

Figure 11.1 illustrates some of the non-technical factors that often need to be incorporated into the risk management process. These factors add layers of complexity to risk management and, while they are essential, they also can make the process become rather messy (Fischhoff, 1985).

Benefit to society–All other factors (for example, size of exposed population, potency) being similar, a chemical that makes vegetables more colourful ought not to be subject to the same standard of acceptable risk as one that is released when processing raw materials for important industries. In general, people are less willing to reduce risks if it also means forgoing benefits (Dwyer and Ricci, 1989).

Litigation – The possibility of litigation by one or more parties makes the task of risk management much more difficult. Litigation may introduce delays and the possibility of additional financial risks into the process (Deisler, 1988).

Economics – The expenditures required to clean up a site, decommission a facility or comply with air and water emission criteria can be substantial to both the company and the economy as a whole. The relationship between expenditures and risk is that expenditures go up as the targeted risk level goes down. There have been cases where decisions were made to achieve relatively low-risk levels (say less than one-in-a-million) because it was easy or inexpensive to do so. Conversely, there have been cases where decisions were made to achieve risks higher than one-in-a-million because it was prohibitively expensive or difficult to achieve lower risks.

11.3.3   Possible Applications for Risk Management

Risk management can be used to address several types of issues:

- Is remedial action required?
- What level of risk is acceptable?

- What is the optimum remedial action?
- What priorities should be assigned to action items?

The first two issues can arise when evaluating situations such as the remediation of a contaminated property or during the assessment of air emissions or liquid discharges. Once a decision has been made that remediation is required, options must be assessed. A preliminary risk assessment of each option can assist in the identification of the optimal alternative.

Assigning priorities to action items is critical when resources such as funding, staffing, technical equipment and technical expertise are scarce. No government agency or company can immediately implement activities for all actual or potential situations that require remediation. Some activities must wait until additional resources can be made available.

An example of a need for prioritization is the clean-up of hazardous waste sites in Canada and the United States which for various reasons have become the responsibility of federal, state and/or provincial agencies. Since the costs of cleaning up such sites tends to be very expensive, there is a vital need to assign priorities in such a way that the sites that pose a clear and immediate threat to human health and the environment are remediated first.

11.4   Risk Communication

11.4.1   Overview

Risk communication often is described as consisting of three main elements: the audience, the message and the communicators. It is an integral part of risk management. It requires training and experience in translating scientific data into clear and understandable language and, therefore, should not merely be passed along to the public relations staff or community affairs department if those skills are lacking.

At the same time, the numbers of occasions that require risk-related information to be shared with the public seems to be ever-increasing. The Ontario Environmental Assessment Act (EAA) and EPA both allow for input from the public. For example, there is a requirement to conduct a public hearing before issuing a C of A for the use, operation, establishment, alteration, enlargement or extension of a waste disposal site (see Section 4.6.3). Although somewhat dated, the MOE has published a guideline for public consultation (MOEE, 1994).

The MOE clearly has indicated that it intends to increase the opportunities that members of the public have to participate in environmental decision-making. The Environmental Bill of Rights, which became law in 1994, created an environmental registry on which are posted all significant environmental proposals and decisions by government including the issuing of approvals and permits. Most proposals are posted along with a time frame for comment. Longer periods are used for decisions of regional or provincial significance.

The MOE *Guideline for Use at Contaminated Sites in Ontario* encourages public communication and states as a principle that the public should be kept informed of the site restoration process.

11.4.2   General Guidance

Risk communication is inherently difficult. Although everyone assesses and manages risk every day, it is a process not easily expressed in words. Environmental risk assessments often make extensive use of assumptions, numerical data, statistics, and mathematical models, all of which can impede the transfer of information from the communicator to the audience. To further complicate matters, many risk assessors have scientific or engineering backgrounds, and studies have shown that their perceptions about risk acceptability and uncertainty often are substantially different than those of members of the general public.

While much can go wrong, there also are a growing number of organizations and agencies that offer guidance on risk communication. (Their guidance often comes from both successful and unsuccessful attempts at risk communication.)

Table 11.1 lists several organizations, individuals and Web sites that provide risk communication guidance.

Table 11.2 presents some general rules and guidelines published by the U.S. Environmental Protection Agencies for effective risk communication. Rules and guidelines like these can be used as a foundation for a risk communication program and lead to a constructive interaction with the community.

Some recurring general themes in the guidance on risk communication are:

1. Be prepared. Plan presentations thoroughly in advance. Rehearse.
2. Be open-minded about what the public may say. Realize in advance that the public may overreact to risks you think are unimportant, or show little interest in those you think are important.
3. Be proactive. Consult with the public. Hold workshops. There are many ways to encourage public involvement and solicit opinions (U.S. EPA, 1999).
4. Create visible connections between the community input and the outcome of the risk management process.

11.4.3   Commonly Experienced Difficulties

There is no way to present risk data that are neutral, only ways that are alarming or reassuring in varying degrees (Sandman, 1986).

Some people have reached the opinion that pollution is morally wrong. They may want to discuss environmental risk in terms of "good and evil" instead of "costs and benefits" (Sandman, 1986).

It is difficult for people to accept the results of risk assessments, risk-benefit calculations, risk-cost ratios and risk comparisons when they are being asked to bear the risks while someone else makes the decision. The process of who decides can be a key factor, and in most cases, much more than the substantive issues.

Consider the example of a town being selected as the future site of a hazardous waste treatment facility. The community, offended at this infringement of local autonomy, prepares to stop the facility by collecting information on the unacceptability of the site and initiates litigation. Both the community's anger and the regulatory assessment/public consultation process encourage community members to overestimate the risk of the proposed facility and to resist any argument that some package of mitigation, compensation and incentives might actually yield a net gain in the community's health and safety, as well as its prosperity (Sandman, 1986).

People will participate more if they exercise some real control over an ultimate decision. In response to this need, regulatory agencies are trying to encourage public participation on various issues. Many previous public participation exercises have been perceived as being too little too late, and often involving only draft decisions. As a result, many members of the public believe that they will not be taken seriously.

11.4.4   The Media

Most people have little contact with the media. When they find themselves at the centre of a story or expected to provide information, they usually prefer that the media would go away and leave them to do their jobs. Since it is highly likely that the media will persist, it is imperative that managers understand how to communicate effectively with the media and to anticipate the sorts of information that are likely to be of greatest interest. While communications often become a source of confusion, the risks of avoiding the media may be far greater than the risks of working with them (Sandman, 1986).

It is important to remember that a reporter's job is news, not education: events, not issues or principles. The news is the "risky" thing that has happened, not the difficult determination of the numerical risk value. A reporter may seek answers to simple direct questions such as:

- What happened?
- How did it happen?
- Who is to blame?
- What are the authorities/company doing?

The media focuses on the politics of risk rather than the science of risk (Sandman, 1986).

A company or organization at the centre of a risk management story needs to make its position known and to realize that the story will be covered, whether or not it arranges to be included. Environmental risk stories often turn into political stories in part because political content is more readily available and understood than technical content.

In responding to the media, keep the following points in mind (Sandman, 1986):

- provide facts; never guess or lie
- if you do not know say so, but get back to the reporter
- remember that journalists have daily deadlines not monthly
- decide on the main points of your story in advance
- stress points consistently and repetitively
- leave out technical terms, jargon and qualifiers whenever possible
- leave in important (essential) terms and qualifiers and explain them

Members of the media may push for a response that is clearly on one side or another of a situation (for example, safe versus unsafe, legal versus illegal.) They do not want to dwell on the nuances of intermediate positions because the length of their news story seldom allows for a lengthy response. Managers, scientists and technical staff often resent the pressure from members of the media to dichotomize and simplify an issue.

11.5   Summary

Risk assessment and risk management are likely to play an increasingly important role in environmental decision-making. The realization that there are insufficient resources to address all environmental concerns will continue to support the need for risk assessment and management.

Risk assessment often is described as consisting of four parts: hazard identification/problem identification, exposure assessment, dose-response or toxicity assessment, and risk characterization.

The information obtained from a risk assessment is only part of the overall package needed to conduct risk management. Risk management is an interactive process of identifying the options available for abating unacceptable risks, evaluating the cost-effectiveness of those options and identifying the preferred course of action for achieving the desired risk reductions.

An integral part of risk management is risk communication. It is imperative that managers understand how to communicate effectively with the media and the public and to anticipate the sorts of information that are likely to be of greatest interest. Risk-related information may need to be presented to the public in response to regulatory requirements or voluntary programs. It is important that the information be communicated effectively.

References

Deisler, P.F. 1988. "The Risk Management-Risk Assessment Interface". *Environ. Sci. and Technol.* 22:1: 15-19.

Dwyer, J.P. and P.F. Ricci. 1989. "Coming to Terms with Acceptable Risks". *Environ. Sci. and Technol.* 23:2: 145-146.

Fischhoff, B. 1985. *Issues in Science and Technology* 2:1: 83-96.

Lioy, P.J. 1990. "Assessing Total Human Exposure to Contaminants — A Multidisciplinary Approach". *Envir. Sci. and Technol.* 24:7: 938-945.

Mackay, D. 1982. "Nature and Origin of Micropollutants". *Wat. Sci. Tech.* 14: 5-14.

National Academy of Science (NAS). 1983. *Risk Assessment in the Federal Government: Managing the Process.* National Academy Press.

Ontario Ministry of Environment and Energy (MOEE). 1997. *Guidance on Site Specific Risk Asssessment for Use at Contaminated Sites in Ontario.* PIBS 3267E01.

Ontario Ministry of Environment and Energy (MOEE). April 1994. *Public Consultation.* Guideline H-5 [formerly 16-09].

Sandman, P.M. 1986. *Explaining Environmental Risk.* U.S. Environmental Protection Agency, Office of Toxic Substances, Washington, D.C.

Travis, C.C. and H.A. Hattemer-Frey. 1987. "Human Exposure to 2,3,7,8-TCDD". *Chemosphere* 16: 2331-2342.

United States Environmental Protection Agency (EPA). 1989. *Risk Assessment Guidance for Superfund. Volume 1. Human Health Evaluation Manual (Part A).* Office of Emergency and Remedial Response, Office of Solid Waste and Emergency Response, U.S. EPA, EPA 540/1-89/002.

United States Environmental Protection Agency (EPA). 1999. *Risk Assessment Guidance for Superfund. Volume 1. Supplement to Part A.* Office of Emergency and Remedial Response, Office of Solid Waste and Emergency Response.

Table 11.1
Sources of Information about Risk Communication

The Agency for Toxic Substances and Disease Registry (ATSDR) has a primer on risk communication that can be obtained at **www.atsdr. cdc.gov/HEC/ primer.html**.

The Canadian Standards Association (CSA) published *Risk Management: Guideline for Decision-Makers*. CAN/CSA-Q850-97. 1997.

The Canadian Chemical Producers' Association (CCPA), in it's document entitled *Responsible Care: A Total Commitment* includes a community right to know policy, which recognizes the need and the right of the public to know the risks associated with the operations and products present in or transported through communities.

The Centre for Risk Communication can be contacted at **www.centerforriskcommunication.com**.

The Harvard Centre for Risk Analysis can be contacted at **www.hcra.harvard.edu**.

The Institute for Risk Research can be contacted at **http://irr.uwaterloo.ca**.

"Risk Analysis" is the journal of the Society of Risk Analysis. The SRA can be contacted at **www.sra.org**.

The U.S. Environmental Protection Agency has published several documents that focus on risk communication. In addition to those cited as references in this chapter, two additional examples are:

- *Superfund Community Involvement Handbook and Toolkit*. 1998.
- *General Guidance for Risk Management Programs (40 CFR Part 68). Office of Solid Waste and Emergency Response. EPA 550-B-00-008. 2000.*

Some frequently cited authors of books and articles about risk communication and one reference for each include:

V. Covello. *Seven Cardinal Rules of Risk Communication*. U.S. Environmental Protection Agency, Office of Policy Analysis, Washington, D.C. 1988 (with E. Allen).

B. Fischhoff. "Risk Perception and Communication Unplugged: Twenty Years of Progress". In *Risk Analysis*, 1995, 15:2: 137-145.

G. Morgan. *Risk Communication: A Mental Models Approach*. Cambridge University Press, 2001 (with B. Fischhoff).

P. Sandman. Various articles can be obtained at **www.psandman.com**.

P. Slovic. *Risk Communication, Risk Statistics, and Risk Comparisons: A Manual for Plant Managers*. Chemical Manufacturers' Association, Washington, D.C. 1988 (with V.T.Covello and P.M. Sandman).

Table 11.2

EPA Rules and Guidelines for Effective Risk Communication

Rule 1. Accept and Involve the Public as a Legitimate Partner
Guideline: Demonstrate your respect for the public and your sincerity by involving the community early, before important decisions are made. Make it clear that you understand the appropriateness of basing decisions about risks on factors other than the magnitude of the risk. Involve all parties that have an interest or a stake in the particular risk in question.

Rule 2. Carefully Plan and Evaluate Performance
Guideline: Begin with clear, explicit objectives such as providing information to the public, motivating individuals to act, stimulating emergency response, or contributing to conflict resolution. Classify the different subgroups among your audience. Aim your communication at specific subgroups. Recruit spokespersons who are good at presentation and interaction. Train your staff, including technical staff, in communication skills, rewarding outstanding performance. Whenever possible, pretest your messages. Carefully evaluate your efforts and learn from your mistakes.

Rule 3. Listen to Your Audience
Guideline: Do not make assumptions about what people know, think, or want done about risks. Take the time to find out what people are thinking: use techniques such as interviews, focus groups, and surveys. Let all parties that have an interest or a stake in the issue be heard. Recognize people's emotions. Let people know that you understand what they said. Recognize the "hidden agendas," symbolic meanings, and broader economic or political considerations that often underlie and complicate the task of risk communication.

Rule 4. Be Honest, Frank, and Open
Guideline: State your credentials, but do not ask or expect to be trusted by the public. If you do not know an answer or are uncertain, say so. Get back to people with answers. Admit mistakes. Disclose risk information as soon as possible (emphasizing any appropriate reservations about reliability). If in doubt, lean toward sharing more information, not less – or people may think you are hiding something. Discuss data uncertainties, strengths and weaknesses – including the ones identified by other credible sources. Identify worst-case estimates as such, and cite ranges of risk estimates when appropriate.

Rule 5. Coordinate and Collaborate with Other Credible Sources
Guideline: Closely coordinate all inter- and intraorganizational communications. Devote effort and resources to the slow, hard work of building bridges with other organizations. Use credible intermediates. Try to issue communications jointly with other trustworthy sources such as credible university scientists, physicians, trusted local officials, and opinion leaders.

Rule 6.   Meet the Needs of the Media

Guideline: Be open with and accessible to reporters. Respect their deadlines. Provide information tailored to the needs of each type of media, such as graphics and other visual aids for television. Provide background material for the media on complex risk issues. Follow up on stories with praise or criticism, as warranted. Try to establish long-term relationships of trust with editors and reporters.

Rule 7.   Speak Clearly and With Compassion

Figure 11.1
Risk Assessment and Risk Management Processes

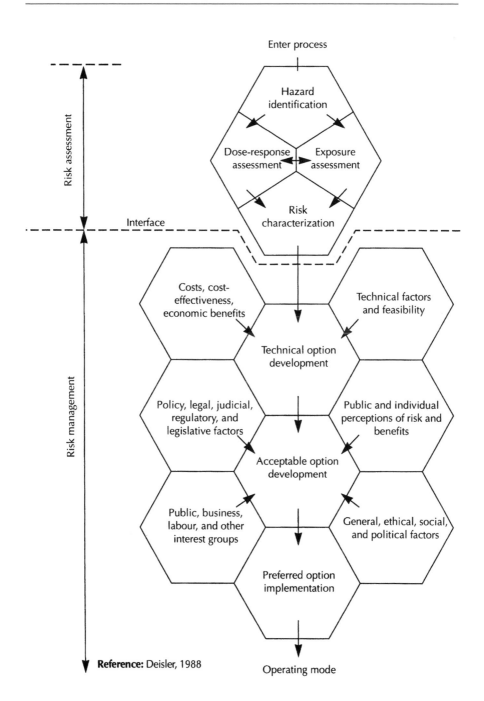

Reference: Deisler, 1988

Figure 11.2

Schematic Plot of Response as a Function of Exposure

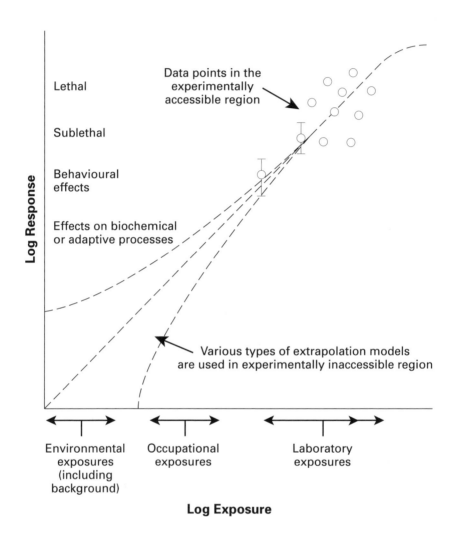

12

Emergency Planning and Spills

12.1 Overview

Despite the attention paid to training, procedures and other precautions, unscheduled releases or discharges of materials to the environment still occur as a result of equipment failure, mishandling, transportation accidents and human error. Approximately 4,000 spills are reported each year to the MOE.

Regulatory aspects of emergency planning and spill response are distributed among the three levels of government. Requirements for federal facilities (research centres, airports, crown land, etc.) and for various types of transportation activities (including airplanes, ships and federal highways) fall under federal jurisdiction. Provincial jurisdiction (in this case, the MOE) extends to any event that can impact the environment. Municipal agencies, such as fire departments and medical officers of health often take on lead roles in responding to emergency situations including spills.

Ontario legislation places far-reaching obligations on companies and individuals to report and clean up spills that can affect the environment. It is a corporation's responsibility to show reasonable care by ensuring that records of inspection and maintenance of equipment are kept, employees receive adequate spill training, and that equipment and procedures are available to respond to spills.

12.2 Federal Acts and Regulations

12.2.1 Canadian Environmental Protection Act

The Canadian Environmental Protection Act, 1999 (CEPA, 1999) prohibits the release of toxic substances into the environment. In the context of CEPA, 1999, a substance is defined as toxic if it can have an immediate or long-term harmful effect on the environment, constitute a danger to the environment on which human life depends, or constitute a danger in Canada to human life or health.

Part 5 of CEPA, 1999 outlines the duty to notify or report if a toxic substance listed in Schedule 1, is released into the environment. The person who owns or has charge of the substance immediately before its release or who causes or contributes to the initial release or increases the likelihood of the initial release, must notify an enforcement officer as soon as possible in the circumstances and make a reasonable effort to notify any member of the public who may be adversely affected by the release or likely release. That person also must provide a written report.

Part 8 of CEPA, 1999 addresses environmental matters related to uncontrolled, unplanned or accidental releases of a toxic substance. Part 8 authorizes the Minister for Environment Canada to require an environmental emergency plan of anyone who manufactures, stores or handles a toxic substance.

In August, 2002, the Minster of the Environment announced proposed Environmental Emergency (E2) regulations. The regulations would have an initial target list of 174 substances and encompass an estimated 1,500 facilities across the country. It is anticipated that the regulations will be finalized in 2003.

12.2.2   Transportation of Dangerous Goods Act and Regulations

As noted in Section 4.9, various materials are defined as dangerous goods by the Transportation of Dangerous Goods Act, 1992 (TDGA, 1992) and associated regulations (TDGR). The term "accidental release" is used in TDGA, 1992 and TDGR to describe events that include spills of dangerous goods.

Anyone who intends to transport or import certain dangerous goods must have an approved emergency response assistance plan (ERAP). Part 7 of TDGR gives recommendations for the contents of an ERAP and describes the process for submitting an ERAP for review and approval by Transport Canada. Part 7 and Schedule XII of TDGR prescribe the dangerous goods and the concentration or quantity that make an ERAP necessary.

In the event that dangerous goods are released in amounts that exceed the quantities shown in Table 12.1, there is a duty to respond immediately. That duty is described in Part 8 of TDGR. It applies to the person who has charge, management or control of dangerous goods at the time that person discovers or is advised of the release or that such a release is imminent.

The parties to be notified of an accidental release include:

- the appropriate authority in the province in which the goods are located (in Ontario, this is the local police)
- the person's employer
- the owner or consignor of the dangerous goods
- the owner, lessee or charterer if a road vehicle is involved
- the Canadian Transport Emergency Centre (CANUTEC) if a railway vehicle is involved
- CANUTEC, a Canadian Coast Guard radio station or a vessel traffic services centre if a ship is involved

- CANUTEC and the nearest regional civil aviation office of the Department of Transport if an aircraft, aerodrome or air cargo facility is involved

The types of information to be reported are described in Part 8 of TDGR and include the location of the accidental release, the shipping name or UN code of the dangerous goods, the quantities of goods involved, the numbers of deaths and injuries that have resulted, and an estimate of the number of people evacuated.

A follow-up report in writing to the Director General of the Department of Transportation must be made within 30 days by the employer of the person who had possession of the goods at the time of the event. The recommended contents of such a report are described in Part 8 of TDGR.

Upon being notified of an accidental release, Transport Canada may send staff to the site. These experts can provide advice, but also can investigate to determine if regulations were followed, and assess the validity of an ERAP.

TDGA, 1992 also sets out various administrative issues such as the powers of inspectors to monitor for compliance, the potential punishments for failure to comply and the recovery of reasonable costs and expenses incurred by the Crown.

Many accidental releases also are classified as spills under Part X of the EPA and therefore also carry requirements for reporting to provincial agencies (see Section 12.4).

12.2.3   Fisheries Act

The Fisheries Act is administered by the Department of Fisheries and Oceans. It has provisions that apply to spills into fish-bearing waters. A spill of a deleterious substance into fish habitat triggers duties for any person who owned or controlled the substance or contributed to causing the dangerous situation. Such people must take all reasonable measures to counteract the adverse effects of the spill. There also is a provision to notify an inspector or authority as prescribed in regulations, but such regulations have not been issued.

Fisheries and Oceans inspectors can direct those responsible for a spill to take appropriate spill response measures, or can take those measures themselves. The Crown then can recover costs and expenses.

12.2.4   Canada Shipping Act

The Canada Shipping Act authorizes the federal government to regulate both planned and accidental discharges from ships of pollutants specified in several associated regulations. It is administered by the Department of Transport.

In general, the master of a ship must report discharges of pollutants named in the Canada Shipping Act regulations without delay to a pollution prevention officer at the Canadian Coast Guard. The types of information to be reported are described in the Pollutant Discharge Reporting Regulations, 1995 (SOR/95-351) and in a guideline published by the Department of Transport.

Larger ships must have arrangements with certified response organizations for oil spills. There are response organizations for eastern Canada, western Canada and the Great Lakes, as well as a co-ordination centre in Ottawa.

12.2.5   Emergency Preparedness Act

The Emergency Preparedness Act requires federal departments to develop and maintain emergency plans relevant to the nature and mandate of the department. For example, Environment Canada has a plan that addresses the identification, assessment and mitigation of environmental hazards (Environment Canada, 1998).

12.3   Related Federal Programs and Facilities

12.3.1   Environmental Emergencies Program

Environment Canada is responsible for developing and maintaining a national spill reporting system and database.

The National Environmental Emergencies Centre (NEEC) has its national headquarters in Ottawa and several regional offices. In Ontario, the regional office in Toronto can be reached at (416) 739-5908 during normal business hours.

Callers can report spills, leaks, fires or other events they believe may impact the environment. The majority of spill reports received at NEEC are sent to regional offices. Environment Canada has information exchange agreements with provincial and territorial counterparts and many reports are received from those sources. (The Ontario Spills Action Centre is described in Section 12.5.2.)

The National Analysis of Trends in Emergencies database was established in 1973. It contains voluntarily reported pollution incidents across Canada. The most recent summary of results covers the period 1984 to 1995 (Environment Canada, 1998).

The National Pollutant Release Inventory (NPRI) contains information on all types of environmental releases, not just spills. NPRI reporting is mandatory for those who manufacture, process or otherwise use numerous substances specified by CEPA, 1999.

There are slight differences in spill reporting requirements among federal, provincial and territorial agencies. Efforts are underway to harmonize these requirements to improve the exchange of spill information.

The Environmental Emergencies Program Web site address is **www.ec.gc.ca/ee-ue**.

12.3.2   National Sensitivity Mapping Program

Environment Canada has the lead responsibility for keeping up-to date-maps of environmentally sensitive areas and seasonal considerations affecting key physical, biological and cultural resources. Such maps can be an important source of information when preparing spill response plans and during the implementation of such plans.

12.3.3   CANUTEC

The Canadian Transport Emergency Centre is operated by Transport Canada to assist responders at emergencies that involve dangerous goods. CANUTEC staff includes scientists specializing in emergency response and providing advice. It maintains extensive records on the chemicals manufactured, stored and transported in Canada as well as lists of specialized equipment suppliers and directories of emergency response groups across Canada.

CANUTEC receives approximately 30,000 calls annually, of which approximately 1,000 require immediate reporting. It offers a 24-hour emergency telephone service for shippers who have registered with CANUTEC. The general telephone number for information or to register is (613) 992-4624. Further information about CANUTEC can be obtained at **www.tc.gc.ca/canutec**.

12.3.4   Federal and International Contingency Plans

Federal agencies play important roles in several federal contingency plans. Two examples are the National Environmental Emergencies Contingency Plan (Environment Canada is the lead agency) and the Marine Spills Contingency Plan (Canadian Coast Guard is the lead agency).

Canada also participates in international contingency plans. Examples are the Canada – United States Joint Inland Pollution Contingency Plan (Environment Canada is the lead agency) and the Canada/U.S. Joint Marine Contingency Plan (Canadian Coast Guard is the lead agency).

12.4　Provincial Acts and Regulations

12.4.1　Environmental Protection Act

Part X of the EPA deals specifically with spills. It establishes a prompt notification requirement and a duty to respond. It applies to spills that have occurred since November 29, 1985. Several key terms are described in s. 91 of EPA:

- **spill** – a discharge of a pollutant into the natural environment from a container, structure or vehicle, which is abnormal in quantity or quality
- **pollutant** – a contaminant other than heat, sound, vibration or radiation and including any substance from which a pollutant is derived
- **substance** – any solid, liquid or gas, or combination thereof
- **owner of a pollutant** – the owner immediately before the first discharge
- **person having control of a pollutant** – the person, employee or agent having charge, management or control immediately before the first discharge

Two other key terms to understanding Part X are "discharge" and "natural environment." Section 1 of the EPA indicates that discharge (when used as a verb) can mean to add, deposit, leak or emit, while natural environment includes the air, land and water, and any combination thereof in Ontario.

The Duty to Report

Section 92 of the EPA describes the duty to report (or notify). This duty applies to every person having control of a spilled pollutant and every person who causes or permits a spill of a pollutant that causes or is likely to cause an adverse effect. The onus is on that person to determine if a spill is causing or is likely to cause an adverse effect. If any of the adverse effects described in s. 1 of EPA is likely, a report must be made immediately to the MOE (see Section 12.5), the municipality, the owner of the pollutant and the person having control of the pollutant. This duty comes into force as soon as a person learns of a spill or should know about a spill. The term "immediately" is interpreted to mean as soon as reasonably practicable in the circumstances. The MOE should be notified as quickly as possible, even if complete information about the event is not available.

The potential for adverse effects to be caused is essential for triggering the duty to report a spill. The EPA definition of adverse effects ranges from impairment of any part of the natural environment, to the loss of enjoyment of the property, to discomfort. All of these concepts rely to some extent on the quantity of material spilled as well as its chemical, physical and toxicological characteristics.

Part X does not specifically address the possibility that small amounts of material may be spilled without causing an adverse effect. Nor does it provide minimum reportable quantities. This lack of direction left it to individuals to decide whether to report small spills. As a result, many small spills were reported. MOE records show that when spill volumes were reported during the 1990s, approximately 25% involved less than 10 litres (MOEE, 1997).

Deciding how to handle small spills is difficult since unwanted results can occur just as easily from choosing to report very small spills as from choosing to report only spills of a certain volume or size. A company may report all spills and feel that it is being a good corporate citizen, however, the associated effort may cause a large portion of the environmental staff's time and the company might gain a reputation of having frequent incidents, particularly if the number of spills is reported to the public without corresponding text about the volumes of material involved.

This condition has been clarified somewhat with the exemptions provided in EPA regulations (see Section 12.4.2).

The Duty to Respond

Section 93 of the EPA describes the duty to respond (or act). This duty requires that the owner of the pollutant and the person having control of the pollutant shall do everything practicable to prevent, eliminate and ameliorate adverse effects and to restore the natural environment. It is interpreted as consisting of three parts:

- stop the spill from continuing
- remove or render harmless the pollutant and everything that has been contaminated, and
- restore the natural environment to a state as if the spill did not occur

The duty to report and the component of the duty to respond that involves stopping the spill from continuing can conflict with one another. In an ideal world, sufficient personnel would be present to stop a spill from continuing and, at the same time, notify appropriate parties. Unfortunately, in many cases there are insufficient personnel to perform both tasks simultaneously.

It is usually during the first few minutes of a spill that the greatest mitigation may be performed such as turning off valves or laying down adsorbent material. Accordingly, the priority should be on performing short duration tasks, getting assistance from others, and notifying the MOE. It is important to avoid the mentality that "If I do a little bit more, I will not have to report it."

The phrase "do everything practicable" makes the scope of this duty somewhat vague, but acknowledges that limitations may exist when attempting to respond and restore the natural environment. At one time, this section indicated that consideration should be given to the technical, physical and financial resources that are or can reasonably be made available. That wording has been removed from s. 93. A clear definition of "practicable" remains elusive.

The restoration or clean-up component of the duty to respond falls on the owner of the pollutant and the person having control of the pollutant. The duty does not fall on the owner of a contaminated site if the owner took reasonable precautions to prevent the contamination, however, a purchaser of contaminated property becomes the "owner of the pollutant" and thereby becomes responsible for remediating a site contaminated by others. This type of undesirable outcome has prompted prospective buyers to perform environmental site assessments (see Section 5.4). It also has put pressure on property owners and potential vendors to perform assessments and remediate properties in advance of offering them for sale.

MOE Responses to Spills

The MOE is most likely to learn of a spill by receiving a report at the Spills Action Centre (SAC) (see Section 12.5). Its role initially is determined by how well MOE guidelines for spill response are being followed (MOEE, 1994). Three levels of response are described in Section 12.5. If MOE staff are sent to investigate a spill, they will be interested in gathering information that can help characterize the spill event, determine the nature and extent of environmental damage and evaluate the adequacy of the clean-up and restoration efforts. All findings, actions and recommendations likely will be documented. Whether the documents are transferred to the Investigation and Enforcement Branch will depend on the circumstances surrounding the spill.

Sections 94 through 97 of the EPA provide the MOE with several administrative tools for responding to spills. If responsible parties fail to clean up a spill site, s. 94 empowers the MOE to initiate clean-up and to sue the responsible parties for the costs. This decision rests with the Minister. Section 97 authorizes the Minister to order clean-up assistance from anyone considered necessary. Ordered persons must act with monetary matters being resolved later. Section 95 enables superior court judges to provide power of entry and the ability to commandeer equipment to anyone who has a duty to respond to a spill or clean-up.

A regional director can apply to the Environmental Clean-Up Fund in emergency situations where the owner of the spilled pollutant cannot be identified or located, has refused, or is unable to take the necessary action (MOE, 2001).

Compensation and Distribution of Costs

Section 99 describes who is financially responsible for costs arising from spills. Direct costs include materials and equipment used to contain, retrieve or dispose of the pollutant or contaminated materials, and the costs to restore the environment, such as replacing damaged or destroyed vegetation. Consequential damages are those caused indirectly by a spill such as food spoiled by a disruption of electricity, the accommodation of temporarily displaced persons and forced closures of businesses. Responsible parties are liable for all the direct costs associated with cleaning up a spill and any consequential damages that result from their fault or negligence during a spill or clean-up.

At one time, applications for compensation could be made to the Environmental Compensation Corporation for loss or damage incurred as a direct result of the spill of a pollutant or neglect or fault in carrying out an order or direction under the Act. This corporation has been dissolved.

Regulatory Requirements to Have Spill Plans

In Ontario, there is no broad regulatory requirement for emergency response plans or spill contingency plans to be in place at facilities, although they are strongly encouraged. Section 18 of the EPA authorizes MOE directors to order persons who own or manage an undertaking or property to prepare contingency plans when the director has reasonable and probable grounds for doing so.

12.4.2   EPA Regulations that Pertain to Spills

In addition to Part X, the following regulations have been issued under the EPA that provide further details about spills and their management in Ontario:

- Classification and Exemption of Spills (O. Reg. 675/98)
- Spills (R.R.O. 1990, Reg. 360)

Regulation 360's name belies its function. It is limited in scope to describing various aspects of claiming compensation from the Crown and the Environmental Compensation Corporation. Until 1998, it also included some sections that exempted several types of spills.

With the issuing of O. Reg. 675/98, the MOE demonstrated the intention to reduce the reporting of spills that are "trivial and frivolous". The regulation describes 11 classes of spills that are exempt from one or more of the duties described in Part X:

1. Class I spills are those that are authorized and approved by the MOE. These are exempt from all regulatory requirements.
2. Class II spills are those that involve water from reservoirs and water mains. These are exempt from all regulatory requirements.
3. Class III spills involve the pollutants produced by the fires from 10 or fewer households. These are exempt from all regulatory requirements.

4. Class IV spills are those that are planned for research or training, to maintain water or waste water treatment systems, and for which the MOE has provided consent. These are exempt from the reporting requirements.

5. Class V spills involve refrigerants in accordance with the O. Reg. 189/94, refrigerants regulation. These are exempt from the reporting requirements.

6. Class VI spills involve less than 100 L of fuels, lubricants or cooling liquids from vehicles. The spill does not cause an adverse effect and is remediated immediately. These are exempt from the requirement to notify the MOE.

7. Class VII spills involve less than 100 L of mineral insulating oil from electrical transformers or capacitors (excluding PCB liquid) owned by a municipal or provincial utility that do not cause an adverse effect and are remediated immediately. These are exempt from the requirements to notify the MOE and the municipality.

8. Class VIII spills involve less than 1,000 L of petroleum product and occur in areas restricted from public access, do not cause an adverse effect, and are remediated immediately. These are exempt from the requirements to notify the MOE and the municipality.

9. Class IX spills involve materials that make the spill subject to the immediate notification requirements of TDGA, 1992 or the provincial Dangerous Goods Transportation Act (see Chapter 4), do not cause an adverse effect, and are remediated immediately. These are exempt from the requirements to notify the MOE and the municipality.

10. Class X spills are those described as "non-reportable" in a spill contingency plan approved by the MOE or that adheres to standard CAN/CSA-Z731-95. These are exempt from the requirements to notify the MOE.

11. Class XI spills are those that are reportable to another provincial or federal agency if there is a one window memorandum of understanding between that agency and the MOE. These need to be reported to the MOE but not immediately.

For Classes V, VII, VIII, IX, X and XI, records of spills shall be kept for at least two years and made available if requested by the MOE. O. Reg. 675/98 describes various types of information that should be included in the records.

O. Reg. 675/98 was amended by O. Reg. 204/01. The amendment clarifies the volumes of materials in Class V spills.

These changes are expected to produce significant reductions in the number of spills reported to the MOE (Environmental Compliance, 2001).

12.4.3   Ontario Water Resources Act

The OWRA has similar notification requirements to those of the EPA. For example, s. 30(2) requires every person that discharges or causes or permits the discharge of any material of any kind, and such discharge is not in the normal course of events, or material of any kind escapes into any waters or on any shore or bank or into any place that may impair the quality of any waters, shall immediately notify the Minister.

An important difference between the EPA and OWRA notification requirements is the use of the term may in the OWRA versus likely in the EPA with respect to adverse effects or impairment. It is less difficult to prove that a contaminant may impair (the OWRA phrase) than it is to prove that a pollutant is likely to cause adverse effects (the EPA wording).

12.4.4   Technical Safety Standards Act

In 2001, seven statutes were replaced by the Technical Safety Standards Act, 2000. These included the former Gasoline Handling Act and regulations that described various requirements for spills of some petroleum products. Under the new Act, there is the Liquid Fuels Handling Regulation (O. Reg. 217/01), the Liquid Fuels Handling Code and the Environmental Management Protocol for Operating Fuel Handling Facilities in Ontario (GA 1/99). All of these plus the statute are administered by the Technical Standards and Safety Authority (TSSA).

O. Reg. 217/01 contains the general notification requirement that a spill or leak of petroleum product into the environment or inside a building shall be reported to the director of the Fuels Safety Division, TSSA.

The Code repeats the notification requirement found in the regulation, indicates that notifying the MOE SAC is sufficient, and directs that investigative and mitigative measures outlined in the protocol be taken to the extent practical.

The protocol requires that any spill of a petroleum product in excess of 100 L at sites restricted from public access (bulk facilities), or 25 L at sites with public access (retail service stations) must be reported immediately to the director of the Fuels Safety Division, TSSA. Lesser spills need not be reported unless the spill would create a hazard to public health or safety, contaminate a fresh water source or waterway, interfere with the rights of any person, or enter a sewer system, underground stream or drainage system.

Where a spill may cause an adverse effect or impact a drinking water supply, the regulatory lead over environmental remediation will be transferred from the TSSA to the MOE even if the facility remains operational.

The investigative measures described in the protocol include recovery of escaped product and determining the extent of the contamination. Depending on the types of impacts, restoration or implementation of a contamination management plan is required. Immediate corrective action is needed to:

- eliminate all liquid phase-separated product evident on the surface or in the subsurface
- eliminate any potential explosion hazards in enclosed spaces
- eliminate the potential for off-site migration of petroleum product and related contaminants.

More information about the TSSA can be obtained at **www.tssa.org**.

12.4.5   Pesticides Act

The Pesticide Act has similar notification requirements to those of the EPA. For example, s. 29 requires every person that discharges or causes or permits the discharge into the environment of a pesticide or substance or thing containing a pesticide that is out of the normal course of events, that causes or is likely to cause an adverse effect, shall forthwith notify the Minister. Such a person also is responsible for taking steps to prevent such releases and to remediate such releases, but the Pesticides Act does not specifically describe emergency plans.

12.5   Related Provincial Programs and Facilties

12.5.1   Province of Ontario Contingency Plan

The MOE maintains the Province of Ontario Contingency Plan for Spills of Oil and other Hazardous Materials that is used to co-ordinate the provincial response to major spills.

12.5.2   Spills Action Centre

The primary role of the SAC is to receive reports of spills (and other urgent environmental issues), and then initiate and help co-ordinate responses to the reports. SAC can be reached 24 hours a day at: 1-800-268-6060.

SAC staff will determine whether the report is to be handled as a spill or as an environmental complaint. If it is a spill, SAC staff may ask for the following information:

- name and address
- telephone number/call back number
- circumstances of the spill
- material spilled (including chemical composition)
- source of spill
- quantity spilled
- weather conditions
- action being taken to clean up the spill
- involvement of other agencies

The SAC works closely with response agencies such as police, fire departments, the Canadian Coast Guard and other reporting or information centres such as CANUTEC. More information about SAC can be obtained at **www.ene.gov.on.ca/spills.htm**.

Depending upon the nature of the spill, SAC can activate three levels of response (MOE, 1998). A Level 1 response is one where SAC determines that it is appropriate to have an environmental officer from the MOE district or area office visit the spill site and check to see that spilled materials are being contained and disposed of by those who have a duty to respond. Most spills warrant this level of response.

A Level 2 response is one where SAC determines that it is appropriate to respond through one of the MOE regional offices. Examples include situations where SAC determines that first responders at the spill need environmental expertise, immediate monitoring, or on-site guidance from MOE staff. This type of response typically happens a few times a year.

A Level 3 response is one where SAC determines that the response requires expertise or resources beyond that found at the regional level. When a populated area may be close to an event that is releasing vapours or fumes, the MOE may mobilize one of its trace atmospheric gas analyzer units to perform air monitoring, including real-time measurements of some chemicals. This type of response typically happens a few times a year.

SAC records all telephone conversations for future reference and as a basis for potential prosecution.

12.5.3   Occurrences Report Information System

All occurrences reported to SAC are recorded on the computerized Occurrences Report Information System. This system facilitates tracking occurrences and generating summary statistics.

Annual summaries of spills prepared by SAC in the 1990s (for example, MOEE, 1993; MOEE, 1997) show some relatively consistent patterns:

1. Approximately 5,000 spills were reported to SAC annually. (This recently has declined to approximately 4,000 spills annually. This decline likely is due to the exemptions provided by O. Reg. 675/98 and companies making greater efforts to prevent spills.)

2. Approximately 60% of reported spills involved oils or fuels. Many of these occurred as a result of transportation accidents or leaks from storage facilities. Chemicals or chemical solutions accounted for approximately 15% of the spills, wastes and wastewaters another 15%, and gaseous emissions about 5%. Other and unknown materials accounted for the remainder.

3. The majority of spills involved small quantities of materials. For those events where quantities were reported, approximately 25%

involved less than 10 L, 60% involved less than 100 L, and 90% involved less than 1,000 L.

6. Approximately 45% of all spills were completely cleaned up, while 22% were partially remediated.

7. Motor vehicles were the most common source of spills, accounting for about 30% of all spills. Spills from other modes of transportation were relatively infrequent. Manufacturing and processing facilities accounted for about 20% of spills.

8. The most frequently cited causes for spills were equipment failure and operator error.

12.6   The Role of Municipalities

While municipalities in Ontario typically have no emergency regulations, municipal agencies such as a fire department or the municipal medical officer of health often are the first to respond to emergency situations including spills and make decisions on important matters such as evacuation. Traditionally, these types of municipal services had emergency response plans for their specific areas of responsibility.

Under the Emergency Planning Act of 1983, municipalities in Ontario may develop emergency plans for disasters (including large spills) to protect the property, health, safety and welfare of inhabitants. In recent years, several cities and regional governments have created broader plans that can co-ordinate and more effectively implement staff and resources.

Section 100 of the EPA acknowledges the right of a municipality to prevent, eliminate and ameliorate the adverse effects caused by a spill and to restore the natural environment if it chooses to do so. Municipalities can call upon the MOE to help assess the environmental aspects of a spill.

12.7   Spill Prevention

Spill prevention is a cornerstone of environmental protection. The following steps can be taken to prevent spills of pollutants into the environment.

1. Process or system modification: In the long run, modifications to processes may be the most cost-effective way to reduce or prevent spills. Modifications can eliminate the need to use certain substances, eliminate the ways in which the substances are handled, or eliminate the use of certain processes. An example of a system modification would be the use of recirculating cooling water as opposed to once-through cooling water. A leak of a chemical into the cooling water would stay within the system and provide an opportunity for corrective action before any of the chemical is been released to the environment.

2. Chemical substitution: There is a growing trend towards replacing hazardous chemicals with non-hazardous chemicals. Wherever practical, chemicals that are spilled frequently or in large volumes should be substituted with products that pose lesser environmental concerns. Parameters that should be reviewed when assessing possible substitutes are toxicity (the concentration at which 50% of test organisms do not survive), persistence (the susceptibility of the substance to degradation) and bioaccumulation (the ability of the substance to accumulate in humans and wildlife). For example, prior to selecting a hydraulic oil, data could be collected concerning operational specifications, toxicity, persistence and costs (both purchase and disposal). If more than one product was able to meet the operational specifications, the next two criteria would be assessed with the goal of reducing the environmental impact of a potential spill.

3. Detection systems: Identify all the indicators that can signify a spill and determine which are the earliest and most reliable detectors. Indicators can include mechanical and electronic meters, gauges and alarms, but also include routine procedures used to inspect and monitor various types of activities or functions. The best indicators of an unplanned or unscheduled loss of a substance are site specific and will change with time as processes and facilities are modified.

4. Planned maintenance: As noted in Section 12.5.3, SAC records show that equipment failure is a frequent contributor to spills whether directly or indirectly.

5. Back-up equipment and systems: Because many spills result from the failure of equipment, some form of back-up or redundancy in equipment and systems is being seen as the only way in which spills can be prevented. It also is looked on by the MOE as an integral part of spill prevention.

6. Containment: Berms, dykes or storage ponds can be used to collect and treat released substances in controlled manners. Adequate containment can prevent releases from reaching the environment and thus prevent the release from being classified as a spill. Containment may allow a facility to continue to operate after a release occurs and thus avoid or reduce down-time.

7. Hazard assessment: Hazardous operations assessment and fault tree analysis are two types of procedures frequently used to assess the appropriateness and effectiveness of equipment, procedures and resources available to prevent and respond to releases, spills and other types of upset conditions. These types of assessment can identify areas of potential vulnerability, deficiencies in design or scheduled maintenance, etc. In addition, a more specific assessment of the risk of spills should be based on a review of spill records and

identify high-risk areas and causes of spills. This helps set priorities for corrective and preventative actions.

8. Employee training: Regular updates of spill notification and response procedures as well as hazardous chemical handling, storage and disposal are important. At some facilities, "spill drills" analogous to fire drills are practised to ensure that all parties know their responsibilities and can carry them out in an effective manner.

12.8 Developing Emergency Response Plans

12.8.1 Initial Response

It is the responsibility of the owner of the facility where the emergency has occurred or the owner of the material being released to initiate emergency response activities. The initial steps or activities of response plans typically focus on safety, stopping the spill from continuing and notification. Accordingly, the sections of emergency plans that address initial response activities often include:

- maps showing evacuation route(s)
- maps indicating location of potential hazards accompanied by brief description of each
- descriptions of alarm systems and locations
- shutdown procedures and valves/switches locations
- contact information for individuals responsible for shutdown
- "call-out" procedures for employees and visitors
- locations of closest residences
- contact information for MOE, municipality, police, fire department, etc.

Contact information should be included for both internal and external parties. Names, titles and telephone numbers (including home telephone numbers for employees) should be available. It is essential that this portion of the response plan be updated frequently.

In the event of an off-site spill or a spill at a place with off-site effects, the initial response should include the immediate dispatching of a company representative to the site to ensure that all action is being taken to mitigate the effect of the spill and that the clean-up and restoration is being done properly and in compliance with the applicable federal and provincial requirements.

12.8.3 Clean-up Procedures

Once the initial steps of an emergency response plan have been implemented, attention will turn to the tasks of containment, recovery and clean-up.

An emergency response plan should contain sections that address:

- classifying unscheduled releases or events
- determining suitable actions
- determining the types and amounts of equipment, supplies and personnel required to implement suitable actions
- contact information for locating sources of equipment, supplies and personnel
- environmental monitoring procedures

As noted in Section 12.4.1, releases to the environment in Ontario must be classified on the basis of whether or not adverse effects are likely to occur. If adverse effects have occurred or are likely to occur, the release is classified as a spill, and the requirements of EPA Part X must be acted upon immediately. If the consensus is that the release is not a spill, internal procedures should be followed.

A large portion of the response plan may be devoted to describing various response options and the decision points for determining when specific options are suitable or necessary. The array of possible actions should reflect the substances involved, the types of events that can occur, the potential locations of spills, the locations of neighbours and sensitive environments, the ways that released materials may move in the environment, the types of adverse effects that could occur, the accessibility of the general area and specific sites of concern, and the proximity of potential sources of assistance.

Each type of action that might be taken to contain, recover or clean up after a release will require specific types and amounts of equipment, supplies and personnel to be implemented successfully. A list of contractors, suppliers and other potential sources of assistance should be kept and updated frequently. Some facilities have clean-up contractors on retainer to ensure a better response time. In addition, a list of registered haulers should also be kept.

Table 12.2 presents equipment that should be accessible to respond to spills. The list includes equipment to contain, recover and clean up spilled material; monitoring equipment; protective and foul weather gear for workers; and various types of useful materials. Relatively simple items such as plywood sheets, rolls of heavy plastic, hay bales, empty drums, portable pumps and generators have proven to be very useful in spill response. A totally encapsulated suit and self-contained breathing apparatus should be available for situations where there is the possibility of hazardous or unknown vapours.

12.8.4   Monitoring and Documentation

For spills reported to the MOE, the MOE may require environmental monitoring during and following clean-up. Monitoring will be needed to assess the types and extents of any adverse effects that have occurred, and to assist in identifying further remedial actions. The monitoring may include collection and chemical analysis of soil, surface water, ground water, air, aquatic organisms, etc. Short-term toxicity tests may be performed (see Chapter 13). Long-term monitoring efforts might investigate changes in the benthic community (bottom dwelling organisms), levels of chemicals in sediment and vegetation, bioaccumulation in aquatic organisms, residual toxicity and contamination of food resources. It may be necessary and beneficial to obtain information about background or ambient conditions to determine whether adverse effects have truly occurred.

The final stage of emergency response is to create a positive paper trail of investigation into the spill, development of recommendations and implementation of the recommendations be undertaken to eliminate or at least mitigate future spills. The "tallying" of spills with no further preventative action can undermine a company's claims of reasonable care or due diligence.

12.9   Similar Types of Preventative Plans

The term "best management practice (BMP) plan" has been used by the MOE for several years in the context of preventing damages to aquatic environments. A BMP plan is generally defined as the processes, procedures, actions and activities used to prevent pollutants or hazardous substances from damaging the aquatic environment. The concept of BMP plans is included in the MOE Model Sewer-Use By-Law, which in turn has been incorporated into many municipal sewer by-laws.

A BMP plan should be based on the activities noted below. There are numerous similarities to those described in Section 12.7 for spill prevention.

1. Risk identification/assessment – It is important to identify those activities or locations that need to be addressed in a plan. All ancillary sources need to be examined to determine the potential risk of toxic pollutant or hazardous substances being discharged to receiving waters.

2. Materials inventory – A listing of all hazardous substances and toxic chemicals used, stored or produced on-site should be prepared. Details of this inventory will be commensurate with quantities present and potential access to receiving waters.

3. Materials compatibility – The plan must provide procedures for ensuring that container materials are compatible with the toxic or hazardous substances to be stored, and with the equipment used to fill and move these materials.

4. Preventive maintenance – Equipment and systems should be tested from time to time to uncover conditions that could lead to malfunction or failure. The elements of a preventative maintenance program

include predetermined schedules for conducting the tests and inspections, written records of the test results and written records of the corrective steps taken (such as adjustment, repair or replacement). Existing preventative maintenance plans should be evaluated by qualified plant personnel.

5. Record keeping – Several types of activities and conditions need to be recorded in a way that allows the information to be retrieved. Record-keeping systems need to capture information concerning preventative maintenance, all unscheduled incidents, response(s) and the corrective actions taken. Record systems should be reviewed using written procedures at specified frequencies to ensure appropriate spill protection and procedures are in place.

6. Incident reporting – A system needs to be in place to ensure that incidents such as spills, leaks or other unusual releases are reported as required by various Acts and regulations. Such systems can help minimize the recurrence of incidents and expedite clean-up, as well as comply with legal requirements.

7. Employee training – All employees need to understand the plan and their roles in it. The program should emphasize plant processes and materials, safety hazards, discharge prevention practices, as well as proper and rapid response procedures. Information meetings should be held at least once a year to highlight individual responsibilities. In addition, any incidents, equipment programs or changes to the plan should be reviewed. Spill or environmental incident drills can be used to improve employees' reactions to situations and incidents. These drills should be a fundamental part of any training program.

8. Housekeeping – A clean, orderly work environment is a subtle but constant reminder of the need for a conscientious approach to environmental management at a facility. This can be accomplished by providing staff with periodic training in good housekeeping techniques examples of which include: neat and orderly storage of bags and drums of chemicals; placing materials such as scraps, cuttings and packaging in proper containers; keeping the workplace floor clean; and ensuring that places have been designated for storing all of the raw and waste materials that are handled.

9. Security – The accidental or intentional entry to a facility resulting in sabotage, theft etc. can be prevented by installing a security system. Existing security systems should be detailed in the plan along with any necessary improvements to ensure no toxic chemicals discharge from unauthorized entry.

A similar and more recent version of the BMP plan is the Storm Water Control Study (SWCS) component of the Municipal, Industrial Strategy for

Abatement (MISA) regulations for both direct and indirect dischargers (see Chapter 3). A SWCS needs to examine the control on site spills, leaks and runoff from raw materials storage and handling areas. The steps of such a study should include:

- identify toxic and hazardous materials to be addressed
- identify potential spill sources
- develop incident reporting procedures
- establish inspections and records procedures
- periodically review the study or plan for possible changes
- co-ordinate incident notification, response and clean-up procedures
- establish training programs for personnel
- assist in the interdepartmental co-ordination of plan implementation

12.10   Sharing Emergency Response Information

12.10.1   Government Reports and Other Forms of Assistance

The *2000 Emergency Response Guidebook* was developed jointly by the federal transportation agencies of Canada, the United States and Mexico for use by fire departments, police and other emergency services personnel who may be the first at the scene of an accidental release of dangerous goods (Transport Canada *et al.*, 2000).

Environment Canada provides emergency response information in several forms. Information about the environmental emergencies program can be found at **www.ec.gc.ca/ee-ue/main/main_e.cfm**. Environment Canada also publishes a Spill Technology Newsletter, and hosts annual spill seminars.

In the 1980s, Environment Canada published various helpful documents such as *Environmental and Technical Information for Problem Spills* (EnviroTIPS) and the *Manual for Spills of Hazardous Materials* (Environment Canada, 1984). Both publications were produced to assist in the design of countermeasures for spills and to assess their impact on the environment. There are EnviroTIPS manuals for more than 50 common and widely used hazardous compounds. The *Manual for Spills of Hazardous Materials* provides qualitative and quantitative information on more than 200 chemicals for those responding to, and planning for, spills of hazardous materials. These documents have not been updated recently.

As part of MOE efforts to foster spill readiness, various training films and reports are available from the SAC.

City, municipal and regional governments should be consulted to determine if they have emergency plans, and to identify ways to co-ordinate efforts or resources. Many municipalities have recognized that they need to play a more active role in responding to spills. Hence, municipal emergency response teams have been formed such as the Emergency Control Group of the Regional Municipality of Halton. Established in 1986, the team is comprised of regional staff from several departments and the plan can draw upon the resources of the police, fire department, public works, medical officer of health, ambulance service, community services, community relations, conservation authorities and hospitals (Region of Halton, 2001). Further information can be obtained at **www.region.halton.on.ca**.

12.10.2   Non-Government Organizations

The Canadian Standards Association (CSA) has prepared a widely referenced guide for emergency planning (CSA, 1995).

The Centre for Occupational Health and Safety provides information on thousands of substances as part of its "CHEMINFO" service. Its databases include information on the physical and chemical properties of substances as well as the toxicological hazards they can pose. Many of the databases can be accessed from its Web site at **www.ccohs.org**. Some databases are available to subscribers only.

The Major Industrial Accidents Council of Canada was a non-profit, multi-stakeholder organization that focussed on emergency prevention, preparedness and response activities as they pertained to the manufacture, storage, use, transportation and disposal of hazardous substances. Accident prevention workshops examined specific substances such as hydrofluoric acid, chlorine, propane and ammonia. The council dissolved in 1999, but much of the information it had gathered and reports it had prepared were transferred to the Canadian Association of Fire Chiefs, which has continued to pursue issues related to emergency preparedness in communities. Publications include guiding principles for industry and municipalities to develop emergency preparedness. Further information can be obtained at **www.cafc.ca.**

12.10.3   Trade Associations and Co-operatives

Several industrial associations and local industries have formed spill response groups to share resources and costs so that a cost-efficient and timely response capability can be achieved.

The Transportation Emergency Assistance Plan was developed by the Canadian Chemical Producers' Association as a Canada-wide emergency-wide response program. The plan allows companies to share resources by pooling data, equipment and expertise. It includes procedure manuals and training. There are response teams at sites across Canada that are available at any time. In addition, there are numerous voluntary response centres. The Transportation Emergency Assistance Plan and CANUTEC will contact one another to share information that can help the other during emergency events. More information about the plan can be found at **www.ccpa.ca**.

Similarly, the Canadian Petroleum Products Institute has established a computerized spill reporting system, training programs for people who transport petroleum products and some generic spill response planning guides.

12.10.4   The Role of the Community

A spill contingency plan should be developed with community involvement. The Canadian Chemical Producers Association document entitled, *Community Awareness and Emergency Response Code of Practice* outlines various forms that community involvement can take. It encourages each member to develop an emergency response plan which:

- identifies situations where company materials or processes can have an impact on a community in the event of an emergency
- is based upon emergency plan framework developed by site management to both address such emergency situations and to assist other authorities in emergency response planning for neighbouring industry and the community
- integrates the company's emergency response planning and organization with those of industrial neighbours and the community into a community emergency response plan
- is communicated regularly, in its key elements, to the community, and in a manner that recognizes its right to know and that gains the community's co-operation and support
- requires active participation, co-operation and co-ordination by company people with local officials and the media during the planning and communication stages

The Canadian Association of Fire Chiefs also has a program for enhancing community safety during emergency events. Further information can be obtained at **www.ptsc-program.org**.

12.11   Summary

Despite the numerous precautions taken, spills still happen. Taking measures to prevent spills and having an effective spill response plan are critical components of environmental protection.

In the event of a spill, individuals and companies are responsible for promptly notifying the appropriate federal or provincial agencies, the municipality, the owner of the pollution and the person who has control of the pollutant. In Ontario, if the spill can cause an adverse effect on the environment, the owner of a pollutant and the person having control of the pollution that is spilled are required to do everything practicable to prevent, eliminate and ameliorate the adverse effects and to restore the natural environment.

References

Canadian Standards Association (CSA). 1995. *Emergency Planning for Industry.* CAN/CSA-Z731-95.

Environment Canada. 1998. *Summary of Spill Events in Canada, 1984-1995.* EPS 5/SP/3. November.

Environment Canada. March 1984. *Manual for Spills of Hazardous Materials.* Environmental Protection Service.

Environmental Compliance Report. 2001. *Draft Guideline Outlines Spill Reporting Exemptions.* Vol. 16, No. 3. March.

Regional Municipality of Halton. 2001. *Emergency Plan.* By-Law 81-01. June.

Ontario Ministry of Environment (MOE). 2001. *Environmental Clean-up Fund.* Guideline G-3. April.

Ontario Ministry of Environment (MOE). 1998. *Spills and Emergencies: The Role of the Ministry of the Environment.* In Brief, October.

Ontario Ministry of Environment and Energy (MOEE). March 1997. *Spills Action Centre Summary Report of 1995 Spills.* ISSN 1192-5078.

Ontario Ministry of Environment and Energy (MOEE). 1994. *Spills Action Centre Operations.* Guideline G-2 [formerly 13-02]. April.

Ontario Ministry of Environment and Energy (MOEE). September 1993. *Province of Ontario Contingency Plan for Spills of Oil and Other Hazardous Materials.*

Transport Canada, U.S. Department of Transportation and Secretariat of Communications and Transportation of Mexico. 2000. *2000 Emergency Response Guidebook.*

Table 12.1

Accidental Release Quantities that Require Immediate Reporting under TDGA

Class and Division	Quantities or Emission Levels
1	>50 kg or any quantity that could pose a danger to public safety
2	any quantity that could pose a danger to public safety or any sustained release of 10 minutes or more
3	>200 L
4	>25 kg
5.1	>50 kg or 50 L
5.2	>1 kg or 1 L
6.1	>5 kg or 5 L
6.2	>1 kg or 1 L or any quantity that could pose a danger to public safety
7	any quantity that could pose a danger to public safety or an emission level greater than that in s. 20 of the Package and Transport of Nuclear Substances Regulations
8	>5 kg or 5 L
9	>25 kg or 25 L

Table 12.2

Potential Equipment for Spill Containment and Clean-Up

- booms and oil skimming devices
- shovels and brooms
- pumps and vacuum equipment
- sorbent materials
- neutralizers
- protective clothing such as respirators, gloves, etc.
- foul weather gear
- material safety data sheets for all chemicals used on-site
- portable air or water monitoring devices/kits
- sample bottles and sample buckets (i.e., for fish toxicity test)
- sand, earth or vermiculite
- plywood sheets

CHAPTER 13

Toxicity Testing

13.1 Overview

Many environmental acts include sections that prohibit releases of substances into the environment that can cause adverse or negative effects. The wording used often is somewhat vague. For example, some prohibit the release of "deleterious substances", others prohibit the discharge of substances in "toxic amounts" and yet others prohibit releases that create "toxic conditions". The federal Fisheries Act has been used to convict persons for discharging a "toxic substance" based on the mere presence of the substance, while the amount of substance involved has been used to assess the level of damage and the extent of fines.

To give scientific meaning to the concept of toxicity, an extensive array of test methods have been developed that use many different test organisms and techniques for interpreting data. Much of this work has focussed on aquatic environments and the characterizing of liquid effluents. As a result, the control of effluent toxicity has been incorporated into water quality regulations, orders, approvals and guidelines for many types of commercial and industrial operations in Canada. Recent advances have been made which likely will lead to terrestrial toxicity tests being incorporated into soil and sediment management policies and regulations.

Table 13.1 lists numerous types of toxicity tests in the Environment Canada Biological Test Method Development Program for effluents, ambient water quality and sediments. Examples of situations when these tests are used and details of tests are provided in this chapter.

13.2 Types of Toxicity Tests

13.2.1 Acute Lethality Tests

The most common use of toxicity tests is to measure acute lethality of an effluent to an aquatic organism. This involves placing a species of fish or invertebrate into a series of containers each holding a different dilution of the effluent or toxicant to be evaluated. One of the containers — the control — contains only dilution water. The test organisms are observed at predetermined time intervals over a relatively short period (typically less than four days). The numbers of organisms that die in each of the dilutions provide the data used to estimate the dilution required for the effluent to be non-lethal. These tests are intended to quantify the overall toxicity of an effluent. They do not identify the substance(s) that cause the toxicity.

Several variations of the basic procedure have been developed to evaluate specific aspects of acute lethality. These variations can be described or categorized according to three characteristics: test duration, test method and organisms tested. The most commonly used test duration is 96 hours, but shorter (typically 24 or 48 hours) or longer durations are also used.

Test methods largely concern how the water-effluent dilutions are managed during a test. In static tests, the solutions are not renewed for the duration of the test. In semi-static or static replacement tests, the solutions are renewed once every 24 hours. In continuous flow (or flow-through) tests, a device called a diluter is used that provides a continuous flow of effluent or toxicant dilution to the test containers. A fourth type of test, the *in situ* or ambient water bioassay test, requires cages to be set at different locations in the receiving water to develop an effluent dilution gradient.

The rainbow trout (*Onchorhyncus mykiss*, previously referred to as *Salmo gairdneri*) is the standard cool-water fish for freshwater pollution studies and research in aquatic toxicology in Ontario and in Canada. A large inventory of toxicological data has been assembled for this species over the last 25 years. Culturing of rainbow trout is well established in Canada and many hatcheries will provide eggs or young fish of appropriate size and quality for toxicity test purposes. Fisheries and Oceans Canada must certify the fish as free of specific pathogens.

The other organism used frequently in acute lethality tests is the water flea (*Daphnia magna*). It has a short life cycle capable of producing young at about two weeks of age and is relatively easy to culture in the laboratory. It is widely distributed in ponds and lakes of intermediate water hardness in Canada and the United States and has been found to be sensitive to a broad range of aquatic contaminants. Because of its small size (about the diameter of a large pencil head), tests using *Daphnia magna* require less than 1% of the volume of test solution required for trout. This greatly reduces the physical requirements for sampling and transporting of effluent samples. *Daphnia magna* are more sensitive and respond more rapidly to some toxicants than fish (Johnson and Finley, 1980) and are easily raised in the lab in large numbers.

Other freshwater species such as the Fathead minnow (*Pimphales promelas*) and bluegill sunfish (*Lepomis macrochirus*) have been used to assess of warm-water environments. The fathead minnow often is used for partial life cycle and chronic studies of fish (see Section 13.2.3). Other organisms, such as sea urchins, bivalves and marine minnow species are used to evaluate marine environments.

In 1990, Environment Canada published universal test methods for the acute lethality test using rainbow trout (EPS 1/RM9) and using *Daphnia* Spp. (EPS 1/RM/11). That same year, Environment Canada also published reference methods for determining acute lethality of effluents to rainbow trout (EPS 1/RM/13) and *Daphnia magna* (EPS 1/RM/14). (Second editions of both documents were published in 2000.) The essential elements of these tests are summarized in Table 13.2.

Since the early 1990s, Environment Canada and the Department of Fisheries and Oceans have undertaken the Environmental Effects Monitoring (EEM) program to assess the adequacy of effluent regulations under the Fisheries Act and to achieve national uniformity in monitoring effects. Under the EEM program, monitoring requirements have been established for specific industrial sectors. As an example, the requirements for the Metal Mining Effluent Regulations are summarized in Table 13.3. Section 13.3 provides additional information on the EEM program. The requirements shown in Table 13.3 illustrate the complex nature of toxicity testing as well as the need to assess the receiving environment and communities of aquatic organisms.

The results of acute lethality tests are expressed in terms of the concentration of the effluent estimated to cause the death of 50% of the test organisms. This is referred to as the 50% lethal concentration (or the LC_{50} value). The LC_{50} value and its 95% confidence limits can be calculated or determined graphically by plotting percentage mortality against effluent concentration. In the past, the term "median tolerance limit" (abbreviated as TL50 or TLm) often was used. It is synonymous with LC_{50}.

If no test organisms die after exposure to undiluted effluent, the result is reported as non-lethal. If less than 50% of the test organisms die in undiluted effluent, the LC_{50} of the effluent is reported as greater than 100% (>100%). If 50% or more organisms die in undiluted effluent, a statistically valid LC_{50} is calculated from the mortality data. Some jurisdictions also require that the 95% confidence limits be reported.

The highest concentration of a test substance in which no mortalities are observed is referred to as the no observed effect-concentration (NOEC).

LC_{50} data (expressed as a percentage of effluent) also can be used to calculate toxic unit (TU) values according to the equation:

Equation 13.1

$$TU = 100 \div LC_{50}$$

This expression can be used to estimate the amount of dilution required to make the effluent non-lethal (LC_{50} >100%). For example, if an effluent has an LC_{50} of 5%, it has a TU value of 20 and requires a 19:1 dilution to become non-lethal. More toxic effluents have higher numbers of toxic units and require greater dilution than less toxic effluents.

The TU value can be used to express the amount of toxicity being discharged per unit of time as the toxicity emission rate (TER) using the equation:

Equation 13.2

$$TER = TU \times Flow$$

The TER can be used to compare quantities of toxicity being discharged from different sewers within a single plant or to compare discharges among plants. The TER concept places equal weight on the toxicity and the flow of the effluent.

13.2.2   Acute Sublethal Tests

While most acute tests are directed toward observing lethal effects, sublethal effects such as organism immobilization and lethargy also may be observed and measured. For example, immobilization often is reported in tests with *Daphnia magna*. The numbers of immobilized organisms can be used to calculate the median effective concentration (EC_{50}) value in the same way that the LC_{50} is calculated.

Sublethal tests have been developed that use bacteria as the test organism. One of these assays, known as the "Microtox," has been used to evaluate various types of effluents. The test establishes an EC_{50} value on the basis of the amount of light produced by the luminescent bacterium *Photobacterium phosphoreum*. The test requires less than one hour to conduct and less than 5 mL of solution. The test has been reported to be less sensitive than tests that use rainbow trout or daphnia (McLeay and Associates Limited, 1987). Environment Canada has not included this test in effluent monitoring procedures, likely because of its lack of sensitivity.

13.2.3   Chronic Tests

Chronic tests take place over longer periods than acute tests, often for one or more stages in the life cycle of the test organism. As controls on acute toxicity become common, regulatory focus shifts to preventing chronic effects. For example, chronic tests involving end points other than lethality are required under Ontario Municipal-Industrial Strategy for Abatement regulations once a discharge has been shown to be non-lethal during acute toxicity testing. The test organisms are exposed during sensitive stages of the life cycle. The types of effects that may be observed include growth, reproduction, hatching, behaviour, and larval survival and development.

Several parameters can be used to express the results of chronic sublethal tests. The lowest concentration at which effects are observed is referred to as the LOEC. The geometric mean of the LOEC and the NOEC is referred to as the threshold effects concentrations (TEC). The test data can be analyzed statistically to determine the concentrations at which inhibition is observed in predetermined percentages of organisms. For example, the concentration at which inhibition is observed in 25% is referred to as the IC_{25}. Confidence intervals can be assigned to IC values using procedures similar to those used to describe LC_{50} values.

Because of the duration and costs involved, chronic toxicity tests are not used as frequently as acute tests to assess effluents. The following are a few examples of chronic toxicity tests (U.S. EPA, 1989).

1. *Ceriodaphnia dubia* survival and growth: Ten animals, each less than 24 hours old, are exposed to different concentrations, each animal being placed in a separate exposure vessel. Each surviving organism is transferred daily into a new test vessel with freshly prepared test solution and food suspension. Once the *C. dubia* has matured and produced young (on about the third day), only the adult is transferred to fresh effluent solution and fed. After seven days, the cumulative number of young produced per female are counted. The test is also conducted in a 10-fold replication. Results are expressed as the TEC, NOEC, LOEC and the IC_{25} with its 95% confidence interval.

2.. *Daphnia sp.* life cycle toxicity tests: Using *Daphnia sp.* which are less than 24 hours old, 10 animals are exposed to different concentrations, each animal being placed in a separate exposure vessel. The test organisms are transferred into fresh test solutions and fed every day. Once the individual daphnids have matured and produced offspring (approximately eight to 14 days), only the adults are transferred to a fresh test solution and fed. The remaining young are examined, counted and recorded. After 21 days, all surviving first-generation daphnids are thoroughly examined, sacrificed and weighed. The chronic LC_{50}, daphnia growth and reproduction data are then evaluated statistically. This test is not used in Canadian jurisdictions but is under consideration.

3. Fathead minnow larval survival and growth: Ten newly hatched fathead minnow larvae are placed into each test vessel. The larvae are fed and after seven days of exposure are sacrificed and weighed. The test is conducted in a four-fold replication. The TEC, NOEC, LOEC, chronic LC_{50} and larval growth IC_{25} are estimated using statistical methods.

4. Fathead minnow embryo-larval survival and teratogenicity: Fifteen eggs fertilized within 24 hours are removed from spawning substrates, washed and placed in test vessels. The test solutions are renewed daily with freshly prepared concentrations. The numbers of surviving, dead and deformed larvae are recorded over a seven-day period. Seven-day embryo and larvae LC_{50} values, TEC, NOEC, LOEC, IC_{25}, as well as teratogenicity endpoints are calculated using statistical methods.

To study marine environments, sea urchins, bivalves and marine minnow species are being used for sublethal toxicity tests. In many instances, corroborative field studies and monitoring of indigenous species are required before the relevance of these tests is appreciated.

13.2.4   Terrestrial Toxicity

Efforts to improve methods used to assess soil quality are considering the potential for soil contaminants to affect terrestrial organisms. There is not yet a consensus as to which organisms should be evaluated, the test protocols that should be used, or how the results might be converted into environmental guidelines, however, all of these aspects are being investigated by organizations including the CCME, U.S. EPA, the International Organization for Standardization, and the Organization for Economic Cooperation and Development.

The deriving of criteria based on terrestrial toxicity likely will begin with identifying the organism(s) potentially at risk from the contamination. Ideally, the selected organism(s) should reflect important characteristics of the ecosystem, be sensitive to the contaminants of concern, and be relevant to the land uses of interest. The available information may limit the extent to which this can be realized.

Once the organism(s) have been identified, the way(s) in which contact occurs with contaminants can be evaluated. These pathways may include direct ingestion of soil, water or other organisms, direct contact with soil, inhalation of air or indirectly via the food chain. Direct contact and/or ingestion are likely to be the pathways of greatest concern for organisms such as soil bacteria, fungi and earthworms. Food chain effects are likely to the pathways of most concern for animals at higher trophic levels that are consumers (that is, herbivores and carnivores). For many species, the information needed to assess pathways is very limited. Numerous assumptions will be needed to estimate intake rates of soil, food items, water, or air, dermal contact rates, amount of time spent in contact with contaminated materials, bioavailability of contaminants from ingested material, etc.

The next step will be to compare the estimated extent of contact to the data from toxicity tests (typically conducted in the laboratory) or observations/measurements made in the field. Soil toxicity information is lacking for many contaminants. Generally, there is more information available for inorganic contaminants than organic contaminants. The largest information base involves terrestrial plants, while a growing area of study involves organisms such as earthworms and isopods (such as wood loose) that are constantly in contact with soil. Few studies have been conducted on birds, with many of the published accounts limited to poultry or other game bird species. Relatively little information is available for mammals other than laboratory rodents. Some information is available for direct soil ingestion by grazing animals including cows and sheep.

The test protocols most commonly used and accepted are short-term (seven or 14 day) mortality tests for earthworms and a five-day seed germination/root elongation test for plants. The results are expressed as LC_{50}, EC_{50}, IC_{25}, NOEC and LOEC values analogous to those used for reporting the results of aquatic toxicity testing. Other tests are being developed that will evaluate end-points other than lethality. Long-term tests also are being developed that will last for at least one reproduction cycle (for soil invertebrates) or one growth cycle (for plants).

To determine soil quality criteria, toxicity test results for a contaminant will be used to create a distribution of concentrations at which effects occur. A point in that distribution, possibly divided by a safety factor, will be used to define a concentration below which adverse effects should not occur. For example, the point corresponding to the 10th percentile (that is, 90% of the effects happen at concentrations greater than the selected point) might be chosen. The magnitude of the safety factor will reflect the quality and quantity of toxicity data, with larger values being used when the data are limited or otherwise suspect. The value produced by dividing the selected point by the safety factor would then be used as one of the pieces of information used to establish a soil quality guideline or criterion.

The preceding discussion indicates that the assessment of soil and sediment toxicity is following much the same path as water quality toxicity assessment, however, several factors make soil and sediment toxicity assessment more complex and may slow the rate at which toxicity is incorporated into setting soil quality criteria. In water, the molecular and soluble fractions of contaminants can be measured directly and related to biological effects with relatively easily derived modifiers such as pH and hardness. In soils, the measured concentrations of contaminants are less important than the percentage that is biologically available, which is influenced by factors such as adsorption, absorption, chelation, and binding to organic compounds as well as the factors that are relevant for water.

These factors (and perhaps more) have made it difficult to draw correlations between measured concentrations of contaminants in soil and the effects observed in biological systems. These findings not only suggest that different approaches should be used to derive water and soil quality criteria, but also may point to limitations in other aspects of environmental management. For example, in Ontario (and in many other jurisdictions), the procedure used to determine if a waste is leachate toxic is to compare concentrations of contaminants in an acidic aqueous extract to predetermined limits (see Section 4.4). While intended to protect human health by controlling the quality of groundwater around disposal sites, the approach may not be protective of aquatic or terrestrial organisms. Given that aquatic biota demonstrate greater sensitivity to ambient contaminant concentrations than do humans, groundwater quality judged acceptable for human exposure will not necessarily provide the same level of safety to other biota.

13.3   Federal Requirements

Aquatic toxicity testing has been incorporated into effluent regulations under the Fisheries Act for several commercial and industrial sectors:

- Meat and Poultry Products Plant Liquid Effluent Regulations (C.R.C. 1978, c. 818)
- Metal Mining Effluent Regulations (SOR/2002-222)
- Port Alberni Pulp and Paper Effluent Regulations (SOR/92-638)
- Pulp and Paper Effluent Regulations (SOR/92-269)

(Toxicity testing is not specified in other effluent regulations for chlor-alkali plants, metal mines [excluding gold mines] or potato processing plants. Those regulations specify maximum discharges based upon the best practicable technology available at the time of promulgation.)

The pulp and paper regulations are highlighted below to show how regulatory expectations are evolving. The regulations for meat and poultry products facilities and the potato processing facilities are identified for repeal. The regulations for meat and poultry products facilities also are based upon the best practicable technology available at the time of promulgation, but they include acute toxicity testing. They stipulate that the undiluted effluent must not be fatal to more than 50% of rainbow trout in a 96-hour, static test.

The two pulp and paper effluent regulations (SOR/92-269 and SOR/92-638) reflect a more extensive approach to regulating effluent toxicity. They require monthly testing of acute lethality using rainbow trout and *Daphnia magna*. The frequency of testing increases if a test fails (that is, the effluent is shown to be toxic). Samples of effluent shall continue to be collected and tested at the increased frequency until they pass three consecutive tests, after which normal monitoring may resume. Every pulp and paper mill operator shall provide the Minister with an interpretative report and supporting data on each environmental effects monitoring study in accordance with the *Aquatic Environmental Effects Monitoring Requirements* (EPS 1/RM/18), as amended from time to time. Details about EEM and the report noted above can be obtained at **www.ec.gc.ca/eem**.

In October, 1997, Environment Canada published *Pulp and Paper Aquatic Environmental Effects Monitoring Requirements - Annex 1 to EEM/1997/1*. The document stipulates that testing is to be conducted during one summer period and one winter period each year over a three-year cycle. The methodologies and reporting requirements for process effluent toxicological testing include "fish early life stage development test", "invertebrate reproduction test" and "plant toxicity test". The document also requires that laboratories accredited by the Canadian Associated for Environmental Analytical Laboratories, Quebec Ministère de l'Environnement et de la Faune or an equivalent level of accreditation be used for the sublethal toxicity testing (Environment Canada, 1997).

In 2000, Environment Canada published a consultation document that examined ways to streamline and improve the two pulp and paper regulations without altering the discharge limits. Some of the proposed amendments included changes to EEM requirements. As of late 2002, neither regulation had been amended.

In 1995, a strategic options process was launched to assess potential options for the management of toxic substances (as defined in CEPA, 1999) in the steel sector. The strategic options process culminated in the development of a strategic options report, which included the following recommendations for both integrated and non-integrated steel mills:

- develop environmental codes of practice
- develop pollution prevention (P2) plans (see Table 1.2 and Section 3.2.1)
- conduct environmental audits (see Chapter 10)

The strategic options report also stated that an EEM program should be developed and implemented where appropriate by each facility in consultation with the appropriate regulatory authorities. This program should be sufficiently comprehensive to enable the facility to measure changes in receiving water quality, aquatic sediments and important aquatic and terrestrial organisms, and assess the need to incorporate changes in operational activities and procedures affecting the receiving environment.

The frequency and duration of this monitoring activity should be assessed, in consultation with the appropriate regulatory authorities, on the basis of test results.

In response to a commitment to update and strengthen the original regulations for metal mining liquid effluent that had been in place since 1978, Environment Canada held a multi-stakeholder consultative workshop in November, 1992. One of the recommendations of the workshop was to assess the known effects of mining in Canada through what was subsequently identified as the AQUAMIN process.

In 1996, the final AQUAMIN report was prepared. It included several recommendations related to water quality. These included: revising the regulations to ensure a consistent national effluent quality requirement at Canadian mines; to set site-specific requirements where necessary to protect local receiving environments; and to require EEM programs to provide reporting and feedback on the effectiveness of protection measures.

The Metal Mining Effluent Regulations (SOR/2002-222) was issued in July, 2002. All sections will be in force as of December, 2002 when the older metal mining liquid effluent regulations are repealed. The new regulations include both acute and chronic toxicity testing.

Acute lethality testing is required once a month using rainbow trout and a grab sample from each final discharge point. Sampling increases if effluent is determined to be acutely toxic and decreases to quarterly after 12 consecutive months of not being acutely lethal.

Chronic toxicity testing (called sublethal toxicity in the regulations) are described in Schedule 5, which is entitled *Environmental Effects Monitoring Studies*. Table 13.3 summarizes the requirements of the EEM studies and includes references to several of the test procedures listed in Table 13.1.

In addition, the EEM studies include biological monitoring studies before and during operations, and in preparation for closing the mine. These include:

- site characterization study:
- assessment of local fish populations and the benthic invertebrate community
- assessment of fish tissue if total mercury in the effluent is equal to or greater than 0.1 μg/L

First study design:

- details of how the fish and benthic populations are to be studied and how the resulting information is necessary to determine if the effluent is having a negative or adverse effect
- details of the mining operation, how and where effluent will be discharged
- estimated concentrations at 250 m from each final discharge point

biological monitoring studies:

- conducted in accordance with the study design
- reports submitted to Environment Canada
- describe how subsequent biological monitoring studies will assess the magnitude and geographic extent of any effects

final biological monitoring study prior to mine closing:

- summary of earlier studies; describe future studies that will be used to characterize effects and the causes of those effects; to be provided not later than six months are proving notice of closing the mine

The combination of toxicity testing and biological monitoring studies required in this relatively recent regulation clearly demonstrate the expectations of regulators and likely signal the types of changes that can be expected as similar regulations are updated or introduced.

13.4   Ontario Requirements

Several provincial acts use toxicity concepts to prohibit specific types of activities. For example, s. 30(1) of the Ontario Water Resources Act prohibits persons from discharging any material into any water that "may impair water quality." Similarly, s. 14 of the Environmental Protection Act states "[no] person shall discharge a contaminant or cause or permit the discharge of a contaminant into the natural environment that causes or may cause an adverse effect." Under the terms of the Canada-Ontario Accord, Ontario has agreed to establish and enforce effluent toxicity testing requirements at least as stringent as federal requirements.

Acute lethality testing is a requirement of the Municipal, Industrial Strategy for Abatement (MISA) regulations (see Section 3.3.2). Dischargers were required to perform acute toxicity testing using rainbow trout and *Daphnia magna* in 100% effluent. The quality of the effluent shall be such that not more than 50% of the test organisms die. This requirement applies to all final effluent and cooling water discharges. If tests indicate non-toxic conditions for 12 consecutive months, the frequency of testing can be reduced to quarterly. The frequency goes back to monthly if the quarterly results indicate that the effluent is toxic.

In addition, some dischargers may be required to perform chronic tests including seven-day reproductive inhibition and survivability tests using *ceriodaphnia*, and seven-day growth inhibition tests using fathead minnows.

In the event that toxicity tests show an effluent to be toxic and the cause of the toxicity is uncertain, the discharger can conduct a Toxicity Identification Evaluation/Toxicity Reduction Evaluation (TIE/TRE), as discussed in Section 13.5.

The document entitled *Water Management: Policies, Guidelines and Provincial Water Quality Objectives* states that one of the overall goals of the MOE is to ensure that surface waters are of a quality that is satisfactory for aquatic life (MOE, 1994). Policy 3 of that document addresses effluent regulations and indicates that bioassay tests may be required to identify discharges deleterious to aquatic organisms. Discharges that produce 96-hour LC_{50} values under static test conditions may require more rigorous biological testing to determine if additional treatment is needed to afford adequate protection to the environment. The testing may include biological responses other than mortality.

The MOE water management document recognizes the concept of a mixing zone around a discharge point, however, mixing zones should not be acutely lethal to important aquatic life or result in conditions that cause sudden fish kills or mortality of organisms passing through the mixing zone or create barriers to the migration of fish and aquatic life.

13.5   Determining the Causes of Effluent Toxicity

13.5.1   Conventional Approach

Traditionally, when an effluent is identified as toxic to aquatic organisms, a sample of the wastewater is analyzed for certain pollutants such as those in the MOE Analytical Test Groups (ATGs) of substances (see Section 3.6.2) or the U.S. EPA priority pollutants. The concentration of each pollutant present in a sample subsequently is compared to toxicity data for the pollutant in the published literature or the rationales for the Canadian Water Quality Guidelines recommended by the CCME (see Section 3.2.7) or the water quality objectives recommended by the MOE (see Section 3.3.5). Unfortunately, determining the source of an effluent's toxicity rarely is so straightforward.

The first problem encountered is one of effluent variability. There is no way to determine whether the toxicity observed over time is consistently caused by a single pollutant or combination of pollutants or a number of different pollutants, each periodically being the cause of the toxicity. Experience has shown that the latter scenario occurs frequently. Further complicating the problem is that the variability in conventional effluent monitoring parameters may not coincide with variability in the effluent of the causative toxicant (Mount and Anderson-Carnahan, 1988).

A second limitation with the conventional approach is the assumption that only the ATGs or priority pollutants cause the toxicity. While some pollutants are regulated largely due to their toxicity to aquatic organisms and/or a high frequency of occurrence in discharges, the ATGs or the priority pollutants list by no means represents the universe of toxic chemicals present in wastewater. Limiting the search to these compounds excludes other chemicals from consideration and limits the chances for successfully identifying the causative toxicant.

Another limitation is that the conventional analytical focus does not account for interactions between toxicants or the influence on toxicity under conditions commonly encountered in effluents such as extreme pH or high-dissolved-solids concentrations.

13.5.2   U.S. EPA Approach to Toxicant Identification

In the United States, the National Pollutant Discharge Elimination System includes toxicity testing requirements. Chemistry and toxicity data pertaining to the discharge are provided by the discharger, the U.S. EPA, and/or the state. The lowest one-week average flow that will likely occur once every 10 years is also provided. Depending on the data, it is then decided if toxicity limits should be incorporated into the permit.

Where toxicity limits have been imposed, the in-stream concentration of the effluent in the receiving water must not exceed the lowest observable effect concentration of the most sensitive test species. For dischargers with greater than a 100:1 dilution at low flow, there must not be acute lethality at the end of the pipe (that is, the LC_{50} must be greater than 100%). For dischargers with less than 100:1 dilution, the dischargers also must comply with a chronic toxicity limit and the effluent concentration that occurs at the edge of the mixing zone at low flow must not cause a chronic effect.

When a discharger is not in compliance, a step-wise investigative process termed a toxicity reduction evaluation (TRE) can be used to determine the measures needed to maintain toxicity at an acceptable level. A key part of a TRE is the toxicity identification evaluation (TIE). Procedures for performing TIEs have been developed by the U.S. EPA National Effluent Toxicity Assessment Centre (NETAC). The NETAC approach is divided into three phases (Burkhard and Ankley, 1989):

Phase I   – identification of physical and chemical nature of toxicant(s)
Phase II   – identify toxicant(s)
Phase III   – confirm suspected toxicant(s)

Figure 13.1 shows the relationships of the NETAC approach to TIE with the overall TRE assessment.

Phase I of the TIE is used to characterize certain physical and chemical properties of the toxicant(s) using a series of relatively simple, low-cost chemical and biological analyses. Each test is designed to remove or render biologically unavailable a specific group of toxicants such as oxidants, cationic metals, volatile compounds, non-polar organic compounds and metal chelates (Mount and Anderson-Carnahan, 1988). Figure 13.2 presents the various Phase I tests that are listed below.

1. The oxidation reduction test is designed to determine whether oxidants and other electrophiles are responsible for effluent toxicity. Examples of oxidants include chlorine, bromine, iodine, ozone and chlorine dioxide. Sodium thiosulfate ($Na_2S_2O_3$) is added in varying ratios to produce reducing agent/total electrophiles solutions and thus reduce any toxicity due to the above compounds.

2. The aeration test is designed to determine whether the toxicants present are volatile. Sparging of samples using air or nitrogen gas is used to remove the volatile substances from solution. Examples of chemicals that may be detected using this method include benzene and hydrogen sulfide.

3. The filtration test involves passing samples of the test solution at three pH values (pH of 3, 7 and 11) through glass fibre filters. The test identifies which groups of compounds can be precipitated under acidic or

alkaline conditions. Data from this test can help define treatment strategies.

4. The pH adjustment test is used to identify the presence of cationic and anionic toxicants. Changing the pH changes the ratio of ionized to un-ionized chemical species and since the later is the more toxic form, the toxicity of the solution will change as the pH is shifted.

5. In the graduated pH test, the pH of the effluent is adjusted within the tolerable range of 6.0 to 8.0 before retesting with aquatic organisms. This test is designed primarily to identify ammonia, hydrogen sulfide and cyanide, but some ionizable pesticides and heavy metals also can be identified.

6. The solid phase extraction (SPE) test is designed to determine the effluent toxicity caused by non-polar organic compounds and metal chelates. The effluent is passed through a small column, which removes non-polar organic compounds. Toxicity reduction suggests the presence of toxic organic compounds.

7. The chelation test is used to determine if the toxicity is caused by cationic toxicants such as heavy metals. Ethylenediamine-tetraacetate ligand (EDTA) is a strong chelating (binding) agent that produces non-toxic complexes with many metals and can reduce the toxicity of solutions containing metal ions.

8. Aquatic organism toxicity tests, performed on the effluent prior to and after the individual characterization treatment, indicate the effectiveness of the treatment and thus provide information on the nature of the toxicant(s). By repeating the series of toxicity characterization tests using samples of a particular effluent collected over a period of time, these screening tests can provide valuable information on the variability associated with the type of compounds causing the toxicity (Mount and Anderson-Carnahan, 1988).

Phase II of the TIE is directed toward identifying toxicants. Various chemical fractionation techniques such as high-performance liquid chromatography, mass spectroscopy and further SPE tests may be used. The results obtained in Phase I are used to select appropriate techniques for Phase II.

For each toxicant that is identified in Phase II, the published literature is searched for LC_{50} values or individual tests of that chemical may be conducted. Concentrations of the toxicant in effluent can be compared with the LC_{50} values. A list of suspected toxicants is then compiled.

Phase III of the TIE is used to confirm the suspected toxicants identified in Phase II. Techniques that can be used in Phase III include correlations, relative species sensitivity, spiking and removal of one toxicant at a time from the effluent. In most instances, several tests are needed to confirm the toxicants. Research is continuing to refine this phase of the overall approach.

Once a TIE is completed, it may be necessary to enter the next step of a TRE that involves the selection and implementation of control methods. In some cases, it is possible that Phase I of the TIE can indicate which treatment methods should remove the causative toxicant(s) from the effluent. Bench-scale studies can be used to evaluate the feasibility of treating effluent toxicity on a large scale. The actual identity of the causative toxicant(s) may not be required; it is only necessary that enough information be available on the toxicant physical/chemical characteristics to predict which treatment options should be studied (Mount and Anderson-Carnahan, 1988). This is analogous to designing a treatment system to handle a specified biochemical oxygen demand (BOD) loading with little or no knowledge of the actual chemicals that comprise the BOD.

In other cases, Phases II and III may be needed to identify the toxicants. If a causative toxicant can be identified in an effluent, the next step of a TRE can be to track a toxicant back through the process line or effluent collection system to its source using chemical analysis (provided that it is not a by-product of other chemicals in the system). Once the source is identified, a source investigation can be conducted. The source control may be in the form of improved spill control, process modification, substitution of raw materials, pretreatment and/or treatment.

13.6   Summary

Toxicity testing is part of a growing number of federal and provincial regulations. Most of these regulations require acute lethality toxicity tests using rainbow trout and *Daphnia magna* as part of the effluent monitoring requirements. Facilities that fail to achieve a non-lethal discharge may be required to perform studies such as a TIE/TRE. Typically, this type of study focuses first on identifying the physical and chemical nature of the toxicant(s), followed by the identification of the toxicant(s) and development of control alternatives. The approach may result in product substitution, process modifications and/or additional treatment.

Both federal and provincial regulations of industrial effluent sources are increasing the amounts and types of toxicity testing. The types of information being required include acute lethality tests; reviews of the receiving environment and populations of fish and aquatic invertebrates; a history of effluent quality; chronic toxicity tests involving early life stage development tests, invertebrate reproduction tests and plant toxicity tests; and chemical characterization of receiving water and sediment.

References

Burkhard, L. P. and Ankley, G. T. 1989. "Identifying Toxicants: NETAC's Toxicity-based Approach". *Environmental Science and Technology*, 12:12: 1438-43.

Environment Canada. 1992a. *Aquatic Environmental Effects Monitoring Requirements.* EPS 1/RM/18.

Environment Canada. 1992b. *Chronic Toxicity Test Using the Cladoceran Ceriodaphnia dubia.* Conservation and Protection, Ottawa, Ontario. EPS 1/RM/21.

Environment Canada. 1992c. *Test of Larval Growth and Survival Using Fathead Minnows. Conservation and Protection.* Ottawa, Ontario. EPS 1/RM/22.

Environment Canada. 1992d. *Algal Growth Inhibition/Simulation Test Using Selenastrum capricornutum.* EPS 1/RM/25.

Environment Canada. 1992e. *Early Life-Stage Test for Toxicity Using Salmonid Fish.* EPS 1/RM/28.

Environment Canada, 1997. *Pulp and Paper Aquatic Environmental Effects Monitoring Requirements: Annex 1 to EEM/1997/1.* October.

Environment Canada. July 2000a. *Reference Method for Determining the Acute Lethality of Effluent to Rainbow Trout.* EPS 1/RM/13. Second Editon

Environment Canada. July 2000a. *Reference Method for Determining the Acute Lethality of Effluent to Daphnia Magna.* EPS 1/RM/14. Second Edition

Environment Canada and Department of Fisheries and Oceans. 1992. *Annex 1: Aquatic Environmental Effects Monitoring Requirements at Pulp and Paper Mills and Off-Site Treatment Facilities Regulated under the Pulp and Paper Regulations of the Fisheries Act.*

Environment Canada and Department of Fisheries and Oceans. 1993. *Technical Guidance Document for Aquatic Environmental Effects Monitoring Related to Federal Fisheries Act Requirements.* Version 1.0, April.

Johnson, W.W. and M.T. Finley. 1980. *Handbook of Acute Toxicity of Chemicals to Fish and Aquatic Invertebrates.* U.S. Fish and Wildlife Service, Publication No. 137.

McLeay and Associates Limited. 1987. *Aquatic Toxicity of Pulp and Paper Mill Effluents: A Review.* Environment Canada Report EPS 4/PF/1.

Mount, D.I. and L. Anderson-Carnahan. February 1988. *Methods for Aquatic Organism Toxicity Reduction Evaluations; Phase I Toxicity Characterization Procedures.* Second Draft - February 1988. U.S. Environmental Protection Agency.

Ontario Ministry of the Environment (MOEE). July 1994. *Water Management: Policies, Guidelines, Provincial Water Quality Objectives of the Ministry of the Environment and Energy.*

Poirier, D.G., G.F. Westlake and S.G. Abernathy. 1988. *Daphnia magna Acute Lethality Toxicity Test Protocol.* Aquatic Toxicity Unit, Aquatic Biology Section, Water Resources Branch, MOE.

United States Environmental Protection Agency (U.S. EPA). February 1989. *Short-Term Method for Estimating the Chronic Toxicity of Effluents and Receiving Water to Fresh Water Organisms*. 2nd ed. U.S. EPA 600/4-89/001, February.

Table 13.1

Environment Canada Test Methods

Test Methods

Acute Lethality Test Using Rainbow Trout	*EPS 1/RM/9, July 1990*
Acute Lethality Test Using Threespine Stickleback (Gasterosteus aculetus)	*EPS 1/RM/10, July 1990*
Acute Lethality Test Using Daphnia spp.	*EPS 1/RM/11, July 1990*
Test of Reproduction and Survival Using the Cladoceran Ceriodaphnia dubia	*EPS 1/RM/21, February 1992*
Test of Larval Growth and Survival Using Fathead Minnows	*EPS 1/RM/22, February 1992*
Toxicity Test Using Luminescent Bacteria (Photobacterium phosphoreum)	*EPS 1/RM/24, October 1992*
Growth Inhibition Test Using the Freshwater Alga Selenastrum capricornutum	*EPS 1/RM/25, November 1992*
Acute Test for Sediment Toxicity Using Marine or Estuarine Amphipods	*EPS 1/RM/26, December 1992 Second Edition*
Fertilization Assay Using Echinoids (Sea Urchins and Sand Dollars)	*EPS 1/RM/27, December 1992 Second Edition*
Early Life Toxicity Test Using Stages of Salmonid Fish (Rainbow Trout, Coho Salmon, or Atlantic Salmon)	*EPS 1/RM/28, December 1998*
Survival and Growth in Sediment Using the Larvae of Freshwater Midge (Chironomus tentans or riparius)	*EPS 1/RM/32, December 1997*
Test for Survival and Growth in Sediment Using the Freshwater Amphipod Hyalella azteca	*EPS 1/RM/33, December 1997*
Test for Measuring the Inhibition of Growth Using the Freshwater Macrophyte, Lemna minor	*EPS 1/RM/37E, March 1999*

Survival and Growth in Sediment Using Estuarine or Marine Polychaete Worms	in preparation
Reference Method for Determining Acute Lethality of Effluents to Rainbow Trout	EPS 1/RM/13, 2000 (Second Edition)
Reference Method for Determining Acute Lethality of Effluents to Daphnia magna	EPS 1/RM/14, 2000 (Second Edition)
Reference Method for Determining Acute Lethality of Sediment to Estuarine or Marine Amphipods	EPS URM/35, 1998
Guidance Document on Control of Toxicity Test Precision Using Reference Toxicants	EPS 1/RM/12, August 1990
Guidance Document on Collection and Preparation of Sediment for Physicochemical Characterization and Biological Testing	EPS 1/RM/29, December 1994
Guidance Document on Measurement of Toxicity Test Precision Using Control Sediments Spiked with Reference Toxicants	EPS 1/RM/30, September 1995
Statistics for the Determination of Toxicity Test Endpoints	in preparation
Guidance Document on Application and Interpretation of Single-species Tests in Environmental Toxicology Testing	EPS 1/2M/34E, December 1999

Source: **http://www.ec.gc.ca/epspubs**.

Table 13.2

Key Elements of Protocols for Acute Lethality Testing

Test Organism: rainbow trout (***Onchorhyncus mykiss*** previously referred to as ***Salmo Gairdneri***)

- 96-hour duration
- individual fish should weigh between 0.5 and 5 g
- the longest fish should not be twice the length of the shortest fish
- 10 fish per test container
- maintain dissolved oxygen concentration of 7 mg/L
- pH between 5.0 and 9.0 (prefer 6.0 and 8.5)
- temperature 15 ± 1 °C
- dilution water must be capable of maintaining healthy fish stocks for at least 10 days
- fish should not be fed within 24-hour period prior to the test or during the test
- observations mandatory at 0.5, 1, 2, 4, 24 hours and every 24 hours thereafter (should be made as often as possible)
- six test concentrations (*i.e.* 10, 20, 30, 40, 65, 100%) and a control solution (100% dilution controls)

Reference: the toxicity monitoring sections of various MISA regulations.

II)   ***Test Organism: water flea*** (***Daphnia magna***)

- 48-hour duration
- minimum of 10 organisms per chamber
- only used when water hardness is greater than 80 mg/L
- pH between 6.0 and 9.0 (prefer 6.5 and 8.5)
- temperature 20 ± 1 °C
- five test concentrations (e.g. 100, 50, 25, 12.5 and 6.3%) and a control solution (100% dilution water)
- observations mandatory at 1, 2, 4, 24, and 48 hours
- dissolved oxygen, pH, hardness, and conductivity must be measured immediately before and after the test is completed

Reference: Poirier et al., 1988.

Table 13.3

Summary of the EEM Studies Required by the Metal Mining Effluents Regulations

In addition to the acute toxicity that is required (see Section 13.3), several types of studies are required as described in Schedule 4 of the regulations. These include effluent characterization, sublethal toxicity testing and water quality monitoring.

Effluent Characterization Study
- effluent is tested four times a year for the parameters prescribed in Part 1 of Schedule 5 (aluminium, cadmium, iron, mercury (not in all cases), molybdenum, ammonia and nitrate)

Sublethal Toxicity Testing
- two times each year for the first three years and then once each year after the third year; this component of the EEM is to include a fish species (larval growth and survival of fathead minnows as well as early life stages of rainbow trout), an invertebrate species (reproduction and survival of *Ceriodaphnia dubia*), a plant species (growth inhibition of *Lemna minor*), and an algal species (growth inhibition of *Selenastrum capricornutum*) for effluents discharged to fresh water; test procedures and test species vary for effluents discharged to marine or estuarine waters
- all tests to be done according to Environment Canada methodologies (see Table 13.1) or U.S. EPA reference methods for some tests of marine organisms

Water Quality Monitoring
- water around effluent discharges is to be tested four times a year for the parameters in Schedule 4
- sampling times should correspond with biological monitoring studies

Figure 13.1

Flowchart for Toxicity Reduction Evaluations

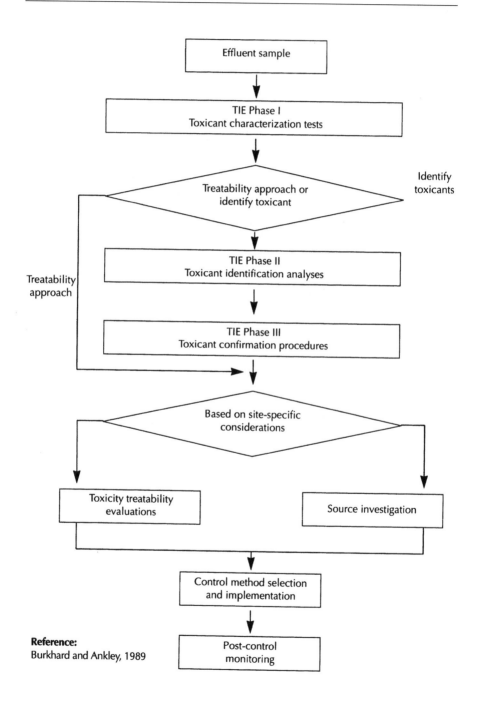

Reference:
Burkhard and Ankley, 1989

Figure 13.2
Toxicity Investigation Evaluations—Phase I Tests

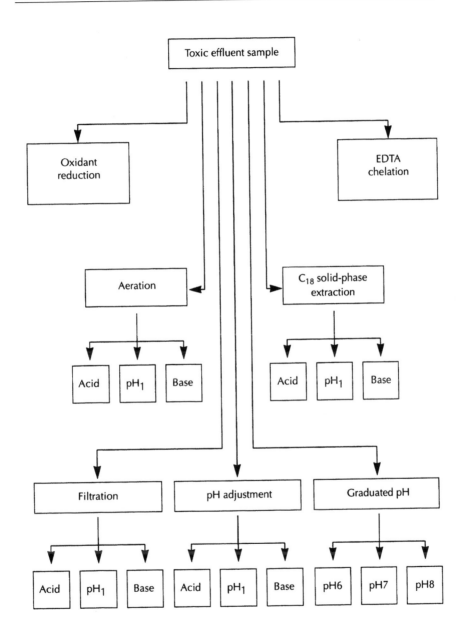

Reference:
Burkhard and Ankley, 1989

List of Abbreviations

AAQC	ambient air quality criterion
ACM	asbestos-containing material
AMPs	administrative monetary penalties
AO	aesthetic objective
API	Air Pollution Index
APPP	Automotive Pollution Prevention Program
AQI	Air Quality Index
ATG	analytical test group
BCF	bioconcentration factor
BMP	best management practice
BOD	biochemical oxygen demanding
CANUTEC	Canadian Transport Emergency Centre
CCME	Canadian Council of Ministers of the Environment
CEAA	Canadian Environmental Auditing Association
CEM	continuous emission monitoring
CEPA, 1999	Canadian Environmental Protection Act, 1999
C of A	Certificate of Approval
CPPR	Canadian Pollution Prevention Roundtable
CSA	Canadian Standards Association
CWQGs	Canadian Water Quality Guidelines
CWS	Canada-wide standards
DGTA, 1992	Dangerous Goods Transportation Act, 1992
DNAPLs	dense, non-aqueous phase liquids
E2	Environmental Emergency
EAA	Environmental Assessment Act
EAAB	Environmental Assessment and Approvals Branch
EBR	Environmental Bill of Rights
EC	European Community
EDTA	Ethylenediaminetetraacetate ligand
EEM	environmental effects monitoring
EMS	Environmental Management System
EPA	Environmental Protection Act
EPS	environmental protection system
ERAP	emergency response assistance plan
ESA	environmental site assessment
ESDM	emission summary and dispersion model
ESSA	extra strength surcharge agreement
HI	hazard index
HQ	hazard quotient
HWIN	Hazardous Waste Information Network
ICI	industrial, commercial and institutional (sector)
IEB	Investigations and Enforcement Branch (of the MOE)
IMAC	interim maximum acceptable concentration
ISO	International Organisation for Standardisation
Leq	energy equivalent sound level
LSB	Legal Services Branch
MAC	maximum acceptable concentration
MDL	method detection limit
MISA	Municipal, Industrial Strategy for Abatement

MOE	Ministry of the Environment
MOEE	Ministry of the Environment and Energy
MPO	manufactured, processed or otherwise used
NAPL	non-aqueous phase liquid
NEEC	National Environmental Emergencies Centre
NEF/NEP	Noise Exposure Forecast/Noise Exposure Projection
NOAEL	no-observable-adverse-effect-level
NOEC	no observed effect-concentration
Nox	nitrogen oxides
NPC	noise pollution control
NPRI	National Pollutant Release Inventory
NTREE	National Round Table on the Environment and Economy
OCWA	Ontario Clean Water Agency
OECD	Organization for Economic Cooperation and Development
OG	operational guideline
OHSA	Occupational Health and Safety Act
OWME	Ontario Waste Materials Exchange
OWRA	Ontario Water Resources Act
PAHs	polycyclic aromatic hydrocarbons
PCBs	polychlorinated biphenyls
PERT	Pilot Emissions Reduction Trading
POI	point of impingement
POPs	persistent organic pollutants
PPE	personal protective equipment
PPV	peak particle velocity
PWQO	provincial water quality objectives
QA/QC	Quality assurance/quality control
REVA	Recognizing and Encouraging Voluntary Action initiative
RfD	reference dose
RMS	root mean square
RSD	risk-specific dose
SAC	Spills Action Centre
SAR	sodium absorption ratio
SDB	Standards Development Branch
SPE	solid phase extraction
SQE	small quantity exemption
SSRA	site-specific risk assessment
STAC	Selected Targets for Air Compliance
SWCS	storm water control study
TCLP	Toxicity Characteristic Leachate Procedure
TDGA, 1992	Transportation of Dangerous Goods Act, 1992
TDGR	Transportation of Dangerous Goods regulations
TEC	threshold effects concentrations
TIE/TRE	Toxicity Identification Evaluation/Toxicity Reduction Evaluation
TKN	total Kjeldahl nitrogen
TPH	total petroleum hydrocarbons
TPHCWG	Total Petroleum Hydrocarbons Criteria Working Group
TSS	total suspended solids
VdB	velocity decibels
VECs	valued ecosystem components
VOCs	volatile organic compounds

Index